PULMONATES

PULMONATES

Volume 1

Functional Anatomy and Physiology

Edited by

Vera Fretter

Department of Zoology, University of Reading

and

J. Peake

British Museum (Natural History), London

1975

ACADEMIC PRESS

London   New York   San Francisco

A Subsidiary of Harcourt Brace Jovanovich, Publishers

ACADEMIC PRESS INC. (LONDON) LTD.
24/28 Oval Road,
London NW1

United States Edition Published by
ACADEMIC PRESS INC.
111 Fifth Avenue,
New York, New York 10003

QL
430.4
P84
v.1

Library of Congress Catalog Card Number: 74-18501
ISBN: 0–12–267501–0

Text set in 11/12pt. Modern Extended, and printed by letterpress
in Great Britain by Page Bros (Norwich) Ltd, London and Norwich

Contributors to Volume 1

H. H. BOER, *Department of Zoology, Free University, Amsterdam, The Netherlands*

C. J. DUNCAN, *Department of Zoology, University of Liverpool, England*

V. FRETTER, *Department of Zoology, University of Reading, England*

F. GHIRETTI, *Institute of Animal Biology, University of Padua, Italy*

A. GHIRETTI-MAGALDI, *Institute of Animal Biology, University of Padua, Italy*

H. D. JONES, *Department of Zoology, Manchester University, England*

J. JOOSSE, *Department of Zoology, Free University, Amsterdam, The Netherlands*

G. A. KERKUT, *Department of Physiology and Biochemistry, Southampton University, England*

J. MACHIN, *Department of Zoology, Toronto University, Canada*

C. P. RAVEN, *Zoological Laboratory, University of Utrecht, The Netherlands*

N. W. RUNHAM, *Department of Zoology, University College of North Wales, Bangor, Wales*

R. J. WALKER, *Department of Physiology and Biochemistry, Southampton University, England*

Preface

The success of the phylum Mollusca is proclaimed by the number of its species; only one other group, the Arthropoda, is in this respect superior. An estimated 100,000 species is made up of approximately 80,000 gastropods, 10,000 bivalves and a total of 5000 in other classes. The gastropods have undergone a rich adaptive radiation: from an undoubted marine origin they have spread to occupy most types of aquatic and terrestrial habitat, an evolution made possible by the remarkable plasticity of the body. In particular, changes related to respiratory function occur in the mantle cavity, and in the reproductive system provision is made for internal fertilization and the production of spawn masses in which embryos can undergo complete development.

About half the species of gastropod are pulmonate, living in fresh water and on land. Although in the British Isles they constitute less than a third of the gastropod species, in more extensive land masses they dominate the molluscan fauna, and everywhere are of economic importance because they are serious horticultural and agricultural pests and because some act as hosts to parasites affecting man and his domestic animals.

Pulmonates are easy to keep in captivity; with few exceptions they can be maintained on a herbivorous diet, eating both fresh and decaying vegetation, and since there is typically no larval stage in the life history successive generations can be kept under observation. For these and other reasons they have become popular animals for laboratory research. Most of the studies described in the chapters of Volume 1 are based on laboratory work; Volume 2 relates to field work. Much recent work on pulmonate physiology has been concerned with two partially overlapping fields, the nervous system and the endocrine organs. The great advances made within these fields rest upon the concentrated pattern of the nervous system with relatively small numbers of large cells in its ganglia, and the resistance of the animals to operative techniques. The pulmonates have provided useful material for work on locomotion, the alimentary canal with its enzyme systems, and have also posed challenging problems in the field of water relationships.

This volume brings together information from various sources related to the functioning of the systems of the pulmonate body and to its embryology. The second volume, edited by Mr J. F. Peake, deals with the ecology of the group, aspects of its economic importance, systematics and evolution.

I am grateful to Dr Elizabeth Andrews, Professor A. Graham, Dr R. H. Nisbet and Professor E. R. Trueman for help in the preparation of the volume.

December 1974.   V. FRETTER

Contents of Volume 1

Volume 2: Systematics, Evolution and Ecology

Contents

Systematics

Ecology

Economic problems

General studies

Introduction

VERA FRETTER

Zoology Department, University of Reading, England

The most perfected terrestrial adaptations are found in vertebrates and arthropods; molluscs which have come on land are still dependent on a moist environment and are inactivated by drought and low temperature. The success of the molluscs relates to some structural changes and considerable physiological adjustments which enable them to take full advantage of favourable conditions, however brief, and then to lapse into inactivity, most pronounced in the hibernation and aestivation of land pulmonates. Pulmonates are essentially land and freshwater molluscs, but a few genera, recognized as amongst the most primitive, live intertidally. They comprise one of the three subclasses of Gastropoda, the Pulmonata, which has two well known superorders: the Basommatophora with one pair of contractile cephalic tentacles bearing eyes at the base; and the Stylommatophora, reaching the peak of pulmonate evolution in the common terrestrial snails and slugs, which have usually two pairs of invaginable tentacles with eyes at the summit of the hinder pair. A third superorder, the Systello-matophora, unites the intertidal onchidiids and the terrestrial veroni-cellids, both slug-like, but with mouth and anus at opposite ends of the body. The other two subclasses, the Prosobranchia and Opisthobranchia, are groupings of marine snails, though a few genera of monotocardian prosobranchs are terrestrial, breathing air, and others live in fresh water where they retain the gill. Land and fresh water are unexploited by opisthobranchs.

Gastropods of the three subclasses display considerable diversity in body form. This is most pronounced in prosobranchs, the archaic group, and least in the pulmonates which, however, have evolved physiological and behavioural innovations in connection with living away from the sea. Perhaps on account of the less appealing external appearance and the habitat of pulmonates their study has been isolated from that of prosobranchs and opisthobranchs. This has led to differences in terminology in the descriptions of various systems of the body. For example, those who have studied the gut of pulmonates have applied terms to different regions which do not conform to those used in other subclasses of gastropods. Thus the buccal cavity is termed the pharynx by some continental malacologists and this term has even been erroneously extended to include the whole buccal mass. Also the anterior and posterior regions of the oesophagus and intestine are referred to by the use of the prefixes pro- and post-; if such terms are to be used pro- and meta- or opistho- would at least avoid coining hybrid words. Again, the anterior pedal gland, opening between propodium and mesopodium, is termed the supra-pedal gland.

Basically the body of the gastropod consists of a head with a terminal mouth, tentacles and eyes, a visceral mass made up of gonad, digestive gland, stomach and loops of intestine, and a large foot with a creeping sole and, typically, a pedal gland discharging along its anterior edge and lubricating the pathway along which the snail creeps. The visceral mass is characteristically wound in a right-handed, helicoid spiral and enclosed in a shell which is secreted by its covering epithelium the mantle or pallium. The shell encloses the mantle cavity which in prosobranchs and those members of the other two subclasses that retain a more primitive organization, lies over the head and acts as a compensation chamber into which the head and foot can be withdrawn. A fold of body wall, the mantle skirt, forms the roof and sides of the mantle cavity and the body wall its floor; the skirt is highly vascular and blood passes from it directly to the heart. Head and foot are withdrawn by the columellar muscle, which arises on the central axis or columella of the shell. The shell is then closed by an operculum in prosobranchs; in the other two subclasses an operculum is present in the more primitive members — the opisthobranchs *Acteon* and pyramidellids and the brackish water pulmonates *Amphibola* and *Salinator*. Under adverse conditions pulmonate snails living on land secrete an isolating sheet of mucus across the shell, the epiphragm, which may become very thick and, in some species, calcified.

Although the mantle cavity retains a more or less standard pattern throughout the gastropods there are considerable differences between the members of the three subclasses reflecting adaptive radiation within the group. In prosobranchs the cavity has a broad anterior opening through which can be seen two gills, left and right (diotocardians) or a single left gill (monotocardians); each is associated with an auricle. The gill of the diotocardian is bipectinate and the heart has a single ventricle receiving blood from two auricles. The gill of the monotocardian is a single series of leaflets hanging from a broad axis which runs in an antero-posterior direction from the mouth of the mantle cavity to its innermost part. The leaflets have supporting rods and ciliary tracts which maintain a flow of water through the cavity from left to right, passing over the osphradium on the left and the hypobranchial gland and anus on the right. The kidney of the monoto-cardian opens to the base of the cavity, typically with no duct, and

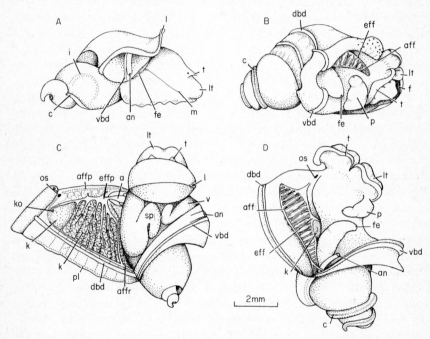

Fig. 1. A, *Chilina dombeyana:* from the right, shell removed. The rectum and ventral ciliated band are at the anterior limit of the exhalant channel of the mantle cavity. B, *Acteon tornatilis:* from the right, shell removed. The mantle edge is reflected, exposing the dorsal and ventral ciliated bands. C, *Chilina:* dorsal view, mantle cavity opened by a cut along anterior and posterior walls of mantle skirt. Reproductive duct, free in haemocoel, seen through dorsal body wall. D, *Acteon:* dorsal view, mantle cavity opened by a cut along the mantle skirt between the dorsal and ventral ciliated bands. Free tip of gill retracted. (A, C, after Haeckel.)

a, auricle; an, anus; aff, afferent branchial; affp, afferent pulmonary; affr, afferent renal; c, pallial caecum; dbd, dorsal ciliated band; eff, efferent branchial; effp, efferent pulmonary; f, foot; fe, female opening; i, intestine; k, kidney; ko, kidney opening; l, mantle-edge lacuna lt, labial tentacle; m, male opening; os, osphradium; p, penis; pl, posterior pallial lacuna; sp, spermoviduct; t, cephalic tentacle; v, vagina; vbd, ventral ciliated band.

the reproductive opening of the female is on the right near the anus. The male opening is on the non-retractile penis behind the right cephalic tentacle.

Opisthobranchs also have a single gill (Fig. 1B), but it differs from that of monotocardians: its attachment to the mantle skirt is narrower and since the mantle cavity is shallower a certain length projects freely from the mouth of the cavity, though it can be withdrawn; there is no skeletal support and only the efferent branchial vessel (eff) runs in the mantle skirt, the afferent (aff) being at the edge of the ctenidial fold. Blood from the efferent branchial flows back to the single auricle of the heart which, in some species, is in the thickness of the mantle skirt. The name opisthobranch refers to the position of the gill with respect to the auricle. In the primitive genus *Acteon* the opening of the mantle cavity is anterior, though twisted to the right, and on the left does not extend beyond the attachment of the gill in the mid line; the gill is anterior to the auricle. In more advanced opisthobranchs the cavity migrates still further along the right side and shallows, so that head and foot can no longer be withdrawn and the gill comes to lie behind the auricle. These changes are accompanied by a reduction in the coiling of the visceral mass, which now rests on the foot, and a reduction of the shell; shell and mantle cavity are lost in the most specialized forms, the sea slugs. In *Acteon* and other shelled opisthobranchs the ciliation of the gill is weak and the flow of water through the mantle cavity is maintained by a dorsal and ventral ciliated band (dbd, vbd) bordering the exhalant channel (Fretter and Graham, 1954). These bands are continuous along a posterior extension of the mantle cavity which forms a long caecum (c) following the coils of the shell in *Acteon*, but the caecum is short when the shell is reduced and the mantle cavity shallower as in *Scaphander* and *Philine*. The haemal spaces in the wall of the caecum collect blood from the visceral haemocoel and distribute it to the kidney.

Pulmonate snails resemble prosobranchs in retaining an anterior cavity which allows head and foot to be withdrawn into the shell, but it communicates with the exterior only by a restricted opening on the right (left in sinistral forms) (Fig. 4D, pn). It functions as a lung: there is a network of capillaries in the roof, which drains into a single auricle (a), and muscles in the floor (the dorsal body wall or diaphragm) by which its volume is controlled. Gill and hypobranchial gland are lost. The lung does not confine the pulmonate to land. Although in some (Onchidiidae, Ellobiidae, Carychiidae, Stylommatophora) the lung is used exclusively for air, in others (*Chilina* (Fig. 1A), *Amphibola*, *Siphonaria* and all limnic Basommatophora) it may be used for holding water or water and air; it is supplemented by pallial gills just outside the pneumostome in planorbids and ancylids. In the first grouping the subdivision of the cavity into a large dorsal chamber, the pulmonary sac, and a small vestibule which communicates with the exterior, is marked by a constriction where the contractile pneumostome is sited. The vestibule is sometimes referred to as the hypopeplar cavity. The reduction or loss of the shell in stylommatophoran slugs and the sinking of the visceral mass into the elongated head-foot is not associated with detorsion but has resulted in a considerable reduction of the lung. The body surface is kept moist by the mucus it secretes and cutaneous respiration is extremely important.

The development of the lung has been investigated in a number of pulmonates, though not the most primitive, and the results of these studies have raised the question as to whether it is homologous with the mantle cavity of other gastropods. It has been suggested that it is an entirely new structure developed along the gastropod line of evolution (Ghose, 1963; Régondaud, 1964). In the embryo of *Lymnaea stagnalis* (L.) the lung develops as an ectodermal invagination, first seen at the age of 7–8 days (Fig. 2A) when there is a cap-shaped shell posteriorly (sh), surrounded by a thickened mantle edge enclosing the pallial groove (Régondaud, 1964). The rudiment of the lung is on the right side (lr) and not related to the pallial groove whereas the developing mantle cavity of prosobranchs is Robert, 1902; (Drummond, 1903; Smith, 1935; Crofts, 1937; Creek, 1951): it is anterior to the groove, immediately anteroventral to the excretory pore (ep) and far from the anal cells (ac) which are mid ventral. In an embryo 11 days old (Fig. 2B) torsion has brought the opening of the rudimentary lung antero-dorsally on the right side and the invagination has deepened and extended dorsally. The adjacent mantle edge (me) has approached the relatively small opening of the invagination and its submarginal tissues have surrounded it to form the boundary of a shallow depression to which are related the osphradium (os), the anus (an) and the kidney aperture, and into the base of which the lung opens. The opening is the pneumostome (pn) and the cavity with which it communicates is, by virtue of its mode of development, homologous with the mantle cavity of prosobranchs. In the development of the terrestrial pulmonates *Helix* (Fol, 1880), *Limax* (Meisenheimer, 1898), *Arion* (Heyder, 1909) and *Achatina* (Ghose, 1963) the ectodermal invagination which forms the lung is also initially independent of the cavity related to the pallial fold which forms a vestibule enclosing the pneumostome (Fig. 2C, pn). The eggs of all these pulmonates are large, their development distorted by the presence of cells filled with yolk, and the formation of such

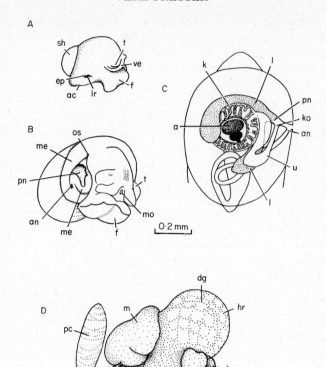

Fig. 2. A, B, *Lymnaea stagnalis* embryo. A, 7–8 days from right side; B, 11 days antero-lateral view. C, *Arion ater rufus* embryo, 28 days, dorsal view of visceral mass. D, *Agriolimax agrestis* (L.) late embryo with podocyst and head vesicle, contractile sacs related to respiratory needs. (A, B, after Régondaud; C, after Heyder; D, after Carrick.) a, auricle; ac, anal cells; an, anus; dg, limit of penetration of digestive gland into vesicle; ep, excretory pore; f, foot; hr, head vesicle; k, kidney; ko, kidney opening; l, lung; ll, lower lip; lr, rudiment of lung; m, mantle; me, mantle edge; mo, mouth; os, osphradium; pc, podocyst; pn, pneumostone; sh, shell; t, tentacle; u, ureter; ve, velum.

structures as podocyst and cephalic vesicle related to development in a terrestrial environment (Fig. 2D). This may account for the unusual development of the mantle cavity. Since torsion is restricted to 90°, or even less in some pulmonates, the opening of the cavity is on the right anteriorly; it is overhung by the pallial margin. In snails the pallial margin (in front of the groove from which the shell is secreted) hypertrophies to form the glandular collar, and this same region in slugs is drawn out into a flap which may be large enough to cover the neck and head when they are not fully extended.

In contrast to these pulmonates the mantle cavity of *Onchidella celtica* (Fretter, 1943) develops before the lung and in the adult has a separate opening to the exterior at the posterior end of the body. In the embryonic veliger stage the cavity is restricted to the right side and is confluent anteriorly with a deep pallial groove. During detorsion, when the visceral mass sinks into the head-foot, it migrates posteriorly and becomes a short narrow tube receiving the anus, kidney opening and vagina. The lung then develops as a separate ectodermal invagination and forms a branching system of spaces reminiscent of the branching lung of athoracophorid slugs. It supplements cutaneous respiration and is used only during periods of activity; at other times the pneumostome is closed. Its homologies are uncertain and the pulmonate affinities of these slugs are revealed by other characters. The terrestrial veronicellids, grouped with the onchidiids, have no shell, mantle cavity or lung. They live in damp tropical and subtropical areas and respire by means of the thin mantle covering the dorsal surface.

Some pulmonates living intertidally have a free veliger larva with a large velum and the trochophore stage is passed through within the egg capsule; they thus resemble monotocardian prosobranchs. They include *Onchidella flavescens* (von Wissel) (Morton and Miller, 1968), *Melampus bidentatus* Say (Russell-Hunter *et al.*, 1972), *Salinator takii* Kuroda (Tanak, 1959), *S. fragilis* (Lamarck), *Siphonaria zelandica* Quoy and Gaimard (Pilkington, personal communication) and some other species of this genus. In *Amphibola crenata* Martyn the velum begins to degenerate within the egg capsule, but if the embryo is accidentally released at the veliger stage the larva will swim rapidly (Farnie, 1924). A heterostrophic shell has been recorded for pulmonate larvae, a feature of some members of prosobranch families (Scalidae, Ianthinidae, Eulimiidae, Aclididae) and of opisthobranchs. It has not been shown to have any functional value and is presumably the result of a mutation. Except for *Salinator fragilis*, in which the mantle cavity is on the right and opens anteriorly, we have no knowledge of the position and extent of the mantle cavity of the pulmonate veliger, or of any changes at metamorphosis. This would be of interest in view of the fact that some regard the lung of pulmonates as being derived from the prosobranch mantle cavity by fusion of the mantle edge with the underlying body wall, except at the pneumostome. Yet in pulmonates with no free veliger there is no evidence of this and the degree of torsion does not bring the opening of the mantle cavity over the animal's back.

Siphonaria, though specialized with respect to its patelliform shape, is probably the most primitive pulmonate with respect to respiratory equipment. It lives intertidally and on some shores replaces the prosobranch limpets, but unlike these limpets it can breathe in water and in air. The mantle cavity, extending over most of the dorsal surface, is subdivided transversely by a gill made up of numerous leaflets with folded walls (Fig. 3B). Afferent and efferent vessels

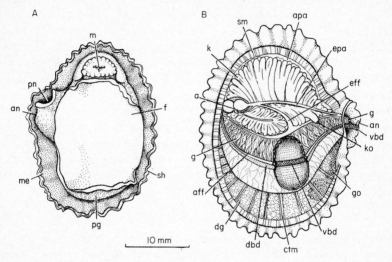

Fig. 3. *Siphonaria* sp. probably *marza* Iredale: A, ventral view of limpet attached to wall of glass aquarium above water level; B, dorsal view, shell removed and also an ovoid area of mantle skirt and underlying gill to expose part of ventral ciliated band.

a, auricle; an, anus; aff, afferent branchial vessel; apa, afferent pallial (= afferent renal) vessel; ctm, strands of connective tissue and muscle; dbd, dorsal ciliated band; dg, digestive gland; eff, efferent branchial vessel; epa, efferent pallial vessel; f, foot; g, gill; go, gonad; k, kidney; ko, kidney opening; m, mouth; me, mantle edge; pg, repugnatorial glands; pn, pneumostome; sh, shell; sm, shell muscle; vbd, ventral ciliated band.

(aff, eff) border the broad axis as in prosobranchs. The right tip of the gill (g) extends to the opening of the mantle cavity where it separates an anterior inhalant from a posterior exhalant respiratory water channel. The ciliation on the gill and underlying body wall is weak and the flow of water through the mantle cavity is maintained by the two ciliated ridges described by Yonge (1952) in *Siphonaria alternata* Say. *Siphonaria* spp. collected in Queensland show that the dorsal ridge, running parallel with the gill axis and immediately posterior, curves ventrally at its inner end and becomes continuous with the ventral one.

They form the anterior boundary of the exhalant channel of the mantle cavity. The dorsal wall of the channel is vascular and the blood flow is to the afferent branchial vessel. The primitive freshwater pulmonate *Chilina dombeyana* d'Orbigny (Haeckel, 1911) has no gill, but similar ciliated bands, which pass into a pallial caecum, as in *Acteon*, and become confluent at its tip (Fig. 1A, vbd, c). Such a caecum does not occur in prosobranchs, nor in higher pulmonates where, however, the ciliated strips are retained in aquatic forms (Harry, 1964). When *Siphonaria* breathes in air the vascularized anterior half of the roof of the mantle cavity, into which the kidney (k) has spread, is the important respiratory surface: only the anterior part of the pneumostome is widely opened (Fig. 3A, pn), the posterior lobes of the kidney all but shield the gill and there are rhythmical pumping movements of the dorsal body wall.

The heart and single kidney of the pulmonate are not posterior to the mantle cavity as in prosobranchs, but in the roof of the pulmonary sac where there is a network of fine blood vessels (Fig. 4C, D). The blood comes from a circular tract of intercommunicating sinuses, the venous circle (vc), which lies along the edge of the sac and receives blood from the visceral haemocoel (circle not complete in *Chilina*). From the lung the blood drains into the efferent pulmonary vein (effp) leading to the auricle (a), most of it directly, though some by way of the kidney (k). The blood supply to the kidney of *Acteon* is also of pallial origin (Fig. 4B, k, c), whereas in monotocardian prosobranchs (Fig. 4A, k) it is essentially venous blood from the visceral haemocoel, and on leaving the kidney it goes to the pallium and then to the gill (aff, eff). Thus in pulmonates and opisthobranchs the pallium receives much of its blood from sources other than the kidney, while in prosobranchs it receives little, and, although in the three subclasses the heart receives oxygenated blood, most of this has been recently monitored by the kidney in prosobranchs, but considerably less in pulmonates and opisthobranchs. The arterial system is best developed in pulmonates: from the anterior and posterior aortae blood is finally channelled into beds of capillaries associated with the various organs of the body. Vessels of $10\mu m$ diameter, or even smaller, on the gut of *Lymnaea stagnalis appressa* Say (Carriker, 1946) end in ostioles which are surrounded by a dense network of muscle fibres capable of occluding the opening and so regulating the flow of blood into the haemocoel. Within the haemocoel there are septa allowing differential pressures to be built up. The transverse septum separates the visceral from the cephalopedal haemocoel as in other gastropods, but the septa of the anterior haemocoel are more complete and have a different disposition

from those described by Nisbet (1953) in *Monodonta lineata* (da Costa).
Kisker (1923) describes two horizontal septa in *Helix pomatia* L. with
ostia controlled by muscle fibres. One, the more dorsal, runs forward
from near the anterior end of the transverse septum beneath the thick
layer of connective tissue of the neck and head, enclosing a space

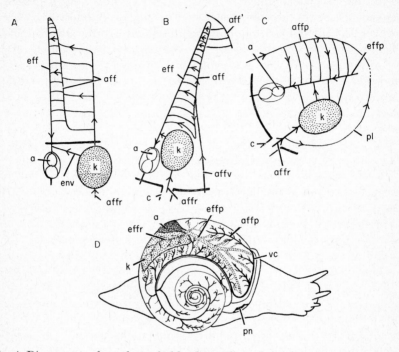

Fig. 4. Diagrams to show the main blood vessels, and the direction of circulation,
in the mantle skirt. A, monotocardian, B, *Acteon tornatilis*, C, *Chilina dombeyana*
in dorsal view; thick line indicates inner limit of mantle cavity. D, *Helix* seen
from the right. (C, after Haeckel; D, after Meisenheimer.)
a, auricle; aff, afferent branchial; aff', marginal pallial vessel homologous to
afferent pallial of *Chilina*; affp, afferent pulmonary; affr, afferent renal; affv,
afferent branchial from visceral haemocoel; c, pallial caecum; eff, efferent branchial;
effp, efferent pulmonary; effr, efferent renal; env, efferent vessel from nephridial
gland; k, kidney; pl, pallial lacunae; pn, pneumostome; vc, venous circle.

exclusively for blood. Anteriorly, before inserting on the head, it
fuses with the second septum which crosses the haemocoel to isolate
the buccal mass and anterior oesophagus from the reproductive ducts
lying above them.

The kidney of the gastropod is a coelomoduct characterized by
deeply folded walls, commonly linked to the pericardial cavity by a

ciliated renopericardial canal and opening into the deeper part of the
mantle cavity. In most basommatophorans and some stylommatophorans it discharges into the pulmonary sac by a papilla, otherwise
in pulmonates there is a long ureter which opens alongside the anus.
The classification of the Stylommatophora is based on the disposition of
the ureter: Orthurethra, with long straight ureter opening near the
pneumostome; Mesurethra, with very short ureter near the side of the
kidney; and the largest order, Sigmurethra, with ascending ureter
alongside the rectum. The siting of the kidney in the mantle skirt is
also a character of the Architaenioglossa which are regarded as an
early offshoot of the monotocardian stock and are freshwater (Vivi-
paridae, Ampullariidae) and terrestrial (Cyclophoridae); there is a long
ureter in members of one genus, *Viviparus*. The initial stage of urine
production in gastropods is by ultrafiltration from the heart into the
pericardial cavity and this filtrate reaches the kidney by way of the
renopericardial duct. In terrestrial pulmonates, however, the filtration
is directly from the renal capillaries into the kidney (Vorwohl, 1961;
Martin *et al.*, 1965) suggesting that the blood pressure in these capillaries
must be high. This does not apply to the terrestrial cyclophorids
(Andrews and Little, 1972). The urine in the pulmonates and the
prosobranch *Viviparus viviparus* (L.) (Little, 1965) is modified in the
kidney sac and ureter by ionic resorption and secretion.

It is assumed that the mantle cavity of the ancestor of the gastropod
was behind the visceral mass and continuous with a pallial groove
running under the margin of the shell. Anus and ducts from kidneys and
gonad presumably opened into the cavity which also housed two gills
and an osphradium associated with each. Studies on the development
of gastropods show that the visceral hump is rotated on the head-foot
bringing the rudiment of the originally posterior mantle cavity anterior.
The twisting is typically in a counter-clockwise direction when the
animal is viewed from above. These molluscs are then said to exhibit
torsion which in prosobranchs is through an angle of 180°. The gut is
now twisted with the anus (Fig. 5A, an) directed forwards, and that part
of the nervous system running from the head to the visceral hump, the
visceral loop, is twisted into a figure of eight so that one half runs
dorsal to the oesophagus, from right pleural (rpl) to supraoesophageal
(= supraintestinal) ganglion (sog), and the other ventrally, from left
pleural to suboesophageal (= subintestinal) ganglion (sbg); the two
parts unite posteriorly dorsal to the gut where they pass to the visceral
ganglia (vg). This streptoneurous condition of the visceral loop is
characteristic of prosobranch gastropods. In monotocardian prosobranchs the nerve ring surrounding the anterior oesophagus has

paired cerebral (cg), pedal and pleural ganglia (rpl). The pleurals innervate the mantle edge and from them arises the visceral loop with three ganglia adjacent to the organs they serve: the supraoesophageal with nerves to the osphradium (os), gill and mantle, the suboesophageal innervating the rectum near the anus, the reproductive duct and mantle,

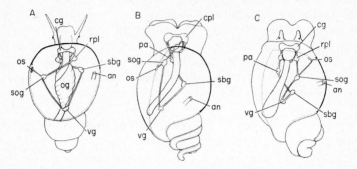

Fig. 5. Diagrams of nervous systems in dorsal view. The extent of the mantle cavity is indicated by a dotted line and its opening by a thick line. A, monotocardian showing the twisting of the mid oesophagus due to torsion; B, *Acteon*; C, *Chilina dombeyana*. (C, after Haeckel.)

an, anus; cg, cerebral ganglion; cpl, cerebropleural ganglion; og, oesophageal gland; os, osphradium; pǝ, left parietal ganglion; rpl, right pleural ganglion; sbg, suboesophageal ganglion; sog, supraoesophageal ganglion; vg, visceral ganglion.

the viscerals near the posterior end of the mantle cavity innervating the heart, kidney, gonad and its duct. In some monotocardians the supra- and suboesophageal ganglia have migrated forwards to the nerve ring (some rissoaceans, calyptraeids, lamellariids); the viscerals have also moved forward in *Cingula*.

The streptoneurous condition is displayed in one genus of opisthobranchs, *Acteon* (Fig. 5B). It gives place to a secondary euthyneury associated with the reversal of torsion in other cephalaspids. In higher opisthobranchs euthyneury involves a shortening of the visceral loop so that all ganglia are concentrated in the nerve ring. A similar concentration is characteristic of pulmonates and for this reason they have been grouped with the opisthobranchs in a major taxon, the Euthyneura and all prosobranchs in the Streptoneura (Spengel, 1881). However, most authorities are unwilling to include the opisthobranchs and pulmonates within a single taxon, though accepting that they are closely related phylogenetically. Their relationship is displayed by another feature of the nervous system — the parietal ganglia which innervate the mantle. Right and left parietals are on the visceral loop of *Acteon* (pa), but in *Chilina* (Fig. 5C), which also retains a ganglionated visceral loop, the right one has apparently fused with the supraoe-sophageal (pa, sog); the fusion of these with other ganglia is of frequent occurrence (Guiart, 1901; Bargmann, 1930) and in higher pulmonates the supra- and suboesophageal ganglia are no longer separate entities. The parietal ganglia may be regarded as a subdivision of the pleurals which in the euthyneuran subclasses rarely give off nerves. Giant nerve cells, offering considerable scope to the neurophysiologist, are another characteristic of members of these two subclasses. In *Archachatina* (*Calachatina*) *marginata* (Swainson) the visceral ganglion has cells of 400 μm diameter, the parietals 300 μm, the pedals 150 μm and the cerebrals 100 μm (Nisbet, 1961).

Torsion of the visceral loop is related to the position of the osphradium and gill which are innervated by the adjacent supraoesophageal ganglion. They are on the left of the body in prosobranchs and *Acteon*, both exhibiting torsion, whereas in other cephalaspids and in aquatic Basommatophora they are on the right, with the result that the visceral loop does not cross the gut dorsally. Thus, although there is a twisted loop in *Chilina* (Haeckel, 1911) this is not the streptoneury of prosobranchs.

Pulmonates, though predominantly herbivorous, eat a variety of foods and some species, such as the worm-eating paraphantid snails and the slug *Testacella*, are specialized carnivores. Unlike prosobranchs and opisthobranchs an elongated proboscis enabling food to be obtained from otherwise inaccessible sites has not been evolved, indeed the methods of feeding and the layout of the alimentary canal are more uniform than in other gastropods. The alimentary canal has the following features. The buccal mass, extremely long in carnivores (Smith, 1970; Crampton, 1973) receives secretion from a single pair of salivary glands; the oesophagus is a simple tube widening posteriorly to form a crop in some species; the stomach which receives the ducts of the digestive gland is wholly or part muscularized in freshwater forms to form a gizzard (lacking in Stylommatophora); the intestine is short in carnivores and long in herbivores. Pulmonates agree with opistho-branchs in lacking the oesophageal pouches which characterize a number of mesogastropod groups. They also agree in having a triturating organ, but whereas this seems to be of gastric origin in pulmonates, in opistho-branchs it is oesophageal.

The radula of gastropods has been a topic of intensive study and, indeed, the breadth of the ribbon, the number of teeth in each radular row and the form of the teeth are of classificatory value. In pulmonates, and the more primitive opisthobranchs, the radular ribbon is broad and the teeth relatively small and simple. The ribbon originates in a radular

sac which in pulmonates and opisthobranchs hardly extends beyond
the posterior limit of the odontophore, whereas in prosobranchs it may
be well over twice the length of the shell (*Patella vulgata* L.), this length
indicating a high rate of wear and replacement. The protein element of
the teeth is tanned and in addition the teeth of prosobranchs, and on
occasion those of euthyneurans, are hardened by inorganic salts. The
numbers of teeth in each row influenced some malacologists in seeking
pulmonate origins in a rhipidoglossan stock (Hubendick, 1945; Morton,
1955). However, apart from this there is no agreement with the
structure of the rhipidoglossan odontophore nor the mode of functioning (Nisbet, 1973). In the typical prosobranch the membrane carrying
the radular teeth runs forward between paired cartilages and as the
odontophore is protracted at each feeding stroke the membrane is
directed forwards over the bending plane at the odontophoral tip and
spreads laterally. The rhipidoglossan radula, possessed by the most
primitive gastropods, brushes rather than scrapes the substratum and
for this no great strength is required, but a broad ribbon with a vast
array of needle-like marginal teeth. When the odontophore protrudes
the teeth erect and swing outward as the ribbon spreads laterally to
cover a broad area of substratum. On retraction over the bending
plane one row of radular teeth after another closes, the marginals
sweeping inwards towards the mid line, brushing any particulate matter
into a median ridge-shaped heap which is carried through the mouth. In
contrast, the odontophore of pulmonates is commonly a scoop with
muscle progressively replacing chondroid tissue and so giving a greater
degree of flexibility under nervous control. When retracted the radular
membrane is not folded away, but still covers the surface of the scoop.
Blood pressure is a more important component in bringing the radula
into the feeding position than in prosobranchs, and greater freedom of
movement is attained. There is only a forward and no lateral movement of the membrane over the bending plane when the teeth are
brought into an erect position to rasp the food or to grip while the
food is cut through by the jaw. The action of the odontophore and jaw
allows large pieces of herbage to be taken by the snail and slug, and these
are swallowed whole by a method unique to pulmonates. A tongue of
tissue projects forward into the concavity of the protracted odontophore, makes contact with the food which has been gathered and on
withdrawal of the odontophore directs it into the oesophagus. This
structure is the collostylar hood: the anterior end of a rod of connective
tissue, the collostyle, which fills the concavity of the short, trough-
shaped radular sac.

Pulmonates, opisthobranchs and a number of prosobranchs, though not the most primitive, are hermaphrodite and the genital ducts open on the right side of the body. The Diotocardia retain the right kidney and discharge the gametes by way of its duct; fertilization is external. All other gastropods have lost the kidney, practice internal fertilization and have glandular ducts producing prostatic fluid for the sperm and nutritive and protective coverings for the eggs. The glandular duct is simplest in the mesogastropod and appears to have originated from the closure of a pallial groove (Johansson, 1948; Fretter and Graham, 1962) leading forwards from the original site of the opening of the lost kidney. It retains this position in the mantle skirt in all mesogastropods, but in pulmonates and opisthobranchs has sunk into the haemocoel where it is free to coil and elaborate. The evolution of a hermaphrodite duct from the female duct of the unisexual monotocardian, such as *Littorina littorea* (L.), necessitates little change for there is already a channel for conducting sperm (ingoing) and another for eggs (Fretter, 1946). The hermaphrodite prosobranch *Valvata piscinalis* (Müll.) has two separate channels (Cleland, 1954) as in many freshwater basommatophorans (Duncan, 1960), and the vas deferens leads to a penis on the right side of the head, which in accordance with all prosobranchs, but unlike that of pulmonates and opisthobranchs (except *Acteon*), is not invaginable. Details of the reproductive ducts of pulmonates are of taxonomic importance, though the significance of differences between related genera is sometimes obscure. The complexity of the ducts is related not only to hermaphroditism, but also the manufacture of spermatophores characteristic of many stylommatophorans, and, in some terrestrial snails, special structures used to stimulate the partner during the elaborate copulatory behaviour — the darts of *Helix*, and the amatorial organ of *Ariophanta ligulata* (Fér.) (Dasen, 1933) armed with a needle and capable of injecting fluid.

The foregoing account attempts to relate some aspects of the functional anatomy of the soft parts of three subclasses of gastropods, which have undoubtedly arisen from diotocardian stock and have attained monotocardian organization. Another very successful group, the Neritacea, approaches this organization in that the right half of the pallial complex is reduced, a pallial genital duct is developed, fertilization is internal and the anus opens at the mouth of the mantle cavity. Neritaceans, however, retain a number of primitive, diotocardian characters and have developed others that have no counterpart in other gastropods (Fretter, 1965). This indicates an independent origin from an archaeogastropod stock and a long period of independent evolution at least since the mid Devonian (Knight *et al.*, 1960). The gastropods in the superfamily Trochacea have soft parts showing trends towards monotocardian organization, especially in the loss of the right gill and osphradium. Their genital ducts, however, are limited to the formation of a urinogenital papilla in females, the eggs are fertilized externally and there is no veliger larva comparable to that of a monotocardian. The spawn masses are of fragile construction as compared with those of the neritids reflecting the simplicity of the reproductive system, which must be an important factor limiting trochids to a marine habitat. The neritaceans and monotocardian gastropods have proved themselves capable of exploiting every type of aquatic and terrestrial niche since they are physiologically more efficient than diotocardians.

There are fossil shells of groups now extinct which indicate that during their evolution they emerged from an archaeogastropod to a submesogastropod level in that one ctenidium was lost. Thus members of the superfamily Murchisoniacea have a shell with a labial emargination, essentially pleurotomarian, implying that the ctenidia were paired, but in some there is an incipient inhalant canal suggesting an inhalant siphon and a single ctenidium. In the superfamily Trochonematacea the right ctenidium was probably present only in more primitive forms. Fossils grouped in these superfamilies were first recorded in the Ordovician. We know nothing of their soft parts except the facts which may be deduced from the shell, nor does the palaeontological record lead us to believe that such forms may have given rise to more recent monobranchiate groups. However, by the Carboniferous divergent families of monotocardian prosobranchs and also opisthobranchs (Knight et al., 1960) and pulmonates were recorded; this early appearance of pulmonates is denied by Knight et al., (loc. cit.) but supported recently by a number of authorities (Morris, Nuttall, Yochelson, personal communication). To the palaeontologist and neontologist the mesogastropods are the most primitive of these three monotocardian groups. Unfortunately the palaeontological record gives no proof of their interrelationship and this can only be conjectured by the malacologist relying on the evidence of soft anatomy.

Two questions arise at this point, firstly, the interrelationship of opisthobranchs and pulmonates; and secondly, their origin from some other gastropod group.

Similarities between opisthobranchs and pulmonates are so marked, especially at the lower levels of each subclass, that they are more likely to indicate origin from similar or even identical ancestral stocks than to have arisen by convergence. Boettger (1954) accepted the view that pulmonates arose from the Acteonidae, a view previously expressed by Pelseneer (1894), whereas Hubendick (1945) and Morton (1955) favoured close, but separate origins from some archaeogastropod group, being influenced, in part, by the number of teeth in each radular row. At whatever level the origin is pitched it is an undoubtedly marine group and the opisthobranchs have continued to exploit this habitat, though only a few of the most primitive pulmonates are still marine. In view of the rarity of animals which have become terrestrial by way of the littoral zone it is much more likely that the adoption of first an estuarine and then a freshwater habitat was the route which pulmonates followed and this is supported by the number of species still living there. The ellobiids would then represent, not so much the direct ancestors of other pulmonates, as a group of equivalent organization which has remained marine, though showing the same trend towards air-breathing and emersion as their more advanced relatives, and acquired some specializations for their habitats.

Though Boettger is right in emphasizing the similarities between opisthobranchs and pulmonates it is not at all probable that the one group arose from the other. This interpretation rests too much on the assumption that groups must all be traceable to a single ancestral stem and neglects the much more acceptable idea that at each level of organization there exists a number of phyletic lines all, in their own way, modifying their structure so as to reach a higher, more efficient level of functioning. The neritaceans, mesogastropods, opisthobranchs and pulmonates are certainly four recent groups of gastropods which are monotocardian in the broad sense; there is, however, no need to suppose that all must be derived from a single stock which crossed the diotocardian-monotocardian boundary, and there is much evidence to suggest that several stocks have done so. The resemblances between opisthobranchs and pulmonates may indicate no more than that they come from some similar area of the promesogastropods, and this could have been a different area from that from which the mesogastropods sprang.

In contrast to this view, affinities between the opisthobranch-pulmonate monotocardian stem and some groups of mesogastropod are also evident and may, indeed, offer, in sum, a better explanation of some features of opisthobranch-pulmonate soft anatomy than a promesogastropod derivation. Boettger (1954) suggested that the opisthobranchs arose from a mesogastropod stock at the level of the Rissoacea. These are small gastropods, Recent ones living in marine, estuarine and freshwater habitats, with a few terrestrial genera. Their small size could have led to the abranchiate condition, also found in primitive opisthobranchs such as the pyramidellids and in pulmonates.

The absence of a gill is compensated for by the development of the ciliated strips on the exhalant side of the mantle cavity characteristic not only of primitive opisthobranchs and pulmonates, but also of a number of prosobranchs (*Omalogyra*, *Rissoella*) related to rissoaceans. Increase in size, necessitating extra respiratory surface and stronger ventilation of the mantle cavity, may explain: 1. the development of the pallial caecum; 2. the re-establishment in opisthobranchs and siphonarians of a gill which, though developed from a vestige of the prosobranch ctenidium, cannot reproduce precisely its structure or mode of function; and 3. the shallowing of the mantle cavity and increasing surface respiration. In the same way the absence in the opisthobranch-pulmonate stock of an oesophageal gland — a structure which persists in a variety of guises throughout the prosobranch series — may be explained on the basis of its origin near rissoaceans or cerithiaceans, both of which have lost the oesophageal glands because they possess a crystalline style. When, with growing size, the adaptive radiation of the group called for new modes of gut function the oesophagus or stomach could then be used as the basis of a crop or gizzard. A mesogastropod origin, too, explains the total absence of a right kidney in the opisthobranch-pulmonate line and the general similarity of their reproductive tract to that of the female mesogastropod.

References

Andrews, E. B. and Little, C. (1972). *J. Zool. Lond.* **168**, 395–422.

Bargmann, H. E. (1930). *J. Linn. Soc. Zool.* **37**, 1–59.

Boettger, C. R. (1954). *Verh. dtsch. zool. Ges., Zool. Anz.* Suppl. **18**, 253–280.

Carriker, M. R. (1946). *Biol. Bull. mar. biol. Lab., Woods Hole* **92**, 88–111.

Cleland, D. M. (1954). *Proc. malac. Soc. Lond.* **30**, 167–203.

Crampton, D. M. (1973). "The functional anatomy of the buccal mass of one opisthobranch and three pulmonate gastropod molluscs". *Ph.D. Thesis, University of Reading.*

Creek, G. A. (1951). *Proc. zool. Soc. Lond.* **121**, 599–640.

Crofts, D. R. (1937). *Phil. Trans. R. Soc.* B **208**, 219–268.

Dasen, D. D. (1933). *Proc. zool. Soc. Lond.* 97–118.

Drummond, I. M. (1903). *Q. Jl microsc. Sci.* **46**, 97–143.

Duncan, C. J. (1960). *Proc. zool. Soc. Lond.* **134**, 601–609.

Farnie, W. (1924). *Q. Jl microsc. Sci.* **68**, 453–469.

Fol, H. (1880). *Arch. Zool. exp. gén.* **8**, 103–232.

Fretter, V. (1943). *J. mar. biol. Ass. U.K.* **25**, 685–720.

Fretter, V. (1946). *J. mar. biol. Ass. U.K.* **26**, 312–351.

Fretter, V. (1965). *J. Zool. Lond.* **147**, 46–74.

Fretter, V. and Graham, A. (1954). *J. mar. biol. Ass. U.K.* **33**, 565–585.

Fretter, V. and Graham, A. (1962). "British Prosobranch Molluscs". London, Ray Society.

Ghose, K. C. (1963). *Proc. malac. Soc. Lond.* **35**, 119–126.

Guiart, J. (1901). *Mém. Soc. zool. Fr.* **14**, 1–219.

Haeckel, W. (1911). *Zool. Jb.* Suppl. **13**, Fauna chilensis 89–136.

Harry, H. W. (1964). *Malacologia* **1**, 355–385.

Heyder, P. (1909). *Z. wiss. Zool.* **93**, 90–156.

Hubendick, B. (1945). *Zool. Bidr. Upps.* **24**, 1–216.

Johansson, J. (1948). *Ark. Zool.* **40**A(15), 1–13.

Kisker, L. G. (1923). *Z. wiss. Zool.* **121**, 64–125.

Knight, J. B., Cox, L. R., Keen, A. M., Batten, R. L., Yochelson, E. L. and Robertson, R. (1960). Systematic descriptions (Archaeogastropoda): *In* "Treatise on Invertebrate Palaeontology" **I**, Mollusca 1: 169–310 (R. C. Moore, Ed.). Lawrence.

Little, C. (1965). *J. exp. Biol.* **43**, 39–54.

Martin, A. W., Stewart, D. M. and Harrison, F. M. (1965). *J. exp. Biol.* **42**, 99–123.

Meisenheimer, J. (1898). *Z. wiss. Zool.* **63**, 573–664.

Morton, J. E. (1955). *Proc. zool. Soc. Lond.* **125**, 127–168.

Morton, J. E. and Miller, M. (1968). "The New Zealand Sea Shore". Collins, London.

Nisbet, R. H. (1953). "The structure and function of the buccal mass in some gastropod molluscs. I. *Monodonta lineata* (Da Costa)." Ph.D. Thesis, University of London.

Nisbet, R. H. (1961). *Proc. roy. Soc.* B **154**, 267–287.

Nisbet, R. H. (1973). *Proc. malac. Soc. Lond.* **41**, 435–468.

Pelseneer, P. (1894). *Mém. cour. Acad. r. Sci. Belg.* **53**, 1–157.

Régondaud, J. (1964). *Bull. Biol. Fr. Belg.* **98**, 433–471.

Robert, A. (1902). *Arch. Zool. exp. gén.* **10**, 267–538.

Russell-Hunter, W. D., Apley, M. L. and Hunter, R. D. (1972). *Biol. Bull. mar. biol. Lab., Woods Hole* **143**, 623–656.

Spengel, J. W. (1881). *Z. wiss. Zool.* **35**, 333–383.

Smith, B. J. (1970). *J. malac. Soc. Aust.* **2**, 13–21.

Smith, F. G. W. (1935). *Phil. Trans. R. Soc.* B **225**, 95–125.

Tanak, Y. (1959). *Venus, Kyoto* **20**, 353–355.

Vorwohl, G. (1961). *Z. vergl. Physiol.* **45**, 12–49.

Yonge, C. M. (1952). *Proc. malac. Soc. Lond.* **29**, 190–199.

Chapter 1

Locomotion

H. D. JONES

Department of Zoology, University of Manchester, England

I. Introduction

Pulmonates, in common with most other gastropods, are primarily adapted for locomotion over and adhesion to relatively hard substrata, and the foot is consequently a muscular organ with a large plantar surface, the sole, with which the animal crawls and remains attached to the substratum. Attachment is principally achieved using the adhesive properties of mucus in conjunction with a large flat sole to give a large area of attachment to the substratum. During locomotion a continuous trail of mucus is laid down at the anterior end of the animal and the sole passes over the mucus using either the cilia on the sole or waves of muscular contraction produced on the sole. Cilia are present

1

on the sole of all gastropods and it is probable that this form of locomotion is the more primitive of the two. In those gastropods that use
pedal muscular waves for locomotion the cilia are not without function
for they probably assist in maintaining the even distribution of mucus
over the sole.

Most pulmonates move by utilizing the properties of the pedal sole:
the aquatic species and some of the smaller terrestrial species use
cilia; the larger terrestrial species use muscular waves. Exceptions to
this, when the musculature of the whole head-foot is used and not just
that of the sole, are the "galloping" of *Helix dupetithouarsi* Deshayes
(Carlson, 1905) and other related species, and the "looping" movements
of *Otina otis* (Turton) and the Ellobiidae (Vlès, 1913; Morton, 1955a, b).

No pulmonate is known to swim, though several early workers
credited freshwater species with this ability. This was a consequence
of the ability of many species to float due to the air in the mantle
cavity.

Many pulmonates, particularly slugs, are adapted to a burrowing
mode of life and, although the mechanism of burrowing of many other
kinds of animals has been investigated, in the pulmonates it is almost
completely unknown. Apparently the only reference to a specialized
mechanism is that of Jousseaume (1909) who merely noted that
Testacella utilizes "annular contractions" of the body to assist in
burrowing. Otherwise progression underground would seem to be a
continuation of normal creeping locomotion.

The possession of a wide flat sole in many slugs seems to be contradictory to their needs as burrowing animals for a more or less circular
cross-section is the ideal for a burrowing animal (Trueman, 1968).
However in some slugs, for example *Milax*, the sole is relatively narrow
and the cross-section roughly circular, but whether or not such slugs are
more efficient burrowers remains to be seen.

II. Anatomy of the muscular system

The gross muscular anatomy of the head-foot is complex, but the
following outline, based on Trappman's (1916) account of the musculature of *Helix pomatia* L., serves to illustrate the muscular anatomy of a
representative animal (Fig. 1).

The foot itself is highly muscular with muscle fibres running in most
directions and sections of the foot of *Helix* are consequently difficult to
interpret. The foot contains longitudinal, transverse and dorsoventral
fibres and muscles inclined at various angles to these three directions

as well as circular muscles running round the head-foot. Most of these muscles are attached to different parts of the body wall at both ends and few can be said to have a definite "origin" or "insertion". The only exceptions to this are the relatively few dorsoventral fibres and some of the roughly longitudinally arranged fibres which are derivatives

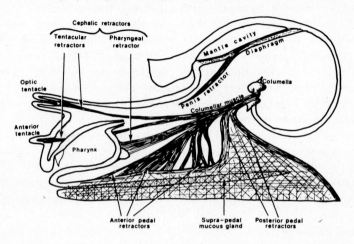

Fig. 1. Diagrammatic longitudinal section of *Helix pomatia* showing the principal muscles of the head and foot. The optic tentacles are shown partly invaginated and the muscles of the body wall are omitted. (After Trappmann, 1916.)

of the columellar muscle and thus originate on the columella. These serve as retractor muscles. Few pedal muscles except the retractors form discrete bundles.

The space between the muscle fibres of the foot is part of the haemocoel but this cavity is barely evident in sections of *Helix*. This may be due to the method of fixation (p. 16). Blood is supplied to the foot through a dorsal median artery and leaves through two lateral veins and a median vein (Schmidt, 1916). The only other major structure in the foot itself is the supra-pedal (= anterior pedal) mucous gland which opens anteriorly below the mouth and is a blind tube running down the median line of the foot. The role of this gland and its secretion is discussed on p. 16. The epithelium of the sole and the body wall contains numerous unicellular mucous glands which apparently secrete several types of mucus (Campion, 1961).

The anterior half of the foot is served with bundles of retractor muscles which at their proximal ends are convergent with the columellar muscle. At their anterior ends their fibres diverge into the foot and the

bundles rapidly lose their identity. The head itself is also served with
retractor muscles which are part of the columellar muscle, while the
single penis retractor muscle originates on the diaphragm (= dorsal
body wall). There is a pharyngeal retractor muscle and four tentacle
retractors, one to each of the paired optic and anterior tentacles.

The body wall of the head and the side and dorsal part of the foot is
highly muscular with longitudinal, circular and obliquely orientated
muscle fibres. Attached to the body wall near the mouth are several
discrete muscles concerned with the operation of the buccal mass.

The diaphragm, the floor of the mantle cavity, is highly muscular.
The muscle fibres are principally longitudinal and transverse (circular)
in orientation. This structure is involved in breathing and in protrusion
of the head-foot from the shell.

Between the epidermis and the various muscle fibres that lie beneath
it there is a layer of connective tissue fibres (Campion, 1961) sometimes
known as the basement membrane. These fibres are probably inelastic
and may serve to define and delimit body shape in much the same way as
the fibres in the body wall of platyhelminths and nemertines (Clark,
1964).

This then is a brief description of the musculature of a representative
shelled pulmonate. Slugs are characterized by the reduction in size of
the visceral mass and the shell, and the columellar muscle as such is
consequently reduced or absent. However much the same arrangement
of retractor muscles is present in the head and anterior foot, the muscles
originating in the vicinity of the vestigial shell. The body wall of slugs
is provided with longitudinal and circular muscle fibres. The foot itself
has, unlike *Helix*, a very distinct arrangement of muscle fibres referred
to below.

III. Muscular locomotion

A. Muscular waves on the sole

1. Types of wave

During locomotion the sole of many gastropods exhibits waves of
muscular contraction passing along, or even across, the sole and Vlès
(1907) has classified these waves according to the relative direction of
motion of the waves and of locomotion. If the pedal waves are moving
in the same direction as the animal they are called direct waves, but if
the waves are moving in the opposite direction to the animal then they
are called retrograde waves. Thus if an animal moving forwards has

waves passing forwards over the sole of the foot, or if an animal moving backwards has waves moving backwards then in both cases the waves are direct. Conversely, if an animal moving forwards has waves moving backwards, or if an animal moving backwards has waves moving forwards then in both cases the waves are retrograde (Fig. 2).

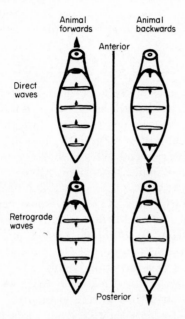

Fig. 2. Diagram to illustrate the difference between direct and retrograde waves in forward and backward locomotion. Large arrows indicate direction of movement of animal; small arrows indicate direction of movement of waves; ovate areas position of snout.

Both direct and retrograde waves may traverse the foot as a single system of waves in which case they are termed monotaxic, or they may pass along the foot as two systems of waves, one on each side of the foot, in which case they are called ditaxic. There is a third arrangement whereby the waves pass along the foot as four separate systems when they are called tetrataxic. Ditaxic waves may be alternate or opposite.

Almost all pulmonates that utilize muscular pedal waves for locomotion use direct monotaxic waves. These muscular waves are more or less confined to the sole, for when a crawling animal is viewed from above no waves are discernible on the foot, except perhaps for slight deflections of the lateral margin. The few exceptions to this are described in section III.B, and they are doubly exceptional in that not only are the waves not confined to the sole, but they are also retrograde.

Vlès's (1907) classification of pedal waves has subsequently been extended by Parker (1911) and Olmsted (1917), but as far as the pulmonates are concerned the classification has not been materially altered. Both Clark (1964) and Morton (1964) give examples of species utilizing the various types of pedal wave and Jones and Trueman (1970) present a discussion on the mechanism of retrograde pedal waves.

2. External appearance and form of the waves

Monotaxic direct waves are seen in their simplest form in *Onchidella celtica* (Forbes and Hanley) in which one wave at a time passes along the sole of the foot, each wave covering the full width of the sole. A wave does not appear at the posterior tip of the foot until the preceding one has passed off the anterior end (Vlès, 1907). *Onchidella* also shows the curious phenomenon of reflection of the waves for when each wave reaches the anterior extremity of the foot it is reflected and passes back down the anterior quarter or so of the sole before being absorbed by the succeeding wave (Vlès). This phenomenon is apparently unique for it has been reported from no other gastropod, nor from chitons. No detailed analysis of the locomotion of *Onchidella* has been carried out and the significance of this reflection is not known.

All other pulmonates having direct monotaxic waves have more than one wave on the sole at a time during locomotion. *Oxychilus alliarius* (Miller.) and *Cochlicopa lubrica* (Müll.), both relatively small animals, have only two or three waves on the sole at one time (personal observation), whereas *Helix pomatia* has eight to ten (Lissmann, 1946) and *Limax maximus* L. eleven to nineteen waves on the sole at any one time (Crozier and Pilz, 1924).

Although direct waves usually traverse the full length of the foot, the situation at the beginning of locomotion is different. When a snail, *Helix aspersa* Müller or *H. pomatia*, starts to move the first wave to appear does so near the anterior end of the foot (Lissmann, 1945). Other waves appear behind this and the propagation of waves spreads posteriorly at a speed greater than the waves pass forwards until the whole length of the foot is covered by waves.

Lissmann has shown that each pedal wave is a region where the sole is longitudinally compressed relative to the rest of the sole. If the sole between waves were to remain at its resting length then because of the appearance of these eight to ten regions of compression on the sole the result would be that the snail would be noticeably shorter during locomotion than when stationary. By observation, this is not so, for snails and slugs are no shorter during locomotion. The reason for this is that the sole is stretched from its resting length between waves and this compensates for the regions of compression which are the waves (Jones, 1973). The compensatory stretching would be impossible to accomplish in such an animal as *Onchidella* where the waves are all propagated at the posterior end of the foot. This indeed may be the reason that *Onchidella* has only one wave, in that it has not evolved a neural mechanism to produce the first wave near the front of foot, and if there were several waves on the sole at once then the animal would be considerably shorter during locomotion.

The lateral extent of the waves varies considerably from species to species. In species of *Helix* the waves cover almost the full width of the sole and there is a relatively narrow strip down each side of the sole not involved in the muscular waves. However, in most slugs, for example, the waves pass down the centre third of the sole only and there is a distinct line of demarcation between this strip and the two lateral strips which play no part in muscular locomotion.

Several species of slug, e.g. *Limax maximus, Agriolimax reticulatus* (Müll.), *Arion ater ater* (L.), can descend from overhanging situations by crawling down a mucous thread. During this form of locomotion the pedal waves are almost semicircular whereas during normal locomotion the waves are only slightly recurved if at all. The mucous thread down which the animals crawl is usually the normal slime track produced by the supra-pedal mucous gland which is not flattened by the weight of the animal (Barr, 1926). However in *Arion* this mucus is supplemented by the mucus of the caudal gland which lies on the dorsal surface of the foot just above the tail (Barr, 1928).

During normal locomotion the waves on the sole usually show as darker areas moving forwards over the lighter background shade of the rest of the sole, but in *Limax* the waves show as lighter areas moving over a darker background (Parker, 1911). The precise shade may well depend on the lighting conditions at the time of observation.

Many earlier workers on pulmonate locomotion considered that the waves on the sole were convexities of the sole, but Parker (1911) showed conclusively that the waves were concavities on the sole by observing the behaviour of air bubbles trapped in the mucus beneath the sole. Olmsted (1917) confirmed this by fixing the open end of a mano-meter tube to a small hole in a glass plate. The animals were allowed to crawl over the plate and when a wave passed over the hole the fluid in the manometer was drawn towards the plate, clearly indicating that each wave was a concavity. Subsequently both van Rijnberk (1919) and Ten Cate (1922), also using manometric techniques, thought that the waves were convexities, but finally Bonse (1935) and Lissmann (1945) using light levers to detect both vertical and forward displacements (Fig. 3) showed that the waves were indeed concavities.

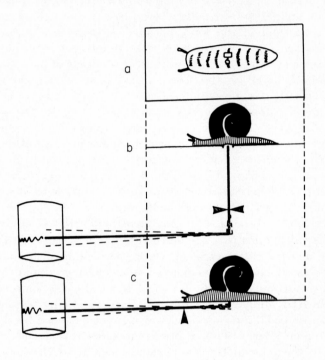

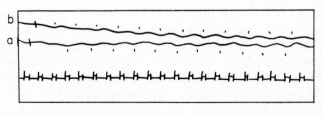

$t = 1\ s \longrightarrow$

Fig. 3. Diagram of the lever apparatus used to record simultaneously the horizontal (b) and vertical (c) displacement of the foot during locomotion; (a) shows the sites from which recordings were made. One such recording is shown below, (a) horizontal displacement; (b) vertical displacement. The marks of coincidence mark the passage of the dark waves. (From Lissmann, 1945.)

3. Mechanics

a. Mechanics

Lissmann (1945, 1946) has extensively investigated the locomotion of *Helix aspersa* and *H. pomatia* and his work remains the only detailed account of the mechanics of locomotion.

The analysis of cine film (Fig. 4) enabled Lissmann to show that any point on the sole remains more or less stationary until a pedal wave

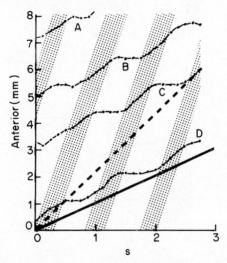

Fig. 4. Graph obtained from cine films of *Helix pomatia* of the movement forwards of four marked points (A, B, C and D) during locomotion. The shaded diagonal lines represent the passage of the darker pedal waves over the sole. Any point moves forward (upwards on the graph) only whilst a dark wave passes over it. The solid line represents the average speed of the animal and the pecked line the average speed of a point in movement. (From Lissmann, 1945.)

approaches. It then begins to move forwards with increasing speed, but always less fast than the wave itself. The wave continues forwards and the point decelerates and finally comes to rest as the wave passes beyond the point. There may be some degree of forward or backward slip evident in the period between waves. Figure 4 illustrates in graphic form the movement of four points on the sole during a short period of time in which some three waves passed.

The lever apparatus already referred to (Fig. 3) enabled Lissmann to show that each wave was a region where the sole was lifted and moved forwards simultaneously.

The mechanics of locomotion are summarized in Fig. 5. The forces involved in locomotion were detected by allowing the animal to crawl across a counter-weighted balance. As the head moves on to the balance it pushes the movable part forwards, then as the foot moves on to it the balance is pushed backwards due to the "positive static reaction" or thrust of this part of the foot. The tail of the foot is passively dragged

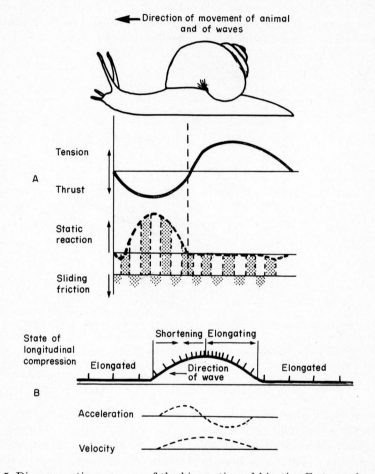

Fig. 5. Diagrammatic summary of the kinematic and kinetic effects as observed on the sole of a snail. A, kinetic effects on the foot as a whole; B, kinematics of an individual locomotory wave. (Modified from Lissmann, 1946.)

off the balance thus exerting, like the head, a "negative static reaction". Figure 6A shows an apparatus similar in principle to Lissmann's except that the movable part of the bridge was attached to a force transducer arranged to measure forces in the anteroposterior direction.

In addition a small hole is present in the centre of the bridge which may be connected to a pressure transducer (cf. the manometric techniques mentioned earlier).

Figure 6B was obtained using *Helix aspersa* crawling across a bridge 1 cm long. The effect of individual pedal waves is minimized due to the

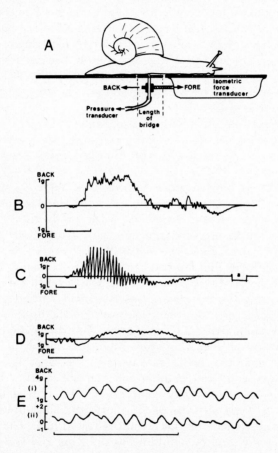

Fig. 6. A, diagram of a "bridge" apparatus for simultaneously recording antero-posterior forces exerted by the sole of a crawling snail or slug and the sub-pedal pressure; B, force recording made by allowing *Helix aspersa* to traverse a bridge 1 cm long; C, as B except that the bridge was 2mm long. The deflection (a) was caused by cutting the mucus remaining on the bridge after the snail had passed. D, force recording of *Arion ater* crawling over a bridge 2mm long; E, recording of (i) force and (ii) sub-pedal pressure made by allowing *Helix aspersa* to crawl over a bridge 5mm long with a hole in the centre. Calibration of the pressure recording is in g/cm². The time scale on all recordings equals 10 s, and each recording reads from left (head) to right (tail).

relatively long bridge. On a shorter bridge, 2 mm long, the effect of
each wave becomes apparent (Fig. 6C). This was also seen using *H.
aspersa*. Both records show that the backward thrust on the substratum
is exerted by the anterior portion of the foot, the posterior half being
dragged off passively. All recordings with snails are consistent in
showing this asymmetrical distribution of forces. The reason lies in
their gross anatomy for lying halfway along the body is the large mass
of the shell and the visceral mass. The front half of the foot must pull
this mass along and thus must exert a larger backthrust on the substratum. The posterior half of the foot merely has itself to propel and
need exert a relatively small backthrust. Indeed Fig. 6C shows that,
in this experiment, the back half of the foot was dragged passively off
the bridge by the front half.

If this explanation were true we might expect slugs, which lack the
heavy shell and demarcated visceral mass, to show a more symmetrical
distribution of forces, and Fig. 6D, obtained from *Arion ater* shows
that this is so.

Figure 6E is a simultaneous recording of force (i) and pressure
beneath the sole (ii), the latter being recorded through the hole in the
bridge. A rise in pressure coincides with production of backward
thrust, but a fall in pressure coincides with a reduction of backward
thrust. As each wave passes the foot is lifted – causing the reduction in
pressure beneath the foot, and moved forwards – causing the reduction
in backward thrust. Between waves the sole is pushed on to the substratum – causing the increase in pressure, and is pulling the body of
the snail forwards – causing the backthrust on the bridge.

Figure 5B summarizes the kinematics of an individual locomotory
wave and it can be seen that each wave is a region of longitudinal
compression. Thus the sole is stationary on the substratum when it is
relatively elongated, and is lifted off the substratum and moving
forwards when it is relatively compressed. This is true for all direct
waves and this type of wave moving forwards must always result in
the forward movement of the animal (Gray, 1968). The fact that a
direct wave is a region of simultaneous uplifting and compression has
frequently been misapprehended though it was clearly illustrated by
Robert (1907) and proved by Lissmann (1945, 1946).

b. The mechanism of a direct wave and the role of the pedal musculature
Simroth (1878, 1879) was able to account for the locomotion of
Helix and *Limax* only by postulating that the longitudinal muscle
fibres in the sole were able to expand actively. This was contested by

Carlson (1905) who proposed that the galloping locomotion of *Helix dupetithouarsi* was produced by the antagonistic action of the transverse and longitudinal muscle fibres of the foot and body wall, exerted on the hydrostatic skeleton of the head-foot. This is probably true for galloping but not for the normal pedal sole locomotion (see section *B.1*). A mechanism proposed by Runham and Hunter (1970) for terrestrial slugs is similar to that proposed by Jones and Trueman (1970) for *Patella*; however, *Patella* has retrograde waves and the mechanism cannot be the same as in terrestrial slugs, where they are direct, for the configuration of the two kinds of wave is different (Trueman, 1975). Trappmann (1916) suggested that the transverse and longitudinal muscles in the foot of *Helix* act antagonistically to produce the waves, but he was under the impression that each wave was a convexity of the sole and since this is not so his mechanism cannot apply. There have been few other attempts to correlate the form of the pedal waves on the sole with the musculature of the foot. There has been a tendency to assume that the pedal waves are produced by a thin layer of longitudinal muscle immediately dorsal to the epithelium of the sole (Morton, 1964). No such layer is visible in the foot of *Helix* (Trappmann, 1916; Elves, 1961) or *Agriolimax* (Jones, 1973).

Jones (1973) investigated the locomotion of *Agriolimax reticulatus* by dropping crawling specimens into liquid nitrogen. This freezes the animal very rapidly and the pedal waves are arrested in mid-stride. A section of a single pedal wave is shown in Fig. 7.

The musculature of the foot has a very distinct arrangement. Immediately ventral to the supra-pedal mucous gland there lies a layer of longitudinal muscle fibres. Ventral to this lies a relatively unrestricted haemocoelic space bounded ventrally by the epithelium of the sole. This space is crossed by two sets of oblique longitudinal muscle fibres and some transverse muscle fibres. The oblique longitudinal muscle fibres are continuous at their dorsal end with the fibres of the longitudinal muscle and at their ventral end attach on to the epithelium of the sole. The oblique fibres are arranged so that one set applies an anteriorly directed upward force to the epithelium and may be called anterior oblique fibres (Fig. 7, AOF). The other set applies a posteriorly directed upwards force to the epithelium and may be called posterior oblique fibres (POF). The forces exerted at the dorsal end of each set are in the opposite direction to those exerted at the epithelial ends. These oblique fibres appear to be the only elements capable of accounting for locomotion and we must look in detail at their disposition to account for the production of the pedal waves and the movement forwards of the body of the slug.

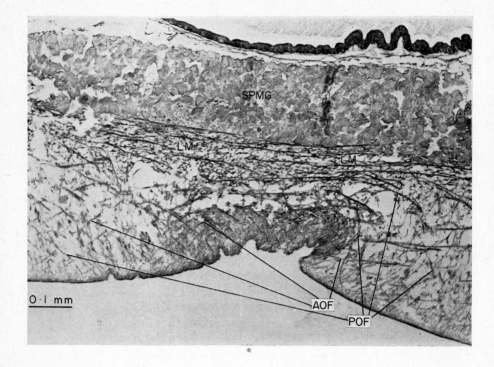

Fig. 7. Part of a longitudinal section through *Agriolimax reticulatus* frozen in liquid nitrogen whilst crawling. AOF, anterior oblique muscles; LM, longitudinal muscles; POF, posterior oblique muscles; SPMG, supra-pedal mucous gland. Anterior is to the right. See text for further information.

Sections show that any anterior oblique fibre is longer when its point of attachment to the epithelium lies immediately in front of a wave than when it lies just behind a wave. The opposite is true of the posterior oblique fibres. This is seen in the section shown in Fig. 7 and is diagrammatically illustrated in Fig. 8. This indicates that the anterior oblique fibres are contracting during the passage of each wave but that the posterior oblique fibres are contracting between the waves. The fibres of each set attach on to the epithelium at varying angles depending on each fibre's position relative to the waves. The shorter each fibre is the nearer its angle of attachment is to a right angle. This arrangement of fibres has certain similarities to the "pinnate" arrangement of the fibres in some arthropod and vertebrate muscles.

The anterior oblique fibres contract lifting the epithelium upwards and forwards relative to both the body of the slug and to the substratum and thus produce each wave on the sole. As the region of contraction of the fibres continues forwards each fibre must then relax and the epithelim is re-applied to the substratum. Once the epithelium

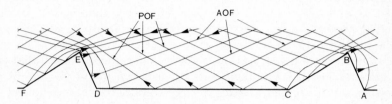

Fig. 8. Diagram to illustrate the method of locomotion of *Agriolimax*. Anterior is to the right. Only the oblique muscle fibres of the sole are illustrated. Muscle fibres contracting bear large arrow heads. The posterior oblique fibres (POF) contract between waves (C – D) pulling the body of the slug forwards and exerting a backward thrust on the substratum. The anterior oblique fibres (AOF) contract to pull the epithelium of the sole forwards and to form each pedal-wave (A – B – C and D – E – F). See the text for further information. (From Jones, 1973.)

is back on the substratum it becomes stationary, probably adhering to the mucous trail produced by the supra-pedal (= anterior pedal) mucous gland. The posterior oblique fibres then contract and since the epithelial end of each fibre is stationary on the substratum the result is that the layer of longitudinal muscle and the body of the slug above move forward. Thus the slug moves forwards relative to the substratum and at the same time the anterior oblique fibres are restored to their original length. Another pedal wave then follows and the sole is again moved forwards by the contraction of the anterior oblique fibres, this in turn causing the posterior oblique fibres to be restored to their original length. Thus the two sets of oblique muscles are mutually antagonistic (Jones, 1973).

During the contraction of the posterior oblique fibres (between successive waves) the body of the slug is being propelled forward and the sole must exert a backward thrust on the substratum. It is this backward thrust which was detected by Lissmann (1945, 1946) using the apparatus shown in Fig. 3 and is evident in the recordings shown in Fig. 6.

The means whereby the sole is replaced on the substratum after each wave has passed is probably a pressure difference between the haemocoel and the space under the sole. There must be a positive pressure in the haemocoel of the ventral part of the foot and, though there is little direct evidence, the pressure in land slugs must be relatively high for a mollusc: Dale (1973) calculated that the pressure in *Helix* is about 8–16 g/cm². Further the waves themselves may well cause an increase in pressure in the haemocoel of the foot due to the increased muscular activity. In addition to this probable high pressure in the foot there is a low pressure beneath the sole as the epithelium is lifted as each wave passes (Fig. 6E). Thus the combination of high internal pressure and low external pressure probably causes the sole to be pushed back on to the substratum after each wave has passed.

Examination of frozen sections confirms the earlier assertion that the sole is stretched from its resting length between waves, for the fibres of either set of oblique muscles diverge towards their attachment on the sole between waves. In slugs which are not crawling the fibres of either set are parallel (Simroth, 1879; Schmidt, 1965; Jones, 1973).

The arrangement of the pedal musculature of *Limax cinereoniger* Wolf (Simroth, 1879) and of *Arionater rufus* (L.) (Schmidt, 1965) is similar to that of *Agriolimax* and it is likely that the same mechanism may apply. In snails, however, the pedal musculature is apparently more complex. There are muscle fibres in *Helix* orientated like the oblique fibres in slugs (Simroth, 1878; Trappmann, 1916), but the haemocoel above the sole appears to be much more restricted than in *Agriolimax*. This may be due to the method of fixation and relaxation, for sections of *Agriolimax* not properly relaxed prior to fixation show the haemocoel occluded. Because of this it is uncertain whether the mechanism described above can apply to snails.

c. The role of mucus

The secretions of the supra-pedal mucous gland and the glands of the sole are of considerable importance during locomotion, indeed locomotion cannot be accomplished without mucus (Barr, 1926; Bonse, 1935). As a snail or slug progresses a trail of mucus is produced by the supra-pedal gland, is flattened as the animal crawls over it and is then left behind as the familiar trail produced by any creeping pulmonate. The mucus produced by the supra-pedal gland is of relatively high viscosity: the mucous produced by the unicellular glands of the sole is, however, of low viscosity. (The references to "high" and "low" viscosity are qualitative only; there are apparently no actual measurements.) Parker (1911) has suggested that the low viscosity mucus produced by the sole fills the cavities under the foot caused by the pedal waves. It may be that as the anterior oblique muscle fibres contract the contents of mucous glands on the sole are squeezed out and that, as the sole is replaced this mucus may be partially resorbed into the glands (Jones, 1973), though this must be against the internal pressure responsible for replacing the sole on the substratum. Otherwise this more fluid mucus would tend to be carried forwards by the pedal waves, somewhat in the manner of a peristaltic pump, and would accumulate at the anterior end of the foot. Almost certainly some of this mucus arrives at the front of the foot where it may serve to dilute the thicker mucus emerging from the supra-pedal gland. The cilia on the sole beat posteriorly and apparently assist in maintaining the even distribution of the mucus (Barr, 1926).

The means whereby the supra-pedal mucus is secreted as rapidly as is necessary during locomotion is not clear. The duct of the gland is ciliated, which may assist in propelling mucus towards the aperture of the gland, but it is difficult to believe that this is the only method of mucus expulsion. The high viscosity of the mucus itself may assist, for once the trail is established then there will be a tendency for the mucus to be drawn out of the gland by its own tenacity (Barr, 1926). The supra-pedal mucous gland in some species, for example *Helix* and *Limax*, is firmly embedded in the musculature of the foot and contractions of the pedal musculature may assist in the ejection of the mucus from the gland, but in other species, for example *Milax* and *Agriolimax* the gland is separate from the musculature and muscular action cannot significantly contribute to the expulsion of the mucus (Barr, 1926). Machin (1964) has shown that the rate of cutaneous mucus production is proportional to the hydrostatic pressure of the blood, and, as has been suggested above, the hydrostatic pressure in the foot may increase during locomotion. This increase may therefore cause an increase in the rate of production of mucus by the supra-pedal gland (Runham and Hunter, 1970).

4. Factors affecting the rate of locomotion

Crozier and Pilz (1924) and Crozier and Federighi (1925a, b) investigated the relationships of step-length (the distance that any point on the foot moves as one wave passes over it), wave frequency and wave speed to the locomotion of *Limax maximus*. The frequency of the waves passing along the foot affects the speed of locomotion for, other things being equal, the larger the number of waves that pass along the foot in a given period of time the faster the animal will be able to go. The relationship is shown in Fig. 9a and is linear in any one animal though varying from animal to animal. The frequency of waves is obviously dependent on the speed that each wave travels over the sole and thus the speeds of the animal and of the waves are directly related (Fig. 9b).

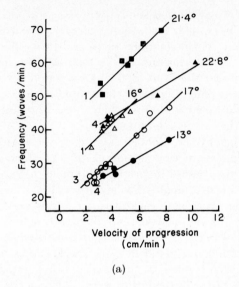

(a)

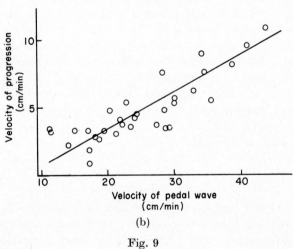

(b)

Fig. 9

The step-length will also affect the speed of locomotion for if the foot moves a greater distance at each step then obviously the animal must move faster. The relationship between step-length and speed of locomotion is linear (Fig. 9c).

In poikilothermic animals such as pulmonates the environmental temperature will have a marked effect on the speed of locomotion (Fig. 9d) and it must be stressed that figures for the speed of loco- motion are of little value without details of this and other environmental

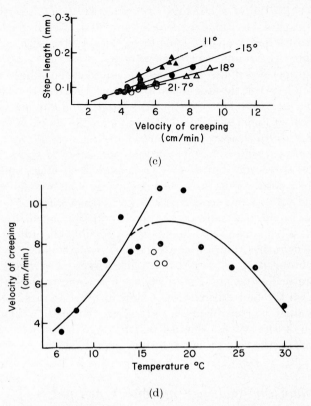

(c)

(d)

Fig. 9. *Limax maximus*. Graphs to show the relationship between the following pairs of variables: a, wave frequency and the velocity of progression for different individuals (numbered 1, 3, 4) and at different temperatures (°C). The slope of the line for animal no. 4 is unusual but consistent at different temperatures. b, velocity of the waves and the velocity of progression; c, step-length and the velocity of creeping at four different temperatures (°C); d, velocity of progression and temperature. (a, b and c from Crozier and Pilz, 1924; d from Crozier and Federighi, 1925a.)

factors, especially humidity. Humidity may well affect the viscosity of the mucus and thus the speed of locomotion, but little work seems to have been carried out to test this.

Temperature has different effects on the frequency of the waves and on the step-length. Frequency is increased with temperature (Fig. 9a) but step-length is decreased (Fig. 9c). An increase in temperature also decreases the viscosity of the mucus and this may explain why the step-length decreases with temperature and why, in Fig. 9d, the rate of locomotion decreases above 15°C.

Jones (1973) noted that slugs of temperate zones are apparently able to adjust the viscosity of the pedal mucus according to the season, and it may be that tropical snails have thicker mucus to compensate for the higher environmental temperatures.

The addition of weights to a creeping slug does not affect the linear relationship between the frequency of the pedal waves and the speed of locomotion, nor that between the step-length and the speed of loco-motion (Crozier and Federighi, 1925b) though the speed of locomotion is altered by adjustment to the speed of the waves. On horizontal surfaces heavy weights may be carried but only for short distances. Taylor (1914) records that *Helix aspersa* can drag fifty times its own weight horizontally but only nine times its own weight vertically. Under light loads slugs usually creep up vertical surfaces faster than they normally do, but under heavy loads adhesion is usually lost. The limiting factor in these experiments was probably the adhesive power of the mucus rather than the muscular power of the foot.

In conclusion it may be said that each wave in any one individual has a more or less constant step-length, a constant number of waves is present on the foot during fast or slow locomotion and the speed of any individual is controlled by adjusting the speed, therefore the frequency, of the waves. Species which move faster generally have more waves on the sole at any one time than do slower species.

B. Locomotion by other muscular methods

1. "Galloping" or "Loping"

Carlson (1905) described a form of muscular locomotion in *Helix dupetithouarsi*, which he called "loping", whereby the animal changes from normal pedal sole locomotion to a more rapid type of locomotion involving the whole of the head-foot (Fig. 10a). To avoid confusion with the "looping" of other species described below, we shall refer to this type of locomotion as "galloping" as it was called by Parker (1911). The snail changes to this form of locomotion only when "endeavouring to escape from an enemy" (Carlson).

When the animal changes to this form of locomotion the head and anterior part of the foot are lifted clear of the ground and pushed forwards, partly by the contraction of the circular muscles in the head-foot and partly as a result of normal locomotion which continues unabated. The head is then flexed downwards, presumably by the contraction of some ventral longitudinal muscle fibres, and re-attaches to the substratum. The anterior third of the head-foot thus forms a large arch 2–4 mm high in the centre (Carlson, 1905). The head then repeats the process and when the snail is in full gallop the foot has $2\frac{1}{2}$–3 such waves. These waves pass backwards relative to the body (but are stationary relative to the substratum) and as the animal is moving forwards the waves are retrograde. As the animal progresses it leaves a dotted slime track instead of a continuous one.

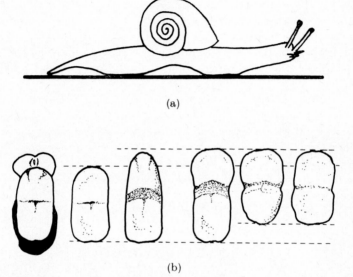

(a)

(b)

Fig. 10. a, galloping of *Helix dupetithouarsi*. (From Carlson, 1905.) b, *Otina otis*. On the left a ventral view of the entire animal (black = shell). To the right, five successive stages of the sole during locomotion. (From Vlès, 1913.)

The mechanism of this form of locomotion is more obvious than that of the normal direct waves. The copious circular and longitudinal muscle fibres of the head-foot are probably mutually antagonistic and the haemocoel acts as the hydrostatic skeleton (Carlson, 1905).

This form of locomotion may be widespread amongst the terrestrial pulmonates for such dotted tracks as are left by galloping snails are not uncommon sights and both Taylor (1914) and Parker (1935) describe galloping in *Helix aspersa*. Robert (1908) describes young *Cepaea hortensis* (Müll.) as exhibiting a similar mode of progression. I have seen such dotted slime tracks in situations where their only apparent source could have been one or more of the commoner species of terrestrial slug though there is no apparent reference to slugs exhibiting such a mechanism.

2. "Looping"

Looping has been described only in *Otina otis* (Vlès, 1913; Morton, 1955a) and the Ellobiidae (Morton, 1955b) amongst the pulmonates though it is found in other gastropods (Morton, 1964).

Otina is a minute pulmonate found in crevices and empty barnacle shells on temperate rocky coasts. The foot is oval in outline and is divided roughly into two by a prominent transverse groove (Fig. 10b). The animal progresses by advancing the anterior portion of the foot and re-attaching it to the substratum and then bringing the rear portion up behind. The wave of muscular contraction can be considered to be passing in the opposite direction to the locomotion and is therefore retrograde.

The mechanism has not been described, though it almost certainly depends on the action of the apparently copious dorsoventral muscles of the foot (Morton, 1955a) on the haemocoel as in the mechanism described by Jones and Trueman (1970) for another species with retrograde waves.

The foot of the Ellobiidae is also divided and locomotion is similar (Morton, 1955b).

IV. Ciliary locomotion

A. Occurrence within the pulmonates

Comparatively few observations on the ciliary locomotion of gastropods have been made despite the relative ease with which the sole of the foot may be observed, for if an animal is held upside-down ciliary movements are readily seen (Kaiser, 1960). However, almost all Basommatophora seem to employ cilia for locomotion, very few (e.g. the Ellobiidae and *Otina*) utilizing muscular waves. Some of the smaller stylommatophorans such as *Discus rotundatus* (Müll.) and *Zonitoides nitidus* (Müll.) (Elves, 1961) also appear to use ciliary locomotion. Larger stylommatophorans use muscular waves though the juveniles probably utilize cilia to some extent.

Most basommatophorans are aquatic, living in fresh water, where the weight of the animal is considerably less than in air. Furthermore the mantle cavity of such animals almost always contains air which further reduces the weight, so that most freshwater pulmonates are at least neutrally buoyant. At certain times, especially just after taking in more air at the surface of the water, they may even be positively buoyant (Henderson, 1963). Thus the load that has to be carried by the foot is minimized and this would seem to be the reason why fairly large animals such as *Lymnaea palustris* (Müll.) (Walter, 1906), *L. stagnalis* (L.) (Kaiser, 1960), *L. peregra* (Müll.) and *Physa fontinalis* (L.) (Elves, 1961) are able successfully to utilize the comparatively weak forces produced by cilia for locomotion.

Clark (1964) has observed that the incidence of ciliary locomotion in gastropods bears no relation to the size of the animal. However it may be that, in pulmonates size is not so important as the effective weight of the animal in the medium in which it lives.

The use of cilia for locomotion by terrestrial snails is probably restricted to the smaller species though there are some anomalies apparent in its occurrence. Elves (1961) examined the locomotion of two terrestrial species, *Discus rotundatus* (<7 mm shell diam.) and *Zonitoides nitidus* (<8 mm shell diam.), and concluded that they both use a ciliary mechanism. Both species are large and heavy as compared to other kinds of animal using ciliary locomotion such as the smaller turbellarians. As these latter increase in size ciliary locomotion is at some (indeterminate) stage supplemented or replaced by the more powerful forces of muscular locomotion (Clark, 1964). It has been suggested above that the size of freshwater snails may not be so important as their effective weight, but these terrestrial species lack the support that is provided by an aquatic habitat; both *Discus* and *Zonitoides* seem rather heavy to utilize a ciliary mechanism successfully. *Oxychilus alliarius*, in the same family (Zonitidae) and of a similar size to *Zonitoides*, utilizes direct monotaxic pedal waves (personal observation); it may be that Elves only examined juveniles in which there may be a greater reliance on cilia. There is clearly a need for a thorough study of the relationship of ciliary locomotion to the size of the animal not only in gastropods but in such animals as Turbellaria as well.

B. Mechanics of ciliary locomotion

The pedal cilia of gastropods have received little attention though other molluscan cilia, especially the ctenidial ones of the Lamellibranchia, have been studied in some detail (e.g. Baba and Hiramoto, 1970).

For a given frequency of beat a longer cilium will have a faster tip-speed. All other things being equal a snail with longer cilia should move faster; a snail with shorter pedal cilia should move more slowly but be able to carry a greater load; freshwater snails with less weight to carry than terrestrial ones may be expected to have longer cilia than terrestrial ones. The results of Elves (1961), bearing in mind the reservations concerning the terrestrial species expressed above, would seem to confirm this. *Discus*, terrestrial, has pedal cilia 4 μm long whereas the aquatic species *Lymnaea peregra* and *Physa fontinalis* have cilia 10 μm and 7 μm long respectively. Kaiser (1960) gives, from several sources, the length of cilia of several aquatic species as from 7–9 μm.

Pelseneer (1935) gives the speed of *L. peregra* as 17·5 cm/min and that of *Physa fontinalis* as 6 cm/min. This is a larger difference than would be expected from Elves's figure for the length of the cilia. This is another region that requires further study.

The speed of locomotion is also dependent on the rate of beating of the cilia but again few details are available. However Kaiser (1960) noted that the cilia of *L. stagnalis* beat at a rate of 50/s. The rate of beating of cilia depends on several physiological and anatomical factors as well as environmental factors such as temperature and viscosity (Sleigh, 1962). Viscosity of the mucus counts as an environmental factor for it is the medium in which the cilia beat. As with locomotion by muscular waves, figures for the speed of ciliary beat and locomotion are of little value without details of the conditions under which they were taken. Though there are no observations on whether the speed of beating of pulmonate pedal cilia may be controlled, the pedal cilia of *Nassarius obsoletus* (Say), a prosobranch that uses ciliary locomotion, do slow their rate of beat as the animal slows (Copeland, 1919). Perhaps the relative pattern of beat may alter in a manner similar to myriapod limbs where the gait and periodicity of the limbs are altered to produce slower or faster movement (See the contribution by Manton in Gray, 1968.)

The relatively weak locomotory forces produced by cilia prohibit most aquatic pulmonates from inhabiting localities subject to fast water currents. A notable exception to this is *Ancylus fluviatilis* Müll., a freshwater limpet. Macan (1963) notes that it occurs on the top of stones in "torrential" streams and it must be able to move against such currents. The shape of the shell must reduce the possibility of the animal being dislodged, but it is difficult to believe that the cilia of the sole are the only method of propulsion.

C. *The control of ciliary locomotion*

The pedal cilia are almost certainly under nervous control, though the pulmonates do not seem to have been investigated from this aspect.

Several prosobranchs have been studied (Copeland, 1919, 1922) and these studies illustrate several general points which possibly apply to pulmonates.

Copeland (1919, 1922) investigated the locomotion of *Nassarius obsoletus*, *N. trivittatus* (Say) and *Polinices duplicatus* (Say). The cilia of *Nassarius* are at rest when the animal is stationary but resume beating immediately the animal is stimulated to move with a piece of dead fish (*Nassarius* is a scavenger). The cilia are unaffected by magnesium sulphate anaesthesia and continue beating when isolated from nervous control. This indicates that the control is inhibitory rather than excitatory. We have already seen that the rate of beating may be controlled in *Nassarius* and the cilia on different parts of the foot are capable of beating independently (Copeland, 1919). This may be a means of steering, for if the cilia on one half of the foot are inhibited then the result must be a turning of the animal.

The pedal cilia, being the only apparent means of propulsion in some species of pulmonates, must at least in these species be under some nervous control, possibly accomplished in a manner similar to that in the prosobranchs Copeland studied.

1. Directional control

No pulmonate using ciliary locomotion has been recorded as being able to reverse, and we can tentatively assume that the cilia are not capable of reversal. Steering in these snails is probably accomplished by the muscular deformation of the foot rather than a change in the direction or rate of ciliary beat, though the latter has not been positively eliminated. We have seen that cilia on one half of the foot may be inhibited and that this may cause a change in direction. Most common freshwater pulmonates seem to steer themselves by swinging the anterior part of the foot laterally, the posterior following passively.

V. Posture, body movements and the hydrostatic skeleton

The main function of a skeleton is, according to Chapman (1958), to provide a means whereby opposing muscles may be antagonized or restored to their relaxed state. Molluscs are essentially soft-bodied. Such skeleton as they have is not used to provide attachment for antagonistic muscles, except in the case of the radular cartilages. The shell in gastropods performs such other skeletal functions as protection and support.

In the absence of rigid skeletal elements gastropods have a fundamental and almost total dependence on the blood acting as a hydrostatic skeleton. Despite this dependence it is surprising how little work has been carried out on the functioning of the musculo-skeletal system, radular movements excepted. Of the work that has been done much of it is of a rather negative nature as the investigator failed to find any other explanation and assumed that the hydrostatic skeleton was involved with little direct evidence of its mode of functioning. Protrusive movements of the odontophore are such an example.

A. Posture

When a snail is crawling on a horizontal surface the shell and its contained viscera must be supported on top of the head-foot. The tentacles also must be maintained in an extended position. The only skeletal support for the tentacles and for the shell and viscera is the blood and it is the blood pressure that enables these to remain erect. Erection of the shell and viscera is initially a result of strong contraction of the columellar muscle but thereafter it relies solely on blood pressure. The anterior lip of the shell and the posterior whorls are, in a crawling snail, resting on the dorsal surface of the head-foot but this imparts no lateral stability to the shell, only longitudinal stability. Thus the tendency for the shell and its contents to fall sideways is resisted solely by the turgidity of the "stalk" of the animal, which relies in turn on the blood pressure.

Aquatic snails live in a medium which is much denser than air and thus receive considerably more support from their environment than do terrestrial snails and we might expect aquatic snails to have lower blood pressures than terrestrial ones. Dale and Jones (in preparation) have measured the blood pressures in some representative freshwater and terrestrial species (Table I). By inserting a cannula into the head between the tentacles, the cephalic pressure applied to the base of the tentacles was measured. While there is some variation amongst the six species the *Helix* spp. have cephalic pressures that are highly significantly different from those of the aquatic species.

The shape of the tentacles of the various snails is also indicative of the higher pressure in terrestrial snails. In these the tentacles are circular in cross-section, whereas the tentacles of freshwater species, especially the Lymnaeidae, are more or less flattened triangular projections of the head. The circular cross-section is a more efficient way of containing high pressures, the freshwater species with a lower blood pressure having less need for tentacles with a circular cross-section

Useful confirmatory evidence was obtained by Dale (1973) who removed the optic tentacles from freshly killed *Helix pomatia* and attached their base to a manometer tube via a hypodermic needle. By varying the height of the tube different pressures could be applied to the base of the tentacle. Experiments carried out with the tentacle in air showed that in order to remain erect it requires a pressure of 16 g/cm², a value which is exceeded by the measured cephalic pressure (Table I). When the experiment is carried out under water a tentacle more or less erects itself with zero (ambient) pressure at its base (Dale, personal communication) clearly showing that aquatic snails have no need for high pressures.

The shell and visceral mass are similarly supported by the pressure in the cephalic sinus for this extends back into the stalk of the animal. It is more difficult experimentally to apply varying pressure to the base of the stalk but from its diameter and the weight of the shell and its contained viscera it can be calculated that the shell and visceral mass exert a downwards force of 8 g/cm² (Dale, 1973). In *Helix* this is comfortably exceeded by the cephalic pressure, though among the aquatic species only *Lymnaea peregra* significantly exceeds this value. *L. peregra* has a strong tendency to make excursions out of the water (hence its common name – wandering snail), more so than the other aquatic species in Table I, and the high blood pressure is a prerequisite that permits this habit.

Viviparus is a prosobranch and is not air-breathing whereas *Lymnaea* spp. and *Planorbarius corneus* (L.) are and their mantle cavity contains

Table I. Blood pressure and ventricle weight of some representative freshwater and terrestrial gastropods. Pressures in g/cm². Adapted from Dale and Jones (in preparation).

Species	Cephalic pressure mean $\pm$ S.D. (n)	Ventricular systolic pressure mean $\pm$ S.D. (n)	Dry ventricle wt as % dry tissue wt of animal (n)
Viviparus viviparus	$8\cdot21 \pm 1\cdot82(14)$	—	—
Lymnaea stagnalis	$6\cdot95 \pm 1\cdot16(10)$	$9\cdot75 \pm 1\cdot76(10)$	$0\cdot10\%(7)$
Lymnaea peregra	$11\cdot65 \pm 2\cdot49(10)$	—	—
Planorbarius corneus	$6\cdot62 \pm 0\cdot54(8)$	—	—
Helix aspersa	$17\cdot15 \pm 2\cdot09(10)$	$25\cdot3 \pm 0\cdot38(10)$	$0\cdot22\%(7)$
Helix pomatia	$17\cdot60 \pm 1\cdot12(5)$	$23\cdot3 \pm 1\cdot08(5)$	—

air. This air further reduces the effective weight of the shell and
visceral mass and lower blood pressures might be expected. From Table
I this seems to be so but with the notable exception of *L. peregra* which
as has been shown requires higher blood pressures for another purpose.

Thus far no mention has been made of the source of the pressure of the
blood. Prosser and Brown (1961) and Chapman (1967) consider that the
somatic musculature has a greater effect on the movements of the blood,
thus the pressure, in an active animal than does the heart, which is
necessary only to circulate blood in a resting animal. If this were so then
the pressure in the haemocoel need only be at a fairly constant standing
level and the heart need only be a relatively weak structure to produce
the low circulatory pressures over the standing pressure. In resting,
terrestrial snails the heart produces pressure gradients of about
19 g/cm² , rising to about 30 g/cm² in active ones (Jones, 1971; Dale,
1974); the pressure of the blood returning to the heart remains fairly
constant at 5 – 6 g/cm². This residual pressure, possibly partly produced
by the somatic musculature and partly the result of the pressure drop
round the circulation, is thus relatively low and any higher pressures
must be produced by the heart. In order to produce the higher pressure
the heart of terrestrial snails must be hypertrophied, more muscular
relative to the freshwater species. Table 1 shows figures for the ventricular weight of *Helix aspersa* and *Lymnaea stagnalis* compared to their
body weight and that of *H. aspersa* is a much higher proportion. From
all this evidence it may be concluded that the heart is producing the relatively high pressures in terrestrial animals and that a hypertrophied
heart may be considered an essential prerequisite to a terrestrial mode
of life among pulmonates.

Slugs lack the heavy shell and demarcated viscera of the snails, but
the tentacles have still to be erected and thus there still needs to be a
high internal pressure. Qualitative comparisons of the ventricle of slugs
with those of terrestrial snails suggest that the slug heart is just as
muscular as the snail heart and thus slugs have a hypertropied heart for
the same reasons as snails.

There are other phenomena linked to the high pressure in terrestrial
pulmonates. If the high pressure blood were to flow through a relatively
open arterial system the pressure would rapidly be dissipated, but in
Helix (Schmidt, 1916) and slugs (Runham and Hunter, 1970) the arterial
system consists of an extensive branching system of tubes, making the
blood system far less open than in other gastropods. The development
of this system, it is suggested, is strongly linked with the need to deliver
high pressure blood to the periphery of the circulation. Blood can still
return to the heart through a more open venous system.

Machin (1964) has shown that cutaneous mucus production is directly linked to the subcutaneous pressure, the higher the pressure on the inside of the skin then the more mucus is produced on the outside. Thus high blood pressure tends to cause terrestrial snails and slugs to dehydrate rapidly, especially in a dry environment. The development of a hypertrophied ventricle has prevented all but a few species, e.g. *Theba pisana* (Müll.), from colonizing arid habitats. Even these few species are active only at night when dew falls or after rain when the local humidity is high and desiccation is kept to a minimum.

B. Protrusion and retraction of the head-foot

Protrusion from the shell, and retraction into the shell, is a basic and characteristic feature of all snails, though the mechanism of protrusion is poorly understood. Trappmann (1916) and Sommerville (1973) described protrusion and retraction in *Helix pomatia*. Brown (1964) investigated the retraction of the marine prosobranch *Bullia,* but was not able to answer many of the questions concerning protrusion. Dale (1974) has recently investigated both protrusion and retraction, though there are still some outstanding problems, particularly concerning protrusion.

Cephalopedal retraction of pulmonate snails is achieved by the columellar muscle and its derivatives, the various pedal and cephalic retractor muscles (Fig. 1), though other factors may be involved. The mechanism of protrusion is more obscure though it too must be due to muscular contraction.

The sequence of retraction of a fully expanded *Helix* may be readily observed and is as follows. By retraction of the tentacular retractor muscles the four tentacles are invaginated from their tips. These continue to contract and the pharyngeal retractor commences contraction and the front of the head is invaginated. The front of the foot is lifted off the substratum and folded down its midline so that opposing sides of the sole touch. Meanwhile the posterior part of the foot remains attached to the substratum. The cephalic retractors continue to contract and the anterior pedal retractors begin to shorten. This causes the whole of the anterior part of the body to be invaginated and drawn into the shell. Simultaneously air is expelled from the mantle cavity but not, as might be expected, through the pneumostome. It is expelled through a median dorsal slit at the front of the mantle (Sommerville, 1973). As it is expelled a yellow-creamy liquid may be secreted and a mass of bubbles forms, presumably repulsive to potential predators. Contraction of the retractor muscles continues and finally the posterior part of the foot is lifted off the substratum by contraction of the posterior retractor muscles, is similarly folded down its midline and is drawn into the shell. Upon complete contraction the aperture of the shell is closed by the soft tissue of the mantle margin. The complete sequence may be interrupted and reversed at various stages or may be continuous.

In *Helix* the retraction of the total volume of the head-foot is compensated for by the expulsion of air from the mantle cavity. However, there must also be a change in the relative volumes of different parts of the haemocoel to allow for the invagination of the anterior portion of the body. Some of the blood is moved back into the posterior extension of the cephalic sinus, the sub-renal sinus, and some moves into the visceral sinus. This was shown by examination of thick sections of active animals which had been injected with neutral red in saline and then dropped into liquid nitrogen at various stages of retraction (Dale, 1974).

In some aquatic snails, for example *Lymnaea stagnalis*, the body is rarely retracted into the shell and in any case is too large to fit into it. On severe stimulation they will retract, but in order to fit into the shell they must first shed some blood. Lever and Bekius (1965) have located the site of the blood outlet as a haemal pore in the mantle cavity near the renal pore. Animals can shed up to 25% of their body weight in this fashion (Bekius, 1972).

The sequence of events during protrusion is the reverse of that during retraction. The rear of the foot emerges first, followed in turn by the anterior part of the foot and the head. Unlike retraction, protrusion takes place in a number of steps or cycles and is not continuous. With the pneumostome open the foot is pushed slowly outwards: Dale (1974) calls this phase 1 of protrusion. Then the pneumostome closes and the foot ceases protruding and slightly retracts. This is phase 2. Phases 1 and 2 are alternated some six or seven times to effect full protrusion.

During phase 1 the diaphragm contracts downwards (Trappmann, 1916; Sommerville, 1973), forcing blood into the head and sucking air into the mantle cavity (Dale, 1974). Thus during extrusion the volume of the head-foot is compensated for by the inhalation of air into the mantle cavity. In *Bullia* the viscera are moved down the shell to compensate for the expansion of the head-foot, the space between the shell and the viscera being occupied by sea water (Brown, 1964). No such movement takes place in *Helix*.

During phase 2 the diaphragm is being raised to its resting position (Sommerville, 1973). When it does this, the pneumostome is closed and a pressure develops within the mantle cavity which forces air into the most posterior part of the cavity (Dale, 1974). The upward displacement of the diaphragm is suggested to be a consequence of the slight retraction of the head-foot that is evident during phase 2, which may cause a high pressure in the cephalic sinus and push the diaphragm upwards.

For the head to evaginate during extrusion there must again be a redistribution of blood within the haemocoel and blood is shifted from the sub-renal sinus and visceral sinus into the head. Brown (personal communication) considers that there are muscles in the viscera of *Bullia* which contract to force the blood out. The viscera of *Helix* are devoid of muscle and a similar mechanism cannot apply. Dale (1974) has shown that heart rate and ventricular systolic pressure increase during protrusion, but the venous return pressure does not. The implication is that the heart is increasing its output to pump up the head, presumably by a redistribution of the aortic blood by constriction of the visceral artery.

VI. References

Baba, S. A. and Hiramoto, Y. (1970). *J. exp. Biol.* **52,** 675–690.

Barr, R. A. (1926). *Q. Jl microsc. Sci.* **70,** 647–667.

Barr, R. A. (1928). *Q. Jl microsc. Sci.* **71,** 503–526.

Bekius, R. (1972). *Neth. J. Zool.* **22,** 1–58.

Bonse, H. (1935). *Zool. Jb.* (*Zool Physiol..*) **54,** 349–384.

Brown, A. C. (1964). *J. exp. Biol.* **41,** 837–854.

Campion, M. M. (1961). *Q. Jl microsc. Sci.* **102,** 195–216.

Carlson, A. J. (1905). *Biol. Bull. mar. biol. Lab., Woods Hole* **8,** 85–92.

Cate, J. ten (1922). *Archs néerl. Physiol.* **7,** 103–111.

Chapman, G. (1958). *Biol. Rev.* **33,** 338–371.

Chapman, G. (1967). "The Body Fluids and their Function". Arnold, London

Clark, R. B. (1964). "Dynamics in Metazoan Evolution". Clarendon Press, Oxford.

Copeland, M. (1919). *Biol. Bull. mar. biol. Lab., Woods Hole* **37,** 126–138.

Copeland, M. (1922). *Biol. Bull. mar. biol. Lab., Woods Hole* **42,** 132–142.

Crozier, W. J. and Federighi, H. (1925a). *J. gen. Physiol.* **7,** 151–169.

Crozier, W. J. and Federighi, H. (1925b). *J. gen. Physiol.* **7,** 415–419.

Crozier, W. J. and Pilz, G. F. (1924). *J. gen. Physiol.* **6,** 711–721.

Dale, B. (1973). *J. exp. Biol.* **59,** 477–490.

Dale, B. (1974). *J. Zool., Lond.* **173,** 427–439.

Elves, M. W. (1961). *Proc. malac. Soc. Lond.* **34,** 346, 355.

Gray, J. (1968). "Animal Locomotion". Weidenfeld and Nicolson, London.

Henderson, A. E. (1963). *Z. vergl. Physiol.* **46,** 467–490.

Jones, H. D. (1971). *Comp. Biochem. Physiol.* **39A,** 289–295.

Jones, H. D. (1973). *J. Zool., Lond.* **171,** 489–498.

Jones, H. D. and Trueman, E. R. (1970). *J. exp. Biol.* **52,** 201–216.

32 H. D. JONES

Jousseaume, F. (1909). *Bull. Soc. zool. Fr.* **34**, 108–115.

Kaiser, P. (1960). *Z. wiss. Zool.* **162**, 368–393.

Lever, J. and Bekius, R. (1965). *Experientia* **21**, 295–396.

Lissmann, H. W. (1945). *J. exp. Biol.* **21**, 58–69.

Lissmann, H. W. (1946). *J. exp. Biol.* **22**, 37–50.

Macan, T. T. (1963). "Freshwater Ecology". Longmans, London.

Machin, J. (1964). *J. exp. Biol.* **41**, 759–769.

Morton, J. E. (1955a). *J. mar. biol. Ass. U.K.* **45**, 113–159.

Morton, J. E. (1955b). *Proc. zool. Soc. Lond.* **125**, 127–168.

Morton, J. E. (1964). *In* "Physiology of Mollusca" (K. M. Wilbur and C. M. Yonge, Eds) **1**, 383–423. Academic Press. New York and London.

Olmsted, J. M. D. (1917). *J. exp. Biol.* **24**, 223–236.

Parker, G. H. (1911). *J. Morph.* **22**, 155–170.

Parker, G. H. (1935). *Biol. Bull. mar. biol. Lab., Woods Hole* **72**, 287–289.

Pelseneer, P. (1935). *Acad. r. Belg. Cl. Sci. Publ. Fond. Ag. Potter* **1**, 1–662.

Prosser, C. L. and Brown, F. A. (1961). "Comparative Animal Physiology", 2nd edition. Saunders, Philadelphia.

van Rijnberk, G. A. (1919). *Archs néerl. Physiol.* **3**, 539.

Robert, A. (1907). *Bull. Soc. zool. Fr.* **32**, 55–62.

Robert, A. (1908). *Bull. Soc. zool. Fr.* **33**, 151–157.

Runham, N. W. and Hunter, P. J. (1970). "Terrestrial Slugs". Hutchinson, London.

Schmidt, G. (1916). *Z. wiss. Zool.* **115**, 201–262.

Schmidt, R. (1965). *Morph. Jb.* **107**, 234–270.

Simroth, H. (1878). *Z. wiss. Zool.* **30** Suppl., 166–224.

Simroth, H. (1879). *Z. wiss. Zool.* **32**, 284–322.

Sleigh, M. A. (1962). "The Biology of Cilia and Flagella". Pergamon Press, Oxford.

Sommerville, B. A. (1973). *J. exp. Biol.* **59**, 275–282.

Taylor, J. W. (1914). "Monograph of the Land and Freshwater Mollusca of the British Isles". **3**, Taylor Brothers, Leeds.

Trappmann, W. (1916). *Z. wiss. Zool.* **115**, 489–585.

Trueman, E. R. (1968). *Nature, Lond.* **218**, 96–98.

Trueman, E. R. (1975). "The Locomotion of Soft-bodied Animals". Arnold, London.

Vlès, F. (1907). *C. r. hebd. Séanc. Acad. Sci., Paris* **145**, 276–278.

Vlès, F. (1913). *Bull. Soc. zool. Fr.* **38**, 242–250.

Walter, H. E. (1906). *Cold Spring Harbor Monographs* **6**, 1–35.

Chapter 2

Respiration

F. Ghiretti and A. Ghiretti-Magaldi

Institute of Animal Biology, University of Padua, Italy

I. Introduction

Respiration is a function common to all living organisms and consists essentially in the uptake of oxygen and the concomitant release of the products of oxidation from the tissues, mainly carbon dioxide and water. Physically, respiration can be considered a process of diffusion through a biological membrane separating the internal medium of the organism from the external environment (Fig. 1). The "respiratory organs", therefore, have not a morphological identity as, for instance, the digestive or secretory organs, nor is there a differentiated "respiratory cell" like the muscle fibre or the nerve cell. In fact any thin layer of epithelial cells separating two media in which a gas is present at different pressures, can function as a "respiratory" organ. Diffusion through the plasma membrane is the only physical factor in unicellular

33

organisms where the diffusion rate depends on the gas concentration on both sides of the membrane. In higher organisms, unless the metabolism is very low, diffusion alone is not sufficient and the efficiency of respiration depends on a continous flowing of the external (air or water) and internal media. These movements are regulated by ciliary or muscular activity and by circulatory currents respectively.

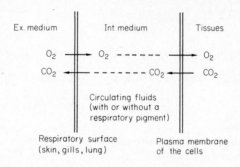

Fig. 1.

Respiratory exchange occurs either through the general body surface or is localized in restricted regions of the body where the rate of exchange per unit area is greater. These regions can be considered parts of the skin that are folded outside or inside the body (Fig. 2); they confer no other advantage to the animal than an increase of the

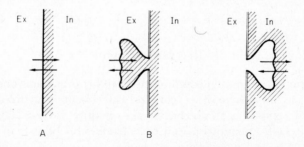

Fig. 2. Types of respiratory systems. A, cutaneous; B, branchial; C, pulmonary.

area for gas exchanges. These foldings are specialized for aquatic or aerial respiration and are called branchiae (or gills) (B) and lungs (C) respectively. Cutaneous respiration is obligatory when no part of the body surface is differentiated as gill or lung or where these organs are largely reduced. It must be remembered, however, that in every case, be the respiration cutaneous or branchial or pulmonary, the gas exchange occurs by diffusion only through the surface which separates the external medium from the blood; the gas is then transported to the tissues by the circulatory system. The primary requirement for a body appendage or evagination to be identified as a gill or ctenidium is the high vascularization of the respiratory surface. The same is true for the lung.

Lungs are usually divided into two types: diffusion lungs and ventilation lungs. In the former the exchanges by diffusion through the separating surface folded into the body are sufficient for supplying the oxygen and for removing the carbon dioxide. This type is most primitive and is found in rather small animals with low metabolism and where the distance to the tissue is short. In ventilation lungs the air is mechanically transported in and out by movements which modify the volume of the cavity. Ventilation lungs are essential for large and active animals where the efficiency of respiration cannot rely upon diffusion only.

II. The "lung" of pulmonate gastropods

The name "Pulmonata" is derived from the Latin *pulmo* = lung and is applicable to those gastropod molluscs which respire atmospheric air, through the surface of the large pallial cavity. In Pulmonata the "lung" is in fact a bag-like cavity with blood vessels disposed as a network in the roof. It should be emphasised that it has no such complexity as the spongy lung of vertebrates, and for this reason we prefer to call it pulmonary cavity or pulmonary sac. In the vast majority of Pulmonata no trace of ctenidia is present. The pulmonary sac is more or less anterior and connected with the outside by a narrow hole, the pneumostome, which rhythmically opens.

A varying degree of adaptation to aerial respiration is found in gastropods. Species of the genus *Littorina* appear progressively adapted to aerial respiration and correspondingly show a greater reduction of the gill.

The volume of the pulmonary sac is variable and changes during the retraction of the body into the shell. Krogh (1941) reports in *Helix* a volume of 5–7 ml when the animal is out and a reduction to less than 0·5 ml when the snail is retracted. In the slugs the volume is smaller: about 0·3 ml in an *Arion* of 10 g weight. As mentioned, the pulmonary cavity is richly supplied with blood vessels and these are disposed along ridges which project unto the sac increasing 3–4 times

c

the available respiratory surface. Pelseneer (1935) reported a value of 107·5 cm² in a *Helix pomatia* L. sample. The ratio respiratory surface to body weight was equal to 8·3, very close to the values calculated for the ctenidia of marine forms (from 7·1 to 9·4) and for the lungs of man (10·9 cm² per g body weight).

All the genera of Pulmonata which inhabit fresh water represent a structurally conservative group. In fact, the pulmonary cavity is typically unchanged and its roof remains vascularized for aerial respiration. The most important feature of freshwater snails is the absence of ctenidia as in the air-breathing forms. As pointed out by Hunter (1964), the evolutionary tendency of freshwater pulmonates contrasts with the general evolutionary trends of marine groups in that life in freshwater involves preadaptations for transition to air breathing. Snails which live in fresh water show in fact an increasing degree of adaptation to aquatic life: from species primarily air breathing and occasionally submerging in water to completely aquatic species where the pulmonary sac is completely absent and new gills are developed. Between these extremes there are species (Physidae, Lymnaeidae) with aquatic habits but retaining the characteristic pulmonary cavity. All forms are capable of cutaneous respiration also. The air dissolved in the water diffuses through the soft and vascular skin and a great deal of oxygen can be carried by the blood to the tissues when the use of the pulmonary sac is discontinued.

III. Respiratory exchange

A. Land pulmonates

Helicidae, Limacidae and other families of Stylommatophora are typical terrestrial and respiration is completely aerial. In these gastropods several modifications have occurred in relation to the breathing mechanism:

1. The restriction of the opening of the mantle cavity to a narrow hole, the pneumostome, with the consequent decrease of evaporation from the respiratory surface.

2. The contraction of the muscular floor of the mantle cavity and the achievement of a pumping mechanism for the renewal of air.

A rhythmic closing and opening of the pneumostome can be observed in air-breathing snails, but it has been questioned whether these movements are related to the pumping mechanism of the pulmonary sac which are required for drawing or alternatively pushing air. According to Wit (1932) the closing of the pneumostome serves only the purpose of reducing evaporation. However, besides temperature and humidity, the movements are controlled by several factors such as the oxygen and carbon dioxide partial pressure. In *Helix* the pneumostome normally remains closed for long periods of time in 20% oxygen and opens when the oxygen content drops to 10% (Maas, 1939). On the contrary, in carbon dioxide at high concentrations (3–5%), the pneumostome is held open for long intervals. Dahr (1924) claimed that in normal circumstances it is the partial pressure of carbon dioxide and not that of oxygen in the pulmonary chamber which controls the opening of the hole.

The respiratory movements of the pulmonary sac have been investigated by many authors, and the pumping mechanism has been matter for debate.

After the classical work of Lazzaro Spallanzani (1803) on the respiration of the garden snail, it was believed that land pulmonates do not display respiratory movements. Pulmonata, therefore, were considered to have a "diffusion lung" where diffusion is regulated by the opening of the pneumostome and no change of volume occurs. In 1924 Dahr found that in *Arion* the pneumostome closes at an oxygen partial pressure of 15 mmHg and that, when the chamber is closed, the pressure inside increases by muscular contractions and reaches values of about 20–30 mmHg. Later, the same author reported that at low oxygen pressure *Helix* and *Arion* show peculiar dyspnoeic reactions during which the pneumostome closes at short intervals. During the closed period, the air is moved inside by muscular contractions (Dahr, 1927). Active respiratory movements have been also described in *Helix* by Ysseling (1930) and by Maas (1939) using a manometer connected with a respiratory chamber containing the animal (Fig. 3). When reporting these data, Krogh (1941) claimed that "these reactions have absolutely no respiratory significance" and that "nothing can be gained by respiratory movements" with regard to gas renewal and transport.

While the problem awaits more experimental data, it must be remembered that, since the internal pressure of oxygen in the lung is certainly different from the tension in the external medium, respiration by diffusion is theoretically possible and can meet the needs of animals living at low metabolic levels. However, on the basis of the anatomical structure of the pulmonary sac, a ventilation mechanism cannot be excluded. As already mentioned, the mantle cavity is covered with a network of ridges containing the pulmonary vessels which converge on the heart. On inflation, the roof stretches and a larger area becomes available for diffusion. When the muscles contract, the arched floor flattens, air is drawn in and, at the limit of contraction, a valve slides across the pneumostome. Conversely, when the muscles relax, the

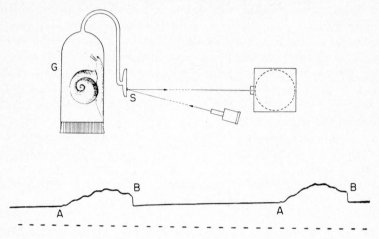

Fig. 3. Respiratory movements in *Helix* as recorded by Maas (1939). The vessel G is connected to a Marey tambour (S). A small mirror on the tambour reflects a beam of light on a photokymograph. At A the pneumostome is closed, at B is open.

cavity decreases in size and the increase of pressure facilitates gas exchange with the blood of the roof vessels. The pneumostome then opens and air is expelled (Meisenheimer, 1912).

Obviously, these breathing movements are not regular and frequent as in vertebrates, although the snail renews the air in the pulmonary chamber at rather short intervals. In the absence of a ventilation mechanism, the air can be changed in the pulmonary sac simply by diffusion and, owing to the large respiratory surface, a slight difference in the oxygen pressure is adequate for gas exchange to occur. In fact, as indicated by Dahr's calculations, values as low as 2 mmHg of pressure difference across the respiratory epithelium are enough for gas exchange to take place.

B. Freshwater pulmonates

As mentioned before, the pulmonate snails which live in fresh water are primarily air-breathing and show various degrees of readaptation to aquatic life. They have no ctenidia and the mantle cavity retains the function of a "lung" which can be used in conditions when the water oxygen content falls significantly. Moreover, the pulmonary cavity serves also as a hydrostatic organ.

In the sinistral Physidae the pulmonary chamber opens on the left side of the body, whereas in the dextral Lymnaeidae it opens on the right. The opening comes to lie at the end of a siphon-like tube, the length of which depends on the animal's proximity to the surface film (Cheatum, 1934).

The capacity of the pulmonary chamber is not proportional to the body weight, since small snails have a greater chamber than larger ones (Cheatum, 1934).

Planorbarius corneus (L.) (the ramshorn) lives in shallow ponds, stagnant pools and waters well stocked with submerged aquatic plants. More often than *Lymnaea*, the ramshorn is found at the bottom where it feeds on decaying material. *Planorbarius* is absent from foul waters. A slip of the mantle, richly vascularized, may be protruded into the water and functions probably as an accessory gill. The mantle cavity is generally filled with air which, at rather long intervals, is replaced by fresh air from the atmosphere (Zaaijer and Wolvekamp, 1958). According to Jordan (1938), the oxygen in the pulmonary cavity is gradually consumed; the pulmonary oxygen tension, therefore, is variable.

Lymnaea peregra (Müll.) can live submerged indefinitely in well oxygenated waters with the pulmonary cavity full of air and pneumostome closed. In waters where the oxygen concentration falls below a definite value this is no longer possible, and the animal has to come to the surface at intervals by flotation or by travelling over submerged plants. When at the surface layer, it directs the body in such a manner as to project the siphonal tube through the surface film and breathe.

The chief type of gas exchange in freshwater pulmonates, however, is cutaneous respiration. Within the genus *Lymnaea* there are species which are known to occur at considerable depths (about 250 m) from where periodic excursions to the surface are impossible and where they can live for long periods of time. Under these conditions, the mantle cavity is transformed into a water lung and remains filled with water throughout life. It is possible that water is rhythmically drawn and expelled from the body by mechanical ventilation and that the animal obtains oxygen through the general surface of the pulmonary sac functioning as a ventilation lung. Diffusion alone would in fact, be, too slow for renewing the water and for providing significant amounts of oxygen.

Living together with snails which have water-filled lungs, are others with a gas bubble in the pulmonary cavity. This bubble is used for cutaneous respiration or for floating or both, and eases the snail's journey to the surface (Cheatum, 1934; Hunter, 1953).

The physiology of the periodic aerial respiration in freshwater pulmonates has been the object of long studies by several investigators (Precht, 1939; Muller, 1943; Fusser and Kruger, 1951; Berg and Ockelmann, 1959; Berg, 1961). Two different stimuli drive the snail to the surface to breathe: the partial pressure of oxygen and the amount of air present in the pulmonary cavity. As mentioned, the latter can also serve a hydrostatic purpose. Precht (1939) followed the excursions of a snail in different conditions and observed that, after respiring pure oxygen, it remained submerged longer than after breathing nitrogen. According to this author, it seems that the critical factor for re-emersion is the volume of gas in the pulmonary chamber. An increase of pressure to, for example, 2 atmospheres, by decreasing the volume of the pulmonary gas, induces the animal to return to the surface. It is interesting that, if the pressure is reduced before the animal reaches the surface, it goes back to deeper water.

Aquatic pulmonates are rather insensitive to carbon dioxide. In low concentration, CO_2 has very little influence upon the frequency of ventilation. At higher concentrations, however, narcotic effects appear.

IV. The transport of oxygen

As stated before, the efficiency of the gas transport depends upon several factors, among them an abundant flow of internal fluids to the respiratory surface. It is essential that the fluids move from the tissues and cells to the respiratory organs and return back to them after oxygenation.

A second factor of no minor importance is the transport capacity of the blood. The amount of oxygen which dissolves in a fluid is determined by the partial pressure (pO_2) and by the absorption coefficient, that is, the quantity of gas which can be held by the fluid at a given pressure and temperature. Water is a very poor transport medium for oxygen, owing to the very low solubility of this gas: 3·4% by volume at 15°C and 160 mmHg. This value is reduced by the presence of dissolved salts as in sea water. Since pO_2 cannot be greater than 160 mm, the oxygen content of the blood can be enhanced by increasing the absorption coefficient, this being obtained by means of substances which chemically combine with oxygen. In order to be used as carriers, such substances must be dissociable: they must have the ability to combine with increasing quantities of oxygen at higher gas tensions and "dissociate" or release it when the partial pressure decreases. This property is expressed in the well known "dissociation curve" which relates the partial pressure of oxygen to the amount of the substance combined with the gas.

The oxygen-carrying substances are called respiratory pigments. Of the four respiratory pigments present in living organisms: haemoglobins, chlorocruorins, haemocyanins and haemerythrins, only haemoglobins and haemocyanins are present in pulmonate gastropods as well as in other molluscs. Their distribution, however, is very irregular among Pulmonata: several species lack any respiratory pigment in the blood, whereas very closely related species living in the same habitat have either haemoglobin or haemocyanin in the blood.

A. Haemoglobin

Among pulmonate gastropods haemoglobin is present only in Planorbidae. In 1869, Ray Lankester first detected the presence of the pigment in the blood of *Planorbarius corneus*. Since then the physiological significance of haemoglobin in this group has been the object of extensive investigations by many authors but rather discordant results were obtained (Read, 1966; Ghiretti and Ghiretti–Magaldi, 1972).

The properties of haemoglobin in *Planorbarius corneus* can be summarized as follows. The pigment is dissolved in the circulating haemolymph and is found in large aggregates of 1 634 000 daltons (Svedberg, 1933; Svedberg and Hedenius, 1934). The oxygen dissociation curve of the blood is practically hyperbolic (Macela and Seliskar, 1925) and shows a very high oxygen affinity: in the absence of CO_2 and at 10°C the p50 is equal to about 1 mm. The increase of temperature and of CO_2 tension shifts the dissociation curve to the right and confers a sigmoidal shape to the curve indicating a positive Bohr effect and haem–haeme interaction. At higher CO_2 tensions, however, the Bohr effect is reversed.

In *Planorbarius*, haemoglobin appears to have no significant function when the partial pressure of oxygen is high. In fact, as long as the air contains as much as 7% of oxygen, the pigment remains fully oxygenated in all parts of the body. On the contrary, the animal dies at tensions lower than 1%. This has been taken as an indication that the pigment becomes important when the oxygen tension of the environment falls to very low levels (Carter, 1931). Since the lowered oxygen tension is a consequence of the behavioural pattern of the animal, the haemoglobin may function at the end of a dive, when the pulmonary pO_2 falls below the critical pressure.

Nevertheless, it seems difficult to ascribe a respiratory function to haemoglobin in the Planorbidae since (1) the animal, by migrating to the surface, is able to avoid the consequences of lowered oxygen tension, and (2) it has been observed that, when the submerged animal remains exposed to oxygen lack for prolonged periods (40–80 min), or is placed in an atmosphere of nitrogen for the same period, the absorption bands of oxyhaemoglobin do not disappear.

Jones (1963), in parallel observations on *Planorbarius corneus* and *Lymnaea stagnalis* (L.), a closely related form which lives in the same habitat and which is believed to lack any respiratory pigment*, found that *Lymnaea* tends to dive for shorter periods than *Planorbarius*. This could suggest that the presence of haemoglobin enables the rams-horn to utilize its oxygen store to a much greater extent and consequently to remain much longer away from the surface. However, the total oxygen uptake of *Planorbarius* is only 60% that of *Lymnaea* (Jones, 1961). When confined in a small volume of water without access to air, both *Lymnaea* and *Planorbarius* are about equally efficient in depleting the dissolved oxygen (Zaaijer and Wolvekamp, 1958).

From these observations it seems that *Planorbarius* derives no benefit from its possession of circulating haemoglobin. It has been claimed that invertebrate haemoglobins with a high oxygen affinity might have a protective function against oxygen poisoning. This hypothesis, first advanced by Manwell (1964), deserves attention but should be substantiated by experimental work.

B. Haemocyanin

The other oxygen-carrying pigment which is found in pulmonate gastropods is haemocyanin. Chemically this respiratory protein is entirely different from haemoglobin, having copper instead of iron and the metal being directly bound to the polypeptide chain. The distribution, chemistry and physiology of haemocyanins have been reviewed in several papers to which the reader is referred (Redfield, 1934, 1950;

---

* The presence of an oxygen-carrying pigment in the haemolymph of *Lymnaea* is doubtful. Cells producing haemocyanin have been observed under the electron microscope in the connective tissue by T. Sminia (*Z. Zellforsch. mikrosk. Anat.* (1972) **130**, 497–526.

Manwell, 1960, 1964; Ghiretti, 1962, 1966b; Redmond, 1968; van Holde and van Bruggen, 1971; Ghiretti and Ghiretti–Magaldi, 1972). Only a brief summary will be given here.

Haemocyanin is present in the circulating haemolymph of gastropods as large aggregates of high molecular weight, up to nine million daltons. One oxygen molecule binds two atoms of copper. *Helix* haemocyanin undergoes reversible dissociation successively into halves, tenths and

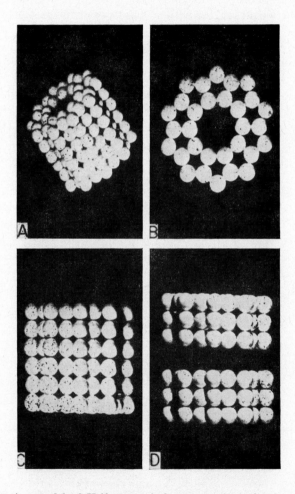

Fig. 4. Tentative model of *Helix pomatia* haemocyanin molecule (according to van Bruggen *et al.*, 1962). A, whole molecule; B, front view; C, side view; D, first dissociation step in halves.

twentieths (Witters and Lontie, 1968). These various states of aggregation appear to be molecularly homogeneous. However, it has long
been known that *Helix* haemocyanin contains two components, α- and
β-haemocyanin, of about the same molecular weight and sedimentation
coefficient, but different dissociation behaviour (Heirwegh *et al.*, 1961;
Lontie *et al.*, 1962; Lontie and Witters, 1966; Wood *et al.*, 1971).

The largest aggregate (which corresponds to 9 million daltons and
which is the main component in the haemolymph of *Helix pomatia* L.)
appears in electron micrographs as circles with 10-fold rotational
symmetry or as rectangles with a subdivision into six parallel rows.
The circles (35 nm diameter) and the rectangles (35 × 38 nm) are
different aspects of the same cylindrical molecule (van Bruggen *et al.*,
1962; Fernandez–Moran *et al.*, 1966; van Bruggen, 1968). Further
studies by Mellema and Klug (1972) have shown that the three-
dimensional structure of the haemocyanin molecule of gastropods can
be described as a hollow cylinder closed at the ends. The end consists
of a "collar" containing a "cap" at the centre (Figs 4, 5).

Fig. 5. Model of the haemocyanin molecule (according to Mellema and Klug, 1972).

The oxygenated pigment is blue and shows typical absorption bands at around 340 nm (the so-called copper–oxygen band) and at 580 nm. Both disappear during deoxygenation. The respiratory function of haemocyanin, that is, the property of the protein of transporting oxygen from the lung to the tissues, has been clearly demonstrated by comparing in *Helix pomatia* the blood taken from the venous circle with that drawn from the heart (Spoek *et al.*, 1964): the first is essentially colourless, the second distinctly blue.

The oxygen capacity of pulmonate haemocyanin is rather low and ranges from 1–5 vols per cent. The oxygen dissociation curve of the oxygenated pigment is sigmoid and the half saturation pressure is greatest at the normal pH value of the blood. Any change of the pH either to the acidic or the alkaline side, increases the affinity of haemocyanin for oxygen (Wolvekamp and Kersten, 1934). See Fig. 6.

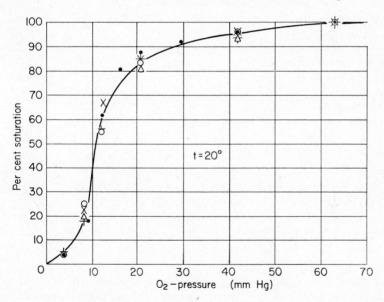

Fig. 6. Oxygen dissociation curve of *Helix pomatia* haemocyanin at different seasons; pH 7·65. ●× June; ⊙△ September; ◯+ October. (From Wolvekamp and Kersten, 1934.)

V. The utilization of oxygen

The oxygen consumption of some pulmonate species, as measured under relatively physiological conditions, is reported in Table I. The

metabolic rate of snails does not show the great variability found in
such groups as Bivalvia and Cephalopoda where the difference ranges
from twice to tenfold. Large variations of the oxygen consumption,
however, are observed in pulmonate species. Also, within the same
species, individual variations exist which can be ascribed to physiological and physico-chemical factors as sex, age, body size, activity,

Table I.   Oxygen consumption of some gastropods. (From Ghiretti, 1966b)

Species	Q_{O_2} $\mu l\ O_2/g$ wet wt/h	temp. (°C)
Cepaea hortensis (Müll.)	80	15
Cepaea vindobonensis (Pfeiffer) (summer)	125	20
Cepaea vindobonensis (winter)	51	20
Helicigona lapicida (L.)	96	15
Agriolimax agrestis (L.)	194	23
Helicella obvia Hartmann	180	23
Helix pomatia L.	20–80	15
Lymnaea stagnalis (L.)	11	—
Haliotis tuberculata L.	24–87	24
Pleurobranchaea meckeli (Blainville)	34–36	24

starvation, season, temperature, oxygen tension, etc. In addition
previous conditions as hibernation, feeding, treatment with nitrogen,
etc. affect the metabolic rate (Fischer, 1931).

Using tissue slices from the various organs of Helix pomatia, Kerkut
and Laverack (1957) found that the brain has the highest Q_{O_2}; it is
followed in order by the liver, gut, mantle, kidney, columellar muscle,
female duct, albumen gland, body wall, dart sac and foot.

In certain pulmonate species the oxygen consumption is directly
proportional to the weight rather than to the surface. In fact in large
specimens of Cepaea and Helix the respiratory rate per unit weight is
not lower than in smaller individuals (Liebsch, 1929). In Biomphalaria
glabrata (Say) and in other aquatic pulmonate snails, however, von
Brand et al. (1948) found a constancy of the oxygen consumption with
reference to the relative surface. As shown in Table II, the oxygen
consumption decreases with increasing size expressed as unit weight.

Like many other molluscs, Pulmonata take up oxygen when active,
but it is not certain whether during hibernation they breathe at all.
In any case the amount of air required to sustain life during the resting

periods must be small. After a period of anaerobiosis, there is a progressive increase in the oxygen consumption (Ghiretti, 1966a). The snail *Strophocheilus oblongus* Müll. can resist total anaerobiosis for up to 48 h without any change in behaviour which remains similar to that of animals breathing normal air (Haeser and De Jorge, 1971). Changes in oxygen uptake with season have been reported several times and very likely they are associated with the physiological activity of the animal or with biochemical changes of the tissues (Wieser *et al.*, 1970). The effect of starvation on the respiratory rate of some pulmonate species is reported in Fig. 7. It is shown that the oxygen uptake decreases

Table II. Influence of size on the oxygen consumption of *Biomphalaria glabrata* at 30°C (mean values). Extremes in parentheses. (From von Brand *et al.*, 1948.)

Weight mg	µl O_2/g wet wt/h
30–40	296 (254–318)
	255 (226–270)
	311 (249–416)
	246 (187–310)
300–400	140 (101–177)
	161 (122–187)
	163 (129–183)
	177 (147–202)

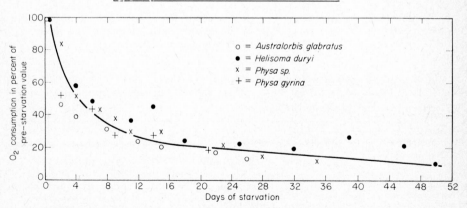

Fig. 7. Effect of starvation on the rate of oxygen consumption of four pulmonate snails. (From von Brand *et al.*, 1948.) (*Australorbis glabratus* = *Biomphalaria glabrata*.)

sharply during the first day and slowly later (von Brand *et al.*, 1948). As in other poikilotherms, the metabolic rate in snails is related to temperature and the relation can be described generally in terms of the well known Krogh's normal curve (Fig. 8). Deviations, however, have been reported for several species (Nopp and Farahat, 1967).

Among species which are capable of maintaining their oxygen consumption relatively unchanged with decreasing oxygen content in the water until a critical point of oxygen supply is reached (Fig. 9), *Biomphalaria* provides a striking example. As reported in Table III,

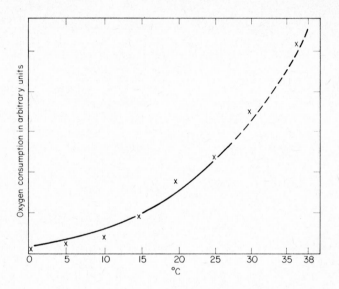

Fig. 8. Effect of the temperature on the oxygen consumption of *Biomphalaria glabrata*. (From: von Brand *et al.*, 1948.)

Table III. Influence of oxygen tension on the oxygen consumption of *Biomphalaria glabrata* at 30°C (mean values). Extremes in parentheses. (From von Brand *et al.*, 1948.)

O_2 mmHg	$\mu l\ O_2$/g wet wt/h	
	30–40 mg	300–400 mg
760	288 (250–334)	144 (97–190)
38	205 (186–221)	138 (99–187)
13	260 (234–279)	156 (110–204)
5	29·6 (8·2–49·2)	12·5 (3·8–17·3)

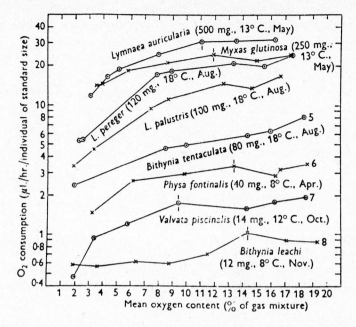

Fig. 9. Effect of the oxygen tension on the metabolic rate of some freshwater snails. (From Berg, 1961.)

even by changing the oxygen tension from 13 to 760 mmHg, the oxygen consumption changes very little.

VI. Concluding remarks

As underlined in this brief review, there are many fundamental problems in the respiration of pulmonate molluscs which remain unsolved. Whereas great progress has been made in the biochemistry of haemocyanin during the last ten years, no attention has been paid to the physiology and biochemistry of haemoglobin in the Planorbidae, where the presence of the pigment remains an odd event in the distribution and the evolution of this respiratory pigment.

Also information on the distribution and the physiological significance of haemocyanin in pulmonate gastropods is very meagre. It is enough to say that even the presence of circulating haemocyanin in *Lymnaea*, a snail which has been the object of many physiological studies, has never been clearly demonstrated. As for the fundamental problem of whether the lung of Pulmonata is a diffusion or ventilation lung, no answer is to be found in the literature of the last forty years.

VII. References

Berg, K. (1961). *Verh. int. Verein. theor. angew. Limnol.* **14,** 1019–1022.

Berg, K. and Ockelmann, K. W. (1959). *J. exp. Biol.* **36,** 690–708.

Carter, G. S. (1931). *Biol. Rev.* **6,** 1–35.

Cheatum, E. P. (1934). *Trans. Am. microsc. Soc.* **53,** 348–407.

Dahr, E. (1924). *Acta Univ. Lund.* **20,** 10, 1–19.

Dahr, E. (1927). *Acta Univ. Lund.* **23,** 10, 1–20.

Fernandez-Moran, H., van Bruggen, E. F. J. and Ohtsuki, M. (1966). *J. molec. Biol.* **16,** 191–207.

Fischer, P. H. (1931). *J. Conch., Paris* **75,** 1–100 and 111–200.

Füsser, H. and Krüger, F. (1951). *Z. vergl. Physiol.* **33,** 14–52.

Ghiretti, F. (1962). *In* "The Oxygenases" (O. Hayaishi, Ed.) pp. 517–553. Academic Press, New York and London.

Ghiretti, F. (1966a). *In* "Physiology of Mollusca" (K. M. Wilbur and C. M. Yonge, Eds) **2,** 175–208. Academic Press, New York and London.

Ghiretti, F. (1966b). *In* "Physiology of Mollusca" (K. M. Wilbur and C. M. Yonge, Eds) Vol. 2, 233–248. Academic Press, New York and London.

Ghiretti, F. and Ghiretti-Magaldi, A. (1972). *In* "Chemical Zoology" (M. Florkin and B. T. Scheer, Eds) Vol. 7, 201–217. Academic Press, New York and London.

Haeser, P. E. and De Jorge, F. B. (1971). *Comp. Biochem. Physiol.* **38B,** 753–757.

Heirwegh, K., Borginon, H. and Lontie, R. (1961). *Biochem. biophys. Acta* **48,** 517–526.

Hunter, W. R. (1953). *Proc. R. Soc. Edin.* B **65,** 143–165.

Hunter, W. R. (1964). *In* "Physiology of Mollusca" (K. M. Wilbur and C. M. Yonge, Eds) Vol. 1, 83–126. Academic Press, New York and London.

Jones, J. D. (1961). *Comp. Biochem. Physiol.* **4,** 1–29.

Jones, J. D. (1963). *In* "Problems in Biology" (G. A. Kerkut, Ed), Vol. 1, pp. 10–89. Pergamon Press, Oxford.

Jordan, H. J. (1938). *Ergeb. Physiol.* **40,** 437–533.

Kerkut, G. A. and Laverack, M. S. (1957). *J. exp. Biol.* **34,** 97–105.

Krogh, A. (1941). "The Comparative Physiology of Respiratory Mechanisms." University of Pennsylvania Press, Philadelphia, Pennsylvania.

Liebsch, W. (1929). *Zool. Jb.* (*Zool. Physiol.*) **46,** 161–208.

Lontie, R. and Witters, R. (1966). *In* "Biochemistry of Copper" (J. Peisach, P. Aisen and W. E. Blumberg, Eds), pp. 1–455. Academic Press, New York and London.

Lontie, R., Brauns, G., Cooreman, H. and Vanclef, A. (1962). *Archs. Biochem. Biophys.* Suppl. **1,** 295–300.

Maas, J. A. (1939). *Z. vergl. Physiol.* **26,** 605–610.

Macela, I. and Seliskar, A. (1925). *J. Physiol., Lond.* **60**, 428–442.

Manwell, C. (1960). *A. Rev. Physiol.* **22**, 191–244.

Manwell, C. (1964). *In* "Oxygen in the animal organisms". (F. Dickens and E. Neil, Eds) pp. 49–119. Oxford, Pergamon Press.

Meisenheimer, J. (1912). "Die Weinbergschnecke, *Helix pomatia*" Monogr. einheimischer Tiere (H. E. Ziegler and R. Waltereck, Eds), Vol. 4. Leipzig Klinkhardt.

Mellema, J. E. and Klug, A. (1972). *Nature, Lond.* **239**, 146–150.

Muller, I. (1943). *Riv. Biol.* **35**, 48–95.

Nopp, H. and Farahat, A. Z. (1967). *Z. vergl. Physiol.* **55**, 103–118.

Pelseneer, P. (1935). *Acad. r. Belg. Cl. Sci. Publ. Fondation de Agathon Potter, I*, 1–662.

Precht, H. (1939). *Z. vergl. Physiol.* **26**, 696–739.

Read, K. R. H. (1966). *In* "Physiology of Mollusca" (K. M. Wilbur and C. M. Yonge, Eds) Vol. 2, pp. 209–232. Academic Press, New York and London.

Redfield, A. C. (1934). *Biol. Rev.* **9**, 175—212.

Redfield, A. C. (1950). *In* "Copper Metabolism" (McElroy and B. Blass, Eds) pp. 174–190. Johns Hopkins Press, Baltimore.

Redmond, J. R. (1968). *In* "Physiology and Biochemistry of Hemocyanins" (F. Ghiretti, Ed) pp. 5–23. Academic Press, New York and London.

Spallanzani, L. (1803). "Mémoires sur la Respiration." Genève.

Spoek, L. G., Bakker, H. and Wolvekamp, H. P. (1964). *Comp. Biochem. Physiol.* **12**, 209–221.

Svedberg, T. (1933). *J. biol. Chem.* **103**, 301–325.

Svedberg, T. and Hedenius, A. (1934). *Biol. Bull. mar. biol. Lab., Woods Hole* **66**, 191–223.

Van Bruggen, E. F. J. (1968). *In* "Physiology and Biochemistry of Hemocyanins" (F. Ghiretti Ed) pp. 37–48. Academic Press, New York and London.

Van Bruggen, E. F. J., Wiebenga, E. H. and Gruber ,M. (1962). *J. molec. Biol.* **4**, 1–7.

Van Holde, K. E. and van Bruggen, E. F. J. (1971). *In* "Subunits in Biological Systems" (S. N. Timasheff and G. D. Fashman, Eds) Part A, pp. 1–35. Dekker, New York.

Von Brand, T., Nolan, M. O. and Mann, E. R. (1948). *Biol. Bull. mar. biol. Lab., Woods Hole* **95**, 199–213.

Wieser, W., Fritz, H. and Reichel, K. (1970). *Z. vergl. Physiol.* **70**, 62–79.

Wit, F. (1932). *Z. vergl. Physiol.* **18**, 116–124.

Witters, R. and Lontie, R. (1968). *In* "Physiology and Biochemistry of Hemocyanins" (F. Ghiretti, Ed) pp. 61–73. Academic Press, New York and London.

Wolvekamp, H. P. and Kersten, H. J. (1934). *Z. vergl. Physiol.* **20**, 702–712.

Wood, E. J., Bannister, W. H., Oliver, C. J., Lontie, R. and Witters, R. (1971). *Comp. Biochem. Physiol.* **40B**, 19–24.

Ysseling, M. A. (1930). *Z. vergl. Physiol.* **13**, 1–60.

Zaaijer, J. J. P. and Wolvekamp, H. P. (1958). *Acta physiol. pharmac. néerl.* **7**, 56–77.

Chapter 3

Alimentary canal

N. W. RUNHAM

Zoology Department, University College of North Wales, Bangor, Wales

I. Introduction

Realization of the importance of aquatic pulmonates as disease vectors and of terrestrial pulmonates as agricultural and horticultural pests, has made a study of their food, feeding and digestion of more than academic interest. It has also been stated that pulmonates are important ecologically as comminuters of dead, dying or fresh plant tissues so increasing the turnover of this material by making it more readily available for bacterial and fungal decay. Such statements are, however, difficult to substantiate. Digestion at first sight appears inefficient, as the faeces often have the same colour as the food and contain large pieces of apparently undigested material, yet utilization of ingested food appears to be very high.

53

Papers published on the digestive system of pulmonates have been well reviewed by Hyman (1967), Owen (1966) and Franc (1968), so I have concentrated on more recent papers. Undoubtedly the greatest contributions in this field have come from Carriker and his co-workers. Although this work was published more than twenty-five years ago no other worker has published a study either of such depth, or with such an integrated approach. The majority of contributions in this field are concerned either with one or at most only a few aspects of the digestive system so that studies of the whole digestive process in one animal are rare. As a result, I have had to rely heavily in this review on the published work of Carriker on *Lymnaea stagnalis* (L.) and Walker on *Agriolimax reticulatus* (Müller).

Dissection readily reveals the gross anatomy of the digestive system (Fig. 1), but a study of the finer details of the anatomy, particularly of the stomach, requires great patience and skill. The mouth leads via a short oral tube into the buccal cavity which has the jaw in its roof, and

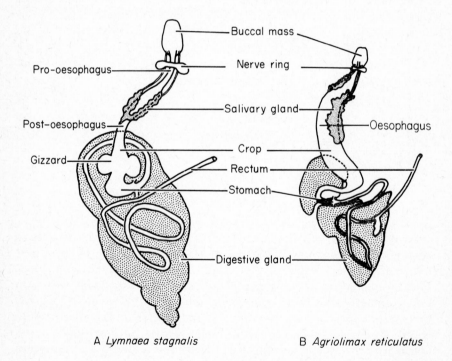

A *Lymnaea stagnalis*   B *Agriolimax reticulatus*

Fig. 1. General anatomy of the pulmonate digestive system. A, an advanced basommatophoran *Lymnaea stagnalis* (after Carriker, 1946 a, b); B, an advanced stylommatophoran *Agriolimax reticulatus* (after Walker, 1969).

projecting from its floor the odontophore supported by the odontophoral cartilage. On part of the surface of the odontophore is the radula. A complex arrangement of muscles is responsible for movement of odontophore and radula and it is the great mass of these surrounding the buccal cavity and odontophore, that is called the buccal mass, which fills a large part of the head region. Paired salivary glands open via ducts into the rear of the buccal cavity. Food passes from the buccal cavity into the oesophagus and then into the greatly dilated crop. Basommatophora have a long oesophagus and short crop while the reverse is found in the Stylommatophora. From the crop food passes into the stomach, consisting in *Lymnaea* of gizzard, the so-called pylorus and style sac, and in *Agriolimax* of anterior stomach and style sac. The presence of a gizzard is a very characteristic basommatophoran feature. Ducts from the stomach lead to the two lobes of the digestive gland which form one of the largest organs in the body. Faeces form in the style sac and they are then consolidated and transported along the intestine and rectum to be voided at the anus which opens close to the pneumostome.

In the following account I shall first discuss food and feeding, and then the anatomy and physiology of each part of the system in turn. As will be seen, our knowledge of comparative anatomy is limited to a few of the pulmonate families and physiological information is derived from only a small number of species.

II. Food

Many accounts exist of the food that pulmonates eat in the wild and particularly in the laboratory (Frömming, 1954, 1956; Mead, 1961; Graham, 1955). However, it almost seems that the more bizarre and the more rarely a food is eaten the more likely it is to be recorded. Very few workers have determined the actual diet of animals in the field. Such studies are much more difficult than first appears. It is not sufficient to record on what plants the animals are discovered as some species have plants on which they like to shelter, although they may rarely consume them, and the reverse is also true. Thus for *Cepaea nemoralis* (L.) Wolda *et al.* (1971) found that *Aegopodium podograria* (Ground Elder) and *Calystegia sepium* (Bindweed), while favoured as plants to rest on, were hardly ever eaten; and *Urtica dioica* (Stinging Nettle) and *Ranunculus ficaria* (Lesser Celandine), the most favoured food plants, were rarely rested upon. The authors suggest that these findings may be related to the fact that the rest plants tend to be glabrous and those avoided for resting were hispid. Other species of snail may not eat the plant on which they are found, but instead consume epiphytic organisms. *Lymnaea peregra* (Müller) was found by Calow (1970) to be almost invariably associated with *Elodea canadensis* (Canadian Pondweed), yet examination of the crop revealed a complete absence of fragments of material from this species; instead, the crop invariably contained filamentous algae and diatoms that were normally epiphytic on the surface of the *Elodea*. Laboratory experiments confirmed this finding and also revealed that filamentous algae were eaten in preference to diatoms and were better assimilated. Determination of what pulmonates eat in the laboratory may bear little relation to what they eat normally. Wolda *et al.* (1971) in early laboratory experiments found that *Aegopodium podograria*, the commonest plant species in the snail's normal habitat, was the most favoured food plant, yet later detailed field observations revealed that this species was only rarely eaten under natural conditions. Pallant (1969) has also found that individual *Agriolimax reticulatus* (Müll.), which were known from examinations of the faeces to have been feeding solely on *Ranunculus ficaria* in a woodland, refused to eat this plant in the laboratory but instead ate another woodland plant.

At the present time it appears that only examination of the gut contents and faeces, followed by laborious examination of the plant fragments found therein, can give meaningful information on the normal diet of pulmonates. Even this approach is not without problems. For example, some plants do not have easily recognizable structures that enable identification of food plants to be made from the minute fragments present. It is possible that only by the development of immunological methods for plant proteins, as has been so successful for the determination of the diet of carnivorous invertebrates, will this be solved. Present methods can also give only a very rough guide to the relative and absolute quantities of the various plant tisuses consumed. Radioactive tracer methods may allow quantification (Rodina, 1957; Malone and Nelson, 1969; Calow and Fletcher, 1972), but this is unlikely to be of wide application under normal field conditions.

Studies that have been published on the diets of animals in the field indicate that pulmonates are selective in their diet (Wolda *et al.*, 1971; Pallant, 1969, 1972; Mason, 1970a). They may also consume only one of several preferred foods plants available in their vicinity (Pallant, 1969). The reasons for this are as yet unknown. Part of the explanation may lie in the limited movement of land pulmonates in their search for food (Moens and Riga, 1966; Hunter, 1968) but it could also imply that individuals may retain a preference for one food species over a period of a few days. The availability of plants varies throughout the year and the composition of the diet also changes (Pallant, 1969; Wolda *et al.*, 1971). Plants also vary in the chemical composition, attractiveness, palatability and physical characteristics of their leaves; thus young, mature, old, dead and decaying leaves are all very different. Our present methods are not sophisticated enough to allow differentiation amongst these various types of leaves once ingested, but a distinction between decaying leaves and the others is possible from the presence or absence of chlorophyll in the ingested fragments. Using this technique *Cepaea nemoralis* and a number of woodland snails have been shown to prefer decaying plant material (Wolda *et al.*, 1971; Mason, 1970a, b). Our present observations are often derived from one population in a restricted habitat in only a small part of the animals' total range. Pallant (1969, 1972) has however compared the very different diets of *Agriolimax reticulatus* in woodland and grassland. Most observations refer to the diet of mature animals, but Wolda *et al.* (1971) found clear evidence that juvenile and mature *Cepaea* had very different diets – *Glechoma hederacea* was mainly eaten by juveniles and *Rumex* and *Chaerophyllum temulum* by mature snails. Snails and slugs on hatching from their eggs might be expected to have a diet different from that of the adult, by virtue of their small size, and because of the different capabilities of the feeding apparatus.

Animal material is very frequently a constituent of the gut contents of herbivorous land snails and slugs. Very small insects may be accidentally consumed along with plant material, but these pulmonates are avid consumers of certain dead animal material, including dead pulmonates, when they have the opportunity. Pallant (1972) records the presence of earthworm chaetae in gut contents of nine *Agriolimax reticulatus* out of one hundred and fifty-two collected from grassland, and arthropod fragments in twenty-eight individuals. Of the seven woodland snails studied by Mason (1970a) only *Oxychilus cellarius* (Müll.) and *O. alliarius* (Miller) apparently consumed significant amounts of animal material. *Lymnaea stagnalis* also consumed dead animal material if available and Bovbjerg (1968) has shown in Y-tube experiments that this snail has a distance chemoreception of animal material but not plant material. Prel'minary experiments indicated that growth and survival were better on a mixed diet and only animals on a mixed diet reproduced.

Many more surveys of the natural diet of molluscs are needed, not only for different species but also for species over the whole of their normal range, at all seasons of the year and at all ages.

Studies of the natural diet are essential basic information for any ecologist studying populations of pulmonates or the effects of molluscs on plant populations. Recently there have been studies on the energy balance of pulmonates. Stern (1970) found that *Arion ater* (L.) during its whole life cycle, under constant laboratory conditions and fed on lettuce, utilized 70% of ingested food. The requirements for respiration amounted to 60% of this total and growth and reproduction 40%. An average of twenty-eight Kcal was ingested during the fifty-two weeks of life resulting in a production of living material equivalent to 5·7 kcal. *Helix pomatia* L. also kept on a diet of lettuce in the laboratory, assimilated 81% of food ingested and 51% of material assimilated was used for growth and reproduction (Turček, 1970). This latter study however, refers only to a fifteen day period during the egg-laying season and the clutches of eggs laid accounted for 60% of the total for growth and reproduction. Stern found that utilization of assimilated nutrients varied throughout the life history and that this could perhaps be related to the reproductive requirements. While these findings are extremely valuable, studies of animals on a more normal diet and under more normal conditions would be welcome. Mason (1970a) has studied assimilation rates of a variety of woodland snails on a variety of different natural food materials (Table I).

The assimilation rates reported above are probably not very accurate. Pulmonates often shred their food, making quantitative collection difficult, and they contribute mucus to the remaining food material during their locomotion over it. Faeces contain, in addition to food passed through the gut, material discharged from the digestive gland which may have been derived from a meal consumed some days previously together with cellular material, mucus secreted by the gut lining during consolidation of the faeces, and in some species (Pallant, 1970) the faeces are accompanied by a string of excreta from the kidney. Two methods to increase the accuracy of these determinations are available, but are of limited application. Artificial foods containing very fine particles of chromic oxide have been prepared by Ridgway and Walker (personal communication). The amount of marker remaining after feeding and that present in the faeces is readily determined, so that from a knowledge of the proportion present in the food, ingestion and assimilation are easily calculated. This food is of course very unnatural, and a method developed by Calow and Fletcher (1972) for use with *Lymnaea peregra* feeding on algae may have wider application. Two radioactive markers were used on the algae, ^{14}C incorporated during photosynthesis and particulate ^{51}Cr, which is not absorbed by the snail, deposited on the outside of the alga. From a knowledge of the ratio between the two isotopes in the ingested food and in the faeces the assimilation was readily determined. While this method was developed for algae there seems to be no reason why this principle cannot be used with a wider range of foods.

As yet the nutritional requirements of pulmonates are virtually unknown. We do not know their requirements for proteins, fats, carbohydrates, vitamins or minerals. Some species are able to live on such a restricted diet as wheat flour. *Limax flavus* L. and *Arion ater rufus* are able to complete their life history on this food (Frömming, 1957), but growth rate and size are reduced for *Limax flavus*. Artificial diets have shown that vitamins appear to be required. *Biomphalaria glabrata* (Say), when reared under axenic conditions and fed on a diet of killed *Escherichia coli* and yeast grew but never laid eggs unless vitamin E was added to the food (Vieira, 1967). Using an artificial diet based on one developed for insects, Ridgway and Walker (personal communication) have shown that in the absence of vitamins of the B group, *Arion ater* has a much reduced growth rate and life expectancy, and shows pronounced cannibalism. A determination of the optimum amount

Table I. Percentage assimilation of food materials by woodland snails at 10°C in the laboratory (Mason, 1970a)

Species	Food	Percentage assimilation (mean $\pm$ S.E.)
Discus rotundatus (Müll.)	Fresh leaf	
	Mercurialis perennis (Dog's Mercury)	40·36 $\pm$ 10·09
	Circaea lutetiana (Enchanter's Nightshade)	46·05 $\pm$ 6·08
	Urtica dioica (Stinging Nettle)	47·70 $\pm$ 8·89
	Leaf litter	
	Carpinus betulus (Hornbeam)	48·67 $\pm$ 6·26
	Castanea sativa (Sweet Chestnut)	45·81 $\pm$ 5·60
	Fagus sylvatica (Beech)	55·06 $\pm$ 3·59
	Quercus robur (Oak)	58·00 $\pm$ 4·95
	Acer pseudoplatanus (Sycamore)	47·49 $\pm$ 3·88
	Earthworm	78·6 $\pm$ 6·73
Hygromia striolata(Prf.)	*Urtica dioica* fresh leaves	52·40 $\pm$ 8·78
	Acer pseudoplatanus litter	38·23 $\pm$ 4·71
Oxychilus cellarius	*Lactuca sativa* (Lettuce)	70·20 $\pm$ 4·40
	Earthworm	86·90 $\pm$ 2·53
Helix aspersa (Müll.)	*Lactuca sativa*	53·50 $\pm$ 6·04

of food to be presented to experimental animals appears to be surprisingly complex. It is apparently not possible to provide food "ad libitum" for *Lymnaea stagnalis* as they show an increased consumption of food (and resultant growth and reproductive activity) with increased amounts of food supplied, up to the point where it is no longer possible to stuff more food into their container (van der Steen *et al.*, 1973).

In addition to the constituents of the diet a knowledge of the factors that determine attractiveness and palatibility of a given food are important, especially with pest species. Both factors depend on the chemical and physical properties of the food. Thus smell and taste together with surface texture and cuticle thickness must be of importance as to whether plants are eaten or not (Grime *et al.*, 1968; Wolda *et al.*, 1971; Kittel, 1956). As yet our knowledge of the touch and chemoreception of pulmonates is very limited. It is to be hoped that the neurophysiological experiments of de Vlieger and his students will be extended to a wider range of chemicals and touch stimuli than the simple sugars and point stimuli already investigated (Jager, 1971; de Vlieger, 1968).

III. The buccal mass

The structures used for feeding are contained in the buccal mass. On dissection the front of the head region is found to be filled by this globular organ. It consists of the radula, jaw, odontophore cartilage and associated muscles. The buccal cavity within which it lies is continuous with the mouth anteriorly, via the oral tube, and with the oesophagus posterodorsally.

Few descriptions exist of the structures around the mouth. In *Lymnaea* there is a solid lip which completely surrounds the mouth and this is extended laterally as the mouth lobes. There is a complex arrangement of glands opening on to the lip surface, and there are many sensory neurones in the epithelium (Zylstra, 1972). In the stylommatophoran *Agriolimax reticulatus* (Walker, 1969) there are several small lips around the mouth: three anterior, five pairs of lateral, one single posterior lip, and lateral to this ring of lips there are the larger mouth lobes. The two most posterior pairs of lateral lips fuse immediately inside the mouth to form lateral folds, which extend back through the buccal cavity, and are continuous with the lateral folds in the oesophagus. In both species much of the surface of the lips, and the linings of the oral tube and buccal cavity, is lined by a cuticle. However, the apparent cuticle over the mouth lobes of *A. reticulatus* is found to be a highly specialized sensory area when examined with the electron microscope (Runham and Cragg, unpublished). This sensory epithelium is connected to a small ganglion which is part of Semper's organ, the other part being a small gland which discharges on to the surface of the mouth lobe (Lane, 1964; Laryea, personal communication).

Within the buccal cavity the cuticle is thickened to form the jaw; and the radula is probably also derived from this cuticle. The jaw varies considerably in morphology and has been well described (Taylor, 1894). Most forms have a single transverse thickening, but a few have 2 lateral jaws and a single median jaw (*Lymnaea*); *Punctum pygmaeum* (Draparnaud) has 19 small plates, and *Testacella* lacks a jaw. The epithelium underlying the jaw consists of tall columnar cells, each of which appears to secrete a rod of material. No details have been published on the process of jaw formation but chemically the jaw consists of chitin and protein together with some calcium, iron, carbonate, phosphate and sulphate (Spek, 1921). If the jaw is removed, with forceps, from *Helix pomatia*, it is regenerated in 12–16 weeks; but the new jaw is smooth and lacks the characteristic ridges (Bierbauer, 1957). In some animals (Runham and Thornton, 1967) the shape of the jaw is apparently determined by the abrading of its surface by the radula. It would therefore have been very interesting if these animals had been examined at longer periods after regeneration. It has been claimed by Taylor (1894) that the number of ridges and the strength of the jaw is directly influenced by the type of feeding; the tougher the food the stronger the jaw.

The oral tube is very short, and when the odontophore is retracted its lumen is I-shaped, because of the lateral folds bulging in from the sides. On entering the buccal cavity these folds become much flatter. Dorsally, and posterior to the jaw, there is a groove leading directly back into the oesophagus – this is the dorsal food channel. Along the upper edge of the lateral folds, just ventral to the channel, there is a narrow band of ciliated epithelium also containing many mucous glands. The mucous glands are possibly comparable to the buccal glands that are found in *Lymnaea stagnalis* and other pulmonates. In *Lymnaea* the glands are unicellular subepidermal mucous glands which are distributed over most of the epidermis of the buccal cavity (Carriker and Bilstad, 1946). No systematic survey seems to have been made of these cells nor of their varied distribution throughout the pulmonates.

Bulging into the buccal cavity from its floor is the large movable odontophore. This structure consists of the odontophore cartilage, muscles, and the radula, which covers most of its surface. It is grooved dorsally and this groove leads posteriorly to the collostylar hood and radular gland. Histologically the cartilage bears little resemblance to vertebrate cartilage, but it is apparently homologous with the highly cartilaginous structures present in, for example, the prosobranchs. Running through the cartilage are many muscle fibres, mainly dorso-ventral in direction, and between the fibres are vesicular connective tissue cells (Loisel, 1893). When the vesicular cells are punctured their contents readily flow out, and Loisel was unable to detect protein or glycogen in this fluid. In the prosobranchs the cartilage cells store glyco-gen (Barry and Munday, 1959), and the cells in pulmonate cartilage closely resemble the vesicular connective tissue cells found elsewhere in the body which function as a glycogen store (Sminia, 1972). A con-nective tissue sheath is present over most of the surface of the cartilage except where muscles are attached. The cartilage is a rubbery bluish structure consisting of two plump rods, which fuse anteriorly where the cartilage is almost completely muscular. The presence of so much muscular tissue implies that the cartilage is highly deformable, and movements of isolated cartilage have been observed (Crampton, 1973). What function the vesicular cells perform is not clear, but the presence of so many fluid-filled cells in a structure with a fibrous limiting sheath suggests that they may function as a hydrostatic skeleton. Loisel indeed likened these cells to those found in the vertebrate notochord.

IV. The radula

The radula consists of a highly flexible membrane to which are attached hard teeth. Superficial examination of the radula shows that the teeth occur in rows – the teeth in any longitudinal row have apparently an identical shape and size, while along a transverse row there are considerable changes in morphology. Usually it is possible to distinguish a central tooth, flanked by lateral teeth, and these in turn are flanked by marginal teeth. There is often a gradual change in tooth form from the centre to the edges of the radula. Each tooth consists of one, or more, pointed cusps on a large flat basal plate. The nomenclature of the teeth cusps is shown in Fig. 2. Due to the considerable variation in arrangement, number, and form of teeth, radular characteristics are often used for specific identification. For convenience a radular formula is used to describe the arrangement of the teeth, thus 45.20.1.20.45, the formula for *Helix aspersa* (Müller), means that there is one central tooth with 20 lateral, and 45 marginal teeth on each side.

Tooth form and number do however vary with the age of the animal. Thus in the radula of *Helix pomatia* studied by Verdcourt (1950) 50 rows, with 29 teeth in each, are present in the radula of the newly hatched snail, while in the mature animal there were 171 rows with 151 teeth in each. There is also a clear change in the form of the teeth, from the teeth with many small cusps, termed echinate, present in the radula of the newly hatched snail, to the mature teeth which have only a few cusps. In a systematic study of the radula of *Lymnaea limosa* (L.) Hubendick (1945) found that the radula increased in size as the animal grew by addition of further longitudinal and transverse rows, together with enlargement of the teeth.

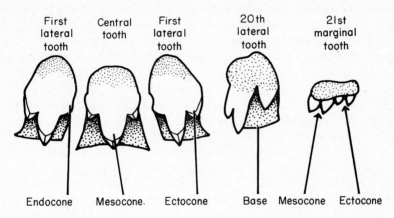

Fig. 2. Morphology of radular teeth. Examples of teeth from the radula of *Helix aspersa* (after Taylor, 1894) illustrating the nomenclature of the cusps. Note the way in which the cusps, first the mesocone and then the ectocone become split towards the edges of the tooth row.

When studying teeth for the identification of a species it is important to consider only unworn teeth. Those teeth at the extreme anterior end of the radula are damaged by wear during feeding (Runham and Thornton, 1967; Oberholzer *et al.*, 1970). Wear is first seen as a blunting of the cusps, and as it becomes more extensive it results in the tooth being reduced to a shapeless nub, or it may even be torn out altogether. Worn teeth are shed at the anterior end of the radula and may be found in the faeces (Carriker, 1943).

Detailed studies of radular form necessitate the examination of large numbers of radulae, and many painstaking measurements. In the recent, and highly sophisticated, study made by Oberholzer *et al.*, (1970) on the difficult *Bulinus natalensis tropicus* group, 1459 radulae were examined from snails representing 84 different populations in South Africa. The authors state that because of the variability of radulae, not less than 5 radulae from a population should be examined.

The results clearly indicated that there was a continuous variation in the form of the teeth over the whole range studied, so this group of snails clearly belongs to one polytypic species. Results collected over a period of three years showed that there was a slight change in the morphology of the teeth during that time which could be related to changes in the habitat of the snails. Of even greater interest was the discovery of some large healthy individuals that completely lacked a buccal mass.

As noted above the teeth in the most anterior part of the radula show signs of wear, whereas more posteriorly they are sharp-pointed and obviously unworn. If the animal is to continue to use the radula for feeding there must be replacement of these worn-out teeth. That continuous replacement does occur has been known for many years, as a result of the comparison of radulae of young and old animals, and from the continuing occurrence of cast teeth in the faeces. By comparison of tooth size, and number of longitudinal and transverse rows of teeth, Sterki (1893) estimated that *Agriolimax laevis* (Müll.) must completely replace its radula 8 times during the animal's life, while in *Mesodon thyroidus* it was 16–18 times. Carriker (1943) collecting the radular teeth occurring in the faeces of *Lymnaea stagnalis* calculated that 0·26 teeth rows are formed every day. The existence of such replacement has more recently been demonstrated experimentally by marking the newly formed teeth by operations, injections of magnesium salts, cold shock (Runham, 1962; Isarankura and Runham, 1968) and by X-rays (Kerth and Krause 1969). These various methods give results that are essentially in agreement (Table II). The rates of replacement are impressive as there may be 100 teeth in each row. It has also been shown with these methods that the rate of replacement is related to the age of the animals and the temperature. Thus newly hatched *Helix aspersa* at 20°C replace 7 rows of teeth a day, while in the mature animal this falls to 3·6 rows a day. Such differences probably reflect the general growth rate of the animal. Adult *H. aspersa* produce no teeth at 0°C, and there is a steady increase in rate of radula production up to the 3·6 rows a day found at 20°C, and this phenomenon may be related to the general level of metabolism.

As the radula moves forward its physical properties clearly change; posteriorly the youngest teeth are very soft while further forward they become considerably hardened. The radula contains chitin and protein and the hardening appears to be the result of a complex series of chemical changes similar to those associated with quinone tanning (Runham, 1963a). Minerals are also present in the radula of *Helix* (Sollas, 1907): silica is found in winter and phosphate in spring. It is not known whether these minerals are important for hardening the teeth, as they are in prosobranchs.

The radula is secreted at its posterior end within the radular gland. Apparently because of the width of the radula and the lack of space within the highly muscular buccal mass, this gland is folded up to form a tubular structure, with its lateral edges nearly meeting dorsally. Filling the centre of the tube there is a rod of connective tissue, the

Table II. Rates of radula replacement in pulmonates

Species	Temperature (°C)	Age	Size (cm)	No. of rows of teeth produced per day
*Lymnaea stagnalis**	20	Mature		2·1
Agriolimax caruanae Pollonera	20	Mature		5·6
*Agriolimax reticulatus**	20	Mature		5·1
*Arion ater**	20	Mature		3·9
Helix aspersa active*	20	Mature		3·5
hibernating†	15	Mature		0·0
active*	15	Mature		2·3
	10	Mature		1·2
	5	Mature		0·4
	20	1 week		7·0
	20	5 week		5·5
	20	17 week		4·3
*Cepaea nemoralis**	20	Mature		3·2
*Achatina fulica**	20	12 month	2·71 × 1·91	2·66
			3·48 × 2·48	2·44
			4·45 × 3·04	2·34
Limax flavus‡	15	48 day		3·1
	15	1–2·5 year		1·4

* Isarankura and Runham, 1968
† Kerth, 1971
‡ Kerth and Krause, 1969

collostyle. This collostyle contains many collagen fibres, some smooth muscle cells, and various types of connective tissue cells scattered in an apparently structureless matrix. Anteriorly the collostyle is covered by a columnar epithelium that secretes the thick cuticle of the collostyle hood, and it is continuous with the epithelia of the oesophagus dorsally and the radular gland ventrally. The radular gland consists of an epithelium beneath (inferior, infra- or subradular epithelium) and above (superior, or supraradular epithelium) the radula, while at the posterior tip of the gland there are specialized cells, the odontoblasts, that secrete much of the radula.

The odontoblasts are large cells and there appears to be a fixed number of them associated with the production of each tooth. In *Lymnaea stagnalis* Hoffman (1932) found that seven odontoblasts were involved in the secretion of the central tooth, eleven for each lateral, and five for each marginal tooth. Cell divisions appear to be rare amongst the odontoblasts, at least in adult animals, so that they are said to be permanent, each longitudinal row of teeth being secreted by one group of cells (Märkel, 1957; Runham, 1963b). Some changes must occur in the odontoblast layer, however, to accommodate the increases in width of the radula, and size of teeth as the animal gets older.

The secretion of the radula involves two components; exceedingly long microvilli projecting from the odontoblasts, and matrix material secreted by both the superior epithelium and the odontoblasts (Runham, unpublished). The characteristic form of the tooth arises from the length and arrangement of the microvilli. Short microvilli are found in the matrix of the cusp while long microvilli extend along the whole length of each tooth base. The front part of the tooth is secreted first and as the tooth lengthens, first the microvilli from the more posterior odontoblasts that secrete the cusps are shed, then when the base has been completed those from the anterior odontoblast are similarly shed. The radular membrane appears to be mainly a product of the inferior epithelium. While the microvilli appear to be shed by the odontoblasts they are no longer visible in teeth from the third row, but their fate is not at the moment clear. The matrix material appears to cement together the microvilli, at least at first, so that the newly secreted tooth can be likened to a paint brush that has had paint left in it to harden. In the older literature there was a considerable controversy about the secretion of the radula by the odontoblasts (Pruvot-Fol, 1926), one view being that the odontoblasts were permanent and that the radula was a simple secretion, while the opposing view was that the radula was formed by "chitinization" of the odontoblasts that therefore had to be replaced. The present observations appear to combine these two theories: the odontoblasts are probably permanent, the radula is partly formed from secretions but the involvement of the microvilli could be interpreted as a superficial "chitinization" of the odontoblasts.

Many cell divisions are found in the epithelia around the edges of the odontoblasts (Runham, 1963b). As the radula elongates so too do the epithelia, by addition of new cells that become closely applied to the newly secreted teeth. Radioactive labelling of the epithelial nuclei, using tritiated thymidine, indicates that as the radula moves forward the epithelia move with it. While the cells are associated with the teeth they produce the series of secretions that harden the teeth (Fretter, 1952; Runham, 1963a).

Anteriorly the cells in the superior epithelium die, thus exposing the teeth they surround, for use in feeding. Not all the cells die at the same level along the epithelium. First about 70% of the cells are lost when they apparently migrate into a position between the teeth, where their nuclei clump and then disintegrate. The remaining cells form an epithelium which appears to secrete a layer of material, possibly mucoid, separating the cells from the teeth. Finally at the junction of the epithelium cells die and disintegrate. Death of the cells is signalled by clumping of pycnotic nuclei, and the presence of much acid phosphatase in the cells and in the overlying collostyle. Production of new superior epithelium posteriorly is therefore balanced by a loss of old cells anteriorly. The basement membrane on which the cells lie is apparently unbroken along its length and is continuous with that underlying the collostyle hood epithelium. In the inferior epithelium cells are similarly produced posteriorly and lost anteriorly where they abut on to the cuticularized epithelium lining the buccal cavity.

While the radula is moving forwards it must be rigidly attached to the underlying epithelium as, during feeding, movement of the radula over the odontophore is a result of the action of muscles that are attached to the basement membrane of the inferior epithelium. Only at the most anterior end of the radula does it become detached, and this is brought about in two ways. In addition to the death of the underlying cells marked changes in the staining properties of the radula, and loss of much of the radula membrane at this point, indicate a chemical attack. As the radula must remain attached firmly to the underlying cells, movement forwards of the radula must result from a movement of the epithelium over its basement membrane. Results supporting this view have been obtained from observations on transplants of radular glands, and from the consequences of experimental removal of the growing tip of the radular gland (Isarankura and Runham, 1968).

Normally the length of the radula is proportional to the size of the animal (Hubendick, 1945). The process leading to the production of new radula must therefore be in dynamic equilibrium with those processes leading to movement forwards and shedding of the radula. Isarankura and Runham (1968) only observed a lengthening of the radula in one series of experiments, which involved keeping *Helix aspersa* in aestivation for very long periods. On occasions a radula three times the

D

normal length resulted. Part of this lengthening was due to the retention of detached radula anteriorly. It appears that the detached radula becomes torn into pieces during feeding, and should animals not feed for a long time, the detached radula accumulates in great scrolls in the buccal cavity. A real lengthening of the radula and radular gland also took place under these conditions. Production of the radula was at the normal rate therefore this lengthening resulted from a slower movement forwards of the radula. Whereas the normal radular gland was straight and closely applied to the collostyle, the longer gland was thrown into folds, as the collostyle had not lengthened.

Examination of radular abnormalities is also interesting (Isarankura, 1966). Some involve a whole longitudinal row of teeth, so that some malfunction of those odontoblasts responsible for the secretion of that row must have occurred. Abnormalities involving one or a few transverse rows are also quite common, and many of these result from such temperature shocks as frosty nights. A common abnormality is the secretion of partial rows of teeth, usually involving only the marginal teeth. As the marginal teeth are much smaller than the lateral teeth, this could be a compensatory secretion by the odontoblasts to avoid excessive shortening of the radula at the lateral edges.

V. Feeding

Anyone who has examined the buccal mass in detail would agree with Carriker's (1946a) comment that the musculature of this structure is the most complex in the animal, and probably in the invertebrates. Carriker has undoubtedly provided a superb description of the arrangement in *Lymnaea stagnalis*, and he comments that it only differs in minor details from the arrangement in *Helix*. Roach (1966) similarly states that *Arion ater* has an arrangement almost identical to that described for *Lymnaea*.

Carriker describes 28 muscles, some single, others double or in larger groups. He divides them into extrinsic – muscles connecting the buccal mass to the body wall, mouth rim, and columella; and intrinsic – muscles which interconnect the jaws, cartilage and radula. Hubendick (1957) interpreted films of *L. stagnalis*, feeding on algae on the wall of an aquarium, in terms of Carriker's anatomical description, and the following summary is based on their work (Fig. 3).

The resting position of the buccal mass will be described, then the 4 phases into which the feeding stroke can be divided:

Resting phase. The mouth is closed. The lips are lateral and are apposed longitudinally along the anteroposterior axis (producing the apparently open position of Fig. 3). The jaw is retracted into the buccal cavity and the odontophore, in its resting position within the cavity, has an almost vertical orientation.

Phase 1. The buccal mass is rotated by muscles through about 90°. Dilator muscles open the mouth and the median jaw moves anteriorly and downwards while the lateral jaws diverge. Within the buccal mass

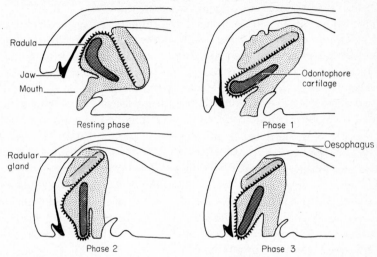

Fig. 3. Sagittal sections through the buccal mass of *Lymnaea stagnalis* illustrating its movements during the feeding cycle (after Hubendick, 1957). The coarse stippling outlines the muscular part of the buccal mass.

the odontophore also rotates through about 90° so that the total turning of the odontophore is 125–170°. Muscles first pull the cartilage from its vertical position to a horizontal or forward tilting position, then the posterior end of the cartilage is moved forward and pressed towards the mouth opening. Contraction of the cartilage muscles acting against the vesicular cells, straightens the cartilage and makes it rigid. The radula is then stretched tight over the surface of the cartilage.

Phase 2. The vertically held odontophore is protracted through the mouth against the substratum. In this position the odontophore is ready for its feeding stroke. The tip of the odontophore where it is applied to the substratum is U-shaped with the opening of the U pointing forwards. In many cases the sides of the U contact the substratum before the central part.

Phase 3. The tip of the odontophore now in contact with the substratum moves forward, so that first the lateral edges of the U and then, at the end of the feeding stroke, the central part is lifted from the substratum. The forwardly directed teeth on the outside of the odontophore are dragged through the food material during this movement. As the odontophore moves forwards the median jaw is pulled back and is held close against the radula at the end of its feeding stroke. With the completion of this movement the mouth begins to close.

Phase 4. Closing of the mouth is completed and the buccal mass and odontophore rotate backwards, mainly by the action of the retractor muscles, until the resting position is reached.

The whole feeding cycle takes 1–2 s. Sugars in the water stimulate the feeding cycle in *Lymnaea*, without there being actual feeding, and different sugars vary in their ability to elicit this response (Jager, 1971).

Hubendick felt that the radula moved very little relative to the cartilage during the feeding stroke. Carriker, however, suggested that the radula slides over the cartilage during feeding, and this was confirmed by experimental studies on the feeding tracks made by *Lymnaea stagnalis* and several other pulmonates (Märkel, 1957). It had been noted by the earliest naturalists that molluscs feeding on surfaces coated with algae left clear feeding marks. By providing slides coated with artificial substrata, of maize meal or fat in agar, Märkel obtained very detailed pictures of these tracks. Even more detailed information can be obtained by examining such feeding tracks in gelatin, with Nomarski Interference contrast optics or with the scanning electron microscope (Runham, unpublished).

From the marks left by individual teeth it can be seen that as the odontophore moves forward so the radula is also moving over its surface. The radular teeth are therefore passing over the substratum at a speed which is the sum of the movement of the odontophore plus that of the radula over the cartilage. As the teeth pass over the edge of the cartilage they swing rapidly through approximately 180°, from the anteriorly directed position on the outside of the odontophore, to the backwardly directed position in the gutter between the cartilages. The edge of the odontophore therefore forms a cutting edge (Runham, 1969). As the teeth pass over the edge of the cartilage they each scrape out a small piece of the substratum and this is clearly visible in the feeding tracks. Backward movement of the radula over the cartilage is produced by contraction of the supralateral radular tensor muscle, shown by Carriker to be the largest muscle in the buccal mass.

From necessity all the studies summarized above refer to feeding on an essentially two-dimensional substratum. Under these conditions the only way that the animal can feed is by rasping. When *Lymnaea* is observed feeding on lettuce, however, it can be seen that "as the radula rasps forward it makes connection with and scrapes past the inner side of the dorsal mandible, much as two jaws would come together, so that the animal . . . actually "bites" off pieces with each rasping stroke" (Carriker, 1946b). Elsewhere Carriker also stated that the main function of the radula, on this type of food, is to cut it up into conveniently sized pieces and that there is no appreciable trituration. If the contents of the crop of *Agriolimax reticulatus* are examined after it has been feeding on lettuce, it is found to contain chunks of lettuce with a characteristic shape rather than the fine particles that would be expected if the food was rasped. On thicker food Carriker felt that "true rasping comes into play." After feeding on carrot, however, the crop of *Agriolimax reticulatus* contains chunks of carrot, similar in form to the pieces of lettuce, so that simple rasping would also seem unlikely with this food. A study of the form of the odontophore and the arrangement of the teeth, at its U-shaped edge, makes it appear more likely that on a three-dimensional food the radula is used rather like a scoop (Runham, 1969). The edge of the U-shaped odontophore is brought against the surface of the food and pushed through an arc down into, forwards, and up through, the food rather as a scoop through ice-cream. During this movement the passage of teeth over the rim of the cartilage makes this area into a cutting edge, and when at the end of this forward stroke the odontophore meets the jaw the piece of food is cut off. Feeding on a two-dimensional food is a modification of this feeding movement – the whole cutting edge is first applied to the substratum but it cannot pass down into the food and so as the radula moves forwards the sides of the U are lifted off, leaving only the centre of the U to rasp the surface. Many reports suggest that either the lateral edges, or the central part of the radula, are most important in feeding in some snails, or on some substrata, but such observations appear to reflect merely the pattern of contact of the odontophore with the unnatural flat surface, rather than indicate any fundamentally different type of feeding activity. During the scooping action of the radula all parts of the radula would seem to be important for feeding. Some pulmonates do appear to feed differently and the reader is referred to the excellent account by Märkel (1957).

Lymnaea peregra (Storey, 1971), *Bulinus tropicus* (Krauss) (Stiglingh and van Eeden, 1970) and probably most other pulmonates, when crawling on a surface, appear to execute random feeding strokes approximately following the line of movement, so sampling the surface for food. When they reach a suitable food material the head is moved from side to side as they feed, so that feeding tracks are made close together and side by side. This movement means that they clear a broad path through the food, and the better favoured the food material the slower they crawl, so that feeding tracks become superimposed.

Such behaviour appears to lead to the most efficient removal of food material.

Most studies on pulmonates refer to the mainly herbivorous forms but there are also some very interesting observations on carnivores. *Testacella* largely feed on earthworms. Their whole body is adapted for movement along the worms' burrows and the feeding apparatus is highly modified for seizing and swallowing the living worm (Webb, 1893). The radular teeth are long, narrow, pointed barbs. When the slug meets its prey the white spoon-shaped odontophore rapidly shoots out of the mouth. The erect teeth, on the outside of the odontophore, penetrate the skin of the worm, and as the radula passes back over the cartilage the worm is dragged into the groove between the cartilages where the many rows of backwardly directed teeth help to hold it firmly. The impaled worm is dragged into the mouth when the odontophore is retracted. Gradually the worm is engulfed and this may take a long time with a large worm.

As would be expected with such an active organ, the arterial and nervous supplies to it are very elaborate (Carriker, 1946a; Bekius, 1972; Walker, 1969). Because of the orderly sequence of contractions involving so many muscles the neurophysiology of the organ must be very interesting, but little has so far been published on this. Preliminary studies on isolated buccal masses indicate that there is a programmed sequence of nerve impulses to the muscles during feeding (Goldschmeding, personal communication; Kater and Fraser-Rowell, 1973).

VI. The salivary glands

The salivary glands are thin sheets of yellowish tissue situated on either side of the post-oesophagus or crop, and are usually joined to each other and to the tract by thin connective tissue bands, blood vessels, and nerves. Their ducts enter laterally on the posterodorsal surface of the buccal mass. The glands are tubulo-acinous, consisting of glandular acini connected via glandular ducts to the main collecting duct.

Studies using the light microscope have indicated variable numbers of cell types, and some workers have suggested that many of these are only stages in the development of a smaller number of cell types. There is fair agreement that ciliated, mucous, serous, and undifferentiated cells are present, but the controversy relates to the number of different types of mucous and serous cells. Most of the difficulties in interpretation relate to the difficulty of fixing this material: the secretion is often badly fixed and its absorption of water during the preparative procedures may destroy the cell. Most recent workers use the infinitely more reliable fixative and embedding procedures for electron microscopy. Using such methods Boer *et al.* (1967) found nine different cell types in the salivary glands of *Lymnaea stagnalis* and Walker ten in those of *Agriolimax reticulatus* (Table III). Both authors believe that these cells are completely different and do not represent stages in cell development. In both species there are two types of mucous cell which produce acid mucopolysaccharides, together with granular, grain, and pseudochromosome cells, that produce proteins combined with neutral polysaccharides. With the exception of the pseudochromosome cell, these cells are present in very large numbers in the gland. The other cell types, which are peculiar to the species, are not common.

The cells discharge their various secretions directly into the lumen of the salivary gland with the single exception of the β cell in *Agriolimax*. This cell type possesses a well-developed intracellular duct system which collects the secretion prior to its discharge. In both species Leydig cells are present between the epithelial cells and in the connective tissue sheath. These cells are probably amoebocytes bringing nutrients to the salivary gland cells.

Secretions produced in the acini are collected by ciliated intralobular ducts. A separate ciliated cell type is present in *Lymnaea* but in *Agriolimax* the ciliated cells are identical in all other respects to the granular secreting cells. The larger interlobular ducts discharge into the salivary duct. All ducts are ciliated in *Lymnaea* but only the intralobular ones in *Agriolimax*. A muscular layer is present in the walls of the ducts in *Agriolimax*; this is thickest in the salivary duct. In *Lymnaea* therefore movement of saliva is due to ciliary action, while in *Agriolimax* saliva, collected by cilia initially, is later moved by peristalsis. This peristalsis is visible in the salivary ducts of the living animal and there appears to be a slight dilatation, which could act as a storage area, near its entry into the buccal mass.

Glandular cells line most of the ducts as well as the acini in *Lymnaea*. The distribution of the various cell types is not random; thus ciliated cells are restricted to the ducts, granular cells are mainly present in the salivary duct, the pseudochromosome cells are mainly found in the interlobular ducts, and mucocytes and some pseudochromosome cells are found in the acini. *Agriolimax* has the granular cells restricted to the walls of the inter- and intra-lobular ducts, while all the other cell types are apparently randomly distributed throughout the acini (Table III).

Table III. Cell types in the salivary glands of *Lymnaea stagnalis* and *Agriolimax reticulatus*

Cell type	*Lymnaea stagnalis* Frequency	*Agriolimax reticulatus* Frequency	Nature of secretion
Basophilic	Interlobular ducts and lobules 2%		Absent
Granular	Secretory duct 90% Interlobular ducts 2·4%	Common in intra- and interlobular ducts	Neutral polysacc-haride and protein Amylase
Pseudo-chromosome	Interlobular ducts 84% Lobules 14%	Very few	Neutral polysacc-haride and protein
Mucocyte 1	Lobules 53%	Common in lobules	Acid mucopolysacc-haride
Mucocyte 2	Lobules 24%	Common in lobules	Acid mucopolysacc-haride
Acidophil	Uncommon		Neutral polysacc-haride and protein
Grain	Lobules of laboratory animals 1%, Field animals 10%	Common in lobules	Neutral polysacc-haride and protein
Mixed	(mixture of properties of pseudochromosome and mucocyte 2 cells) Lobules 6·2%		
Mixed 1		(Mixture of properties of grain cell and β cell) Rare	
Mixed 2		(Mixture of properties of grain cell and mucocyte 2) Rare	
Ciliated			
Leydig	Common near connective tissue sheath	Common near connective tissue sheath	Protein
α		Uncommon	Acid and neutral polysaccharides and protein
β		Uncommon in intralobular ducts	Unidentified

The data for *Lymnaea stagnalis* are derived from Boer *et al.* (1967) and those for *Agriolimax reticulatus* from Walker (1970b).

The salivary ducts in *Agriolimax* and *Helix* are highly specialized structures (Walker, 1970b; Quattrini, 1967). Lining the muscular duct is an epithelium of cells bearing well-developed microvilli, highly elaborate infoldings of the plasma membrane at both basal and luminal surfaces, complex interdigitations between the cells, well-developed junctional complexes at the luminal junctions of the cells and very many large mitochondria sandwiched between the many folds of the plasma membrane. All of these features are typical of cells engaged in water transport. Why this type of epithelium is present in the ducts of these animals, and not in *Lymnaea*, is unknown, as is also the direction of the water flow, if this is indeed its function.

Analysis of the salivary glands clearly indicates the presence of amylase together with small amounts of a trypsin-like protease (Carriker, 1946b; Boer *et al.*, 1967; Walker, 1970b). Boer *et al.* (1967) were able to subdivide the glands in *Lymnaea* into three fractions; the salivary ducts, the interlobular ducts and the acini with their interlobular ducts. On analysis they found that amylase activity was almost entirely restricted to the salivary gland ducts, hence concluded that amylase was produced by the granular cells.

As yet nothing appears to be known of the volume or nature of the saliva. All analyses reported refer to tissue extracts. Blain (1957) in a light microscope study of the glands in *Helix pomatia* reported that the proportion of the various cell types changed following feeding on different types of food. This phenomenon was interpreted as evidence for the view that there were cycles in the development of these cells, but as this now seems unlikely these interesting experiments need to be repeated using modern techniques.

The function of the saliva is believed to be lubrication during feeding, assistance with the removal of food from the radula, and with its passage back into the oesophagus. Extracellular digestion is also started by the amylase, and trypsin where present, that are found in the saliva. It is not known if saliva is poured on to the surface of the food, but in *Otina* (Morton, 1955a), *Oxychilus* (Rigby, 1963) and *Succinea* (Rigby, 1965), it is claimed that the pedal mucus is used in this way. The buccal glands, and in some species accessory salivary glands around the entry of the salivary ducts into the buccal mass (Argaud and Bounoure, 1910), also appear to function in lubrication. How and when these various secretions are produced during the feeding cycle is unknown.

The blood supply to the salivary glands is well developed (Carriker, 1946a; Walker, 1969). The salivary artery has many fine branches which supply all parts of the gland and ducts, and also contribute to the vascularization of the crop or oesophagus. Nerves to the glands have a double origin – anteriorly there are paired salivary-gland nerves arising from the buccal ganglia, that pass along the salivary ducts to break up into many fine branches supplying the gland, while the posterior salivary-gland nerve, a branch of the intestinal nerve arising from the visceral ganglion, subdivides extensively over the gland. As the buccal ganglia appear to programme the complex feeding movements it is possible that passage of saliva into the buccal cavity is integrated, via the anterior salivary gland nerve, with these movements.

VII. The oesophagus

Opening from the posterodorsal aspect of the buccal mass, the oesophagus passes posteriorly to the stomach. Along its length it can be subdivided into regions mainly by changes in its diameter. In Basommatophora the part of the gut between buccal mass and stomach is divisible into pro- and post-oesophagus, and crop, while in the Stylommatophora there are two parts termed oesophagus and crop. The crop has the largest diameter, and is much longer than the oesophagus in the Stylommatophora, but is very short in Basommatophora.

In *Agriolimax reticulatus* the cuticularized dorsal food groove and ciliated lateral folds, in the roof of the buccal cavity, continue into the oesophagus. The ventral part of the oesophagus is continuous with the collostylar hood. Six longitudinal ridges are present in the oesophagus; two lateral ridges continuous with the lateral folds, a single ventral and three dorsal ridges. Cilia are present at the anterior end of the lateral ridges and at the posterior end of all the ridges. Over most of the surface of the oesophagus there is a cuticular lining. Mucous cells are interspersed with the ciliated cells. As the lumen enlarges at its junction with the crop, the oesophagus loses the ridges and cuticle. The lining epithelium of the crop has scattered patches of ciliated cells and many uniformly distributed mucous cells. Most of the cells in this region bear a brush border and contain many apical granules. Beneath both regions there are layers of muscle cells – an outer circular layer and an inner longitudinal, or oblique layer.

Amongst the pulmonates there seems to be considerable variation in the amount of ciliation in this region of the gut. *Lymnaea stagnalis* has a number of ridges along both parts of the oesophagus and these together with the smooth crop are richly ciliated. The minute *Otina otis* Turton and ellobiids studied by Morton (1955a, b) have this whole region of the gut very densely ciliated, while in *Oxychilus cellarius* and

Succinea putris (L.) (Rigby, 1963, 1965) cilia are apparently absent. The reasons for such variation are unknown, but may reflect the relative importance of cilia and muscles for transport and mixing of the food material.

Histochemically the granules in the epithelial cells of the crop wall have been shown to contain non-specific esterase (Walker, 1969), one of the enzymes associated with lysosomes. Acid phosphatase, the enzyme most frequently used as a marker for lysosomes, was absent from the crop wall of *Agriolimax* but has been demonstrated there in *Arion ater* (Bowen, 1970). The presence of many microvilli, and some pits between them, may indicate uptake of food materials. When the crop walls of fed and starved animals are compared there is a vast difference in the number of granules present. In the starved animals granules are present only in the cytoplasm above the nucleus, while in the fed animal they fill most of the apical cytoplasm. So many granules may be present after a meal that the cell becomes swollen and the microvilli are lost. In starved animals the granules consist of a dense body surrounded by a clear fibrous zone, while after a meal the granules are heterogeneous, and there are many that apparently contain lipid.

The crop lumen in Stylommatophora is frequently filled with a reddish brown liquid known as crop juice. It contains a rich soup of enzymes and is sold commercially, because of its value in digesting such a wide range of materials. Food entering the crop remains in this fluid for some hours. The crop juice is active against a wide range of polymers, e.g. carbohydrates, proteins and lipids (Table IV).

Most work in this field relates to the effect of whole crop juice on a variety of substrata. As indicated by Evans and Jones (1962a) carbohydrases are not very substrate-specific so that one enzyme moiety may be capable of hydrolysing several different substrata. Future work using purified extracts from the juice may clarify this point.

Much research has centred on the presence of cellulase and chitinase in the juice, because for many years it was believed that these enzymes could not be produced by animals, only by bacteria, and also because they allow much greater efficiency of digestion in a herbivore. Jeuniaux (1954, 1963) has isolated a large number of bacteria from the crop juice that are capable of digesting cellulose and chitin. However, Strasdine and Whitaker (1963) were able to show that the digestive gland in *Helix pomatia*, which was almost free of bacteria, contained these enzymes and Parnas (1961), administering various antibiotics and fungicides to *Levantina hierosolyma* (Boiss.), found that cellulase activity persisted only in the digestive gland. They also observed that in aestivating snails cellulase was similarly present only in the digestive

gland. The above authors conclude that these enzymes are produced
within the digestive gland of the snail, as well as by the bacteria.
Cellulose digestion is not a simple process because of the very complex
nature of this polymer and there are several stages in its breakdown

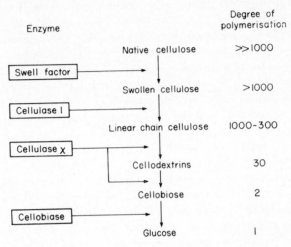

	Enzyme	Degree of polymerisation
	Native cellulose	>>1000
Swell factor		
	Swollen cellulose	>1000
Cellulase I		
	Linear chain cellulose	1000–300
Cellulase x		
	Cellodextrins	30
	Cellobiose	2
Cellobiase		
	Glucose	1

(Koopmans, 1970). Most published work on cellulose digestion by
pulmonates is concerned with the shorter polymers, but the animals
normally eat native cellulose in the form of plant cell walls. Koopmans'
(1970) careful work has shown that in *Helix pomatia* there is one enzyme
capable of attacking native cellulose, together with two separate
enzymes that are capable of digesting the degraded celluloses. It is to be
expected that more than one enzyme could be involved in the breakdown of the similarly complex chitin molecule, where again the
published work refers to digestion of degraded chitins. In view of the
complexity of cellulose, and probably chitin digestion, it would appear
that most of the early work on bacterial *vs* snail origin of the enzymes
will have to be repeated, in order to establish where the different
enzymes are produced.

Peristalsis has been observed in the oesophagus of most of the
pulmonates that have been studied. In *Arion ater* Roach (1968) found
complex peristaltic rhythms. The oesophagus was found to have small
contractions, 3/min; the front part of the crop had stronger contractions
lasting 1·2 min, characterized by a maintained contraction, with fast
relaxation and contraction between successive waves; and the hind crop
showed large changes in muscular tone lasting 3 minutes each. Roach
suggests that the oesophagus rapidly transports food from the buccal
cavity back to the crop where it is mixed with crop juice. Then the slow

peristaltic waves, coupled with the differences in activity between the two ends of the crop, ensure that the contents are passed to and fro and thoroughly mixed. In *Achatina fulica* Bowdich, there appears to be a sphincter dividing the crop into two halves (Ghose, 1963). Its function is unknown, but *Achatina* differs from most other Stylommatophora in that one of the lobes of the digestive glands opens into the so-called posterior crop.

An extensive network of nerves arising from the paired dorsal gastric nerves passes down the oesophagus and crop to the stomach (Carriker, 1946a; Walker, 1969). Studies on isolated oesophagus and crop indicate that peristalsis is controlled by the plexus in the wall rather than by the central nervous system (Roach, 1968; Minker and Koltai, 1961). Arteries in this region are also complex and extensive. Anteriorly, paired dorsobuccal arteries mainly supply the oesophagus while extensions of the salivary artery supply the mid regions and a posterior crop artery, arising from the minor hepatic artery, supplies the region near the stomach.

In *Lymnaea* the posterior oesophagus and crop appear to function largely as a temporary store for food, prior to its passage into the gizzard, and little extracellular digestion appears to take place. In *Agriolimax* and other Stylommatophora much extracellular digestion probably results from the activity of the crop juice. Some small molecular weight materials must therefore be released and, with such a thin-walled structure, absorption would seem likely. This has apparently been confirmed only for *Agriolimax reticulatus* (Walker, 1972). Foods mixed with labelled glucose, galactose, glycine, palmitic acid, or with Thorotrast (colloidal thorium dioxide) were fed to the animals. The fine particles of Thorotrast were not taken up, but the soluble materials were absorbed. Glucose, galactose and glycine were apparently taken up by membrane transport, as they appeared to be diffused throughout the general cytoplasm, but palmitic acid was found in large amounts in the cytoplasmic granules. It seems likely, considering the evidence given above, that the cells take up lipid by pinocytosis. The label was present in the granules for up to 5 days after the meal. The crop appears to be a important area for lipid uptake, and this may be related to the presence of large amounts of esterase in the cytoplasmic granules.

The oesophagus and crop function as an area for food storage, extracellular digestion and absorption. There appear to be no quantitative data available on the proportion of food breakdown and absorption that take place in this region. In *Agriolimax reticulatus* radio-opaque material fed to the animals is found by X-rays to appear almost immediately after feeding in the crop, and it first moves into the stomach after 20–40 min. Several hours are needed for all of the meal to pass into the stomach and very fine particulate material may be present for several days.

VIII. The stomach

The crop leads directly into the stomach. The digestive tract executes a sharp bend at this point and receives the ducts from the two lobes of the digestive gland. Considerable variation in the anatomy of the stomach is found throughout the pulmonates but, perhaps because of small size, it has not been as well studied as in other gastropods and lamellibranchs. There is great similarity between the stomachs of primitive lamellibranchs and gastropods, and the pulmonate stomach can readily be derived from a somewhat similar structure.

In order to understand the pulmonate stomach it is necessary first to describe the basic structure of a primitive gastropod stomach. The primitive stomach (Graham, 1949) is divisible into ciliated and cuticularized areas (Fig. 4). The cuticle is restricted to the relatively small area of the gastric shield against which the tip of the protostyle, a rod of incipient faecal material, is rotated. Part of the stomach wall, together with the caecum, forms a sorting area. This consists of parallel folds which are richly ciliated; the cilia on the tops of the folds transport small particles to the two digestive gland openings, while those in the grooves carry the large particles to a rejection tract, and thence to the style sac. The style sac is the last part of the stomach, and opens into the intestine. It is subdivided into dorsal and ventral channels by two large folds in the wall, the major and minor typhlosoles. The larger dorsal channel is lined with a richly ciliated epithelium which rotates the protostyle. The ventral channel is the intestinal groove, through which waste material from the digestive gland and large food particles pass to the intestine. Starting at the tip of the caecum, the major typhlosole passes down the caecum, across one side of the sorting area and along the length of the style sac. The minor typhlosole is only present in the style sac. Food arrives in the stomach as a steady stream of fine particles, embedded in a string of mucus, and this is wound on to the tip of the protostyle. During the rotation of the protostyle the particles are to some extent triturated against the wall of the gastric shield, but they also become dislodged from the food string because the low acidity in the stomach reduces the viscosity of the mucus. Rotation of the protostyle also aids in sweeping particles over the surface of the sorting area. The accepted food particles are transported

through the digestive gland ducts to the gland, while waste materials
are transported in the opposite direction, back to the stomach, via
ciliated grooves in the walls of the ducts. The fine particles are digested
by intracellular digestion in the gland. Faeces are formed from the
various waste materials which are conveyed to the end of the intestinal

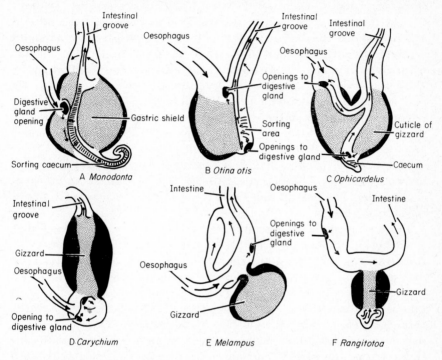

Fig. 4. Structure of the stomach. A, a primitive prosobranch (after Fretter and
Graham, 1962); B, *Otina otis* (after Morton, 1955a); C, a primitive ellobiid (after
Morton, 1955c); D, E, F, advanced ellobiids (after Morton, 1955b). The cuticularized areas are shaded and the muscular layer is solid black. Arrows indicate
ciliary currents.

groove where they are plastered on to the end of the protostyle, and
this is pinched off at intervals to pass into the intestine. The protostyle
therefore has several functions; it helps wind in the food string, it is
involved in trituration of food material, it assists sweeping of material
over the sorting area, and it is involved in faeces formation. This
primitive stomach is clearly adapted for the treatment of food that
arrives in a steady slow stream of fine particles.

   Morton (1955a) believes that *Otina otis* (Turton) is one of the most
primitive pulmonates. This species feeds on fine particles that are collec-

ted by the radula and passed into a stomach (Fig. 4b) that is little modified from the primitive type described above. A protostyle is present and the food material, mainly diatoms, filamentous algae and detritus collected from the rock surface, become incorporated into it. It rotates, sweeping material over the ciliated sorting area and strong muscular contractions of the stomach wall help both to triturate the food material against the protostyle, and to squeeze fluid and fine particles into the digestive glands. Faeces are nipped off from the protostyle by peristaltic contractions at the entrance to the intestine. Waste material is conveyed by grooves from the digestive gland openings into an intestinal groove along the style sac. As indicated by Morton this stomach shows the first stages in the development of a gizzard and there is a greater reliance on muscular, rather than ciliary movement, but it still retains most of the features of the primitive stomach. The arrangement of grooves from the digestive gland openings into the intestinal groove resembles more closely that found in higher pulmonates. It is not known what type of particles pass into the digestive glands and the only enzyme that has been demonstrated is a protease, apparently free in the stomach lumen.

Morton (1955b, c) has also studied a number of the primitive basommatophoran Ellobiidae, and here there is an amazing variety of stomach form. Why this group should have experimented so extensively with this organ is not clear, although its present unusual position in this respect may only reflect the paucity of work on other pulmonate families. A primitive ellobiid stomach such as that in *Ophicardelus* (Fig. 4c) is easily derived from that found in *Otina*. There is a greater development of the muscles and cuticle to form a large gizzard, and there is a well developed Y-arrangement of grooves from the digestive gland openings into the intestinal groove. In *Carychium* (Fig. 4d) only the intestinal end of the gizzard is developed, while in other genera such as *Melampus* (4e) and *Rangitotoa* (4f) the gizzard has become variously cut off from the stomach as a diverticulum. Apparently the posterior caecum forms an expandable process in *Rangitotoa*, and some other forms, that allows the movement of food material in and out of this gizzard. With such an interesting range of structure a detailed study of function – the types of food entering the stomach, their fate in the stomach, the enzymes present and what material passes into the digestive gland – should prove a fascinating study.

Of the other Basommatophora, detailed information is available only for *Lymnaea stagnalis* (Carriker, 1946a, b). The Planorbidae are said to closely resemble *Lymnaea* (Graham, 1949) as also are the Physidae

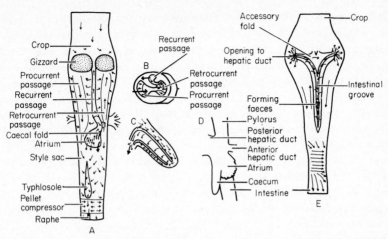

Fig. 5. Structure of the stomachs of *Lymnaea stagnalis* A, B, C (after Carriker, 1946b) and *Agriolimax reticulatus* E (after Walker, 1969). A, stomach, cut ventrally and opened out, showing ciliary currents; B, section through the pylorus; C, ciliary currents around the folds in the caecum; D, dorsal view of the stomach region; E, stomach, cut ventrally and opened out, showing ciliary currents.

and Ancylidae (Rigby, 1965), but *Amphibola* appears to be very different (Sinclair in Morton, 1955c). *L. stagnalis* (Fig. 5) has a very large, highly muscular, bilobed gizzard normally containing sand grains. Behind the gizzard there is a thin-walled chamber commonly known as the pylorus despite the lack of resemblance to the pylorus of vertebrates, with an atrium at one side. A caecum with a sorting area opens from the atrium and there is also a ciliated sorting area leading to two digestive gland openings. This latter area is termed the hepatic vestibule. Along the pylorus (Fig. 5b) there are three well developed folds in the wall, two dorsal and a single ventral, which subdivide it into three channels – retrocurrent, procurrent and recurrent passages. The typhlosoles are apparently represented by a vestige of one fold at the intestinal end of the style sac (termed by Carriker prointestine) and one of the folds in the caecum.

Little comparative anatomy of the stomach of stylommatophorans has been published, and the lack of clear divisions between crop and stomach make comparisons difficult. In *Oxychilus cellarius* the stomach is simple, a gizzard and sorting area being absent. The two separate digestive gland openings open to large highly ridged and heavily ciliated ducts leading to the glands. Rigby (1963) terms these ducts hepatic vestibules, although whether these two structures are merely enlarged hepatic ducts which normally have their own well developed ciliated ridges, or are homologous with the single hepatic vestibule of *Lymnaea* is open to debate. A Y-shaped groove bounded by typhlosoles conveys waste material from the digestive glands to the style sac. Passing anteriorly from the digestive gland openings there are crop folds that resemble the dorsal folds along the pyloric chamber of *Lymnaea*. The function of the crop folds in *Oxychilus* is not clear, but their presence raises the problem of the origin of the crop. It is normally assumed that the crop is a dilatation of the posterior oesophagus, but the situation in *Oxychilus* could be interpreted as an incorporation of the pyloric stomach into that region. Studies on the embryology of *Achatina fulica* (Ghose, 1962) indicate that the oesophagus arises from the stomodeum, while the crop develops as a diverticulum from the stomach. Is it therefore possible that the anterior stomach of the Basommatophora forms the stylommatophoran crop? Much more information on the comparative anatomy and embryology of the stomach in the pulmonates is needed to settle this question. More advanced Stylommatophora lack the crop folds.

In *Agriolimax reticulatus* (Fig. 5) the crop passes directly into the stomach where there are separate digestive gland openings connected via channels to the intestinal groove. The arrangement of typhlosoles, with the development of the anterior accessory fold, is similar to the situation in *Oxychilus*.

Functionally the stomachs in Basommatophora and Stylommatophora appear very different although information is available only for *Lymnaea stagnalis* (Carriker, 1946b) and *Agriolimax reticulatus* (Walker, 1969). In *Lymnaea stagnalis* the particles of food, both large and small, pass into the posterior oesophagus and crop where they are stored temporarily. Sand grains are essential for trituration in this species (Carriker, 1946b), and also in *L. peregra* (Storey, 1970), and they are found mainly in the gizzard but also in the crop and retrocurrent passage in the pylorus. The contents of the gizzard are subject to a slow rotation and kneading movement, material squeezed out is passed back by the muscular action of the crop, and strong ciliary movements in the retrocurrent passage. Trituration of the food, aided by secretions from the salivary glands, results in a gradual comminution of the food particles. Fine particulate and soluble materials entering the procurrent passage can pass through the fine slit between the dorsal folds and into the dorsal recurrent channel. Larger particles are trapped in mucus and returned to the gizzard via the retrocurrent passage. As emphasized by Carriker the dorsal folds act as a filter ensuring that no large particles enter the atrium. A mechanism of this kind is essential with a triturating gizzard and the necessity for returning large particles to the gizzard for further treatment. At intervals some of the contents of the gizzard pass into the style sac for faeces formation. In the hepatic vestibule sorting of the particulate matter ensures that only the finest particles enter the hepatic ducts, while the larger particles pass to the caecum. In the caecum these particles become entrapped in mucus, and pass in a stream to the style sac. There is a continuous string of such material, termed the caecal string, which is an easily recognizable constituent of the faeces. Carriker believes that pulsations of the stomach region, rather than ciliary currents, are responsible for most of the movement of material into the digestive glands. At intervals such pulsations stop and waste material leaves the digestive gland, passing via the hepatic vestibule into the style sac. Faeces are divisible, because of colour differences, into the long thin caecal string, a thicker discontinuous liver (digestive gland) string, alternating with a thick gizzard string. In this stomach there therefore appears to be considerable emphasis on muscular movements and trituration which results in mechanical breakdown of the food with a little help from enzymes. There is also an efficient sorting of particles, so that only the smallest enter the digestive glands, and segregation of the food still being comminuted from that being sorted, and the waste material.

Agriolimax in common with other Stylommatophora has apparently no method of triturating food once it has entered the oesophagus. On entering the crop it meets the very powerful crop juice and the mixture is then churned around before being released into the stomach. The first material is released from the crop after 20–40 min, but it may be several hours before the majority of the food has passed into the stomach, and very small amounts may be present in the crop for more than 3 days. Food material remains in the stomach only for a short time, suitably fine particles and soluble material passing into the digestive glands by pulsations of the stomach and digestive glands (Walker, 1969; Roach, 1968). The ciliated folds around the digestive gland openings serve to exclude all large particles from the ducts. Barium sulphate with a particle size of 1–3 μm was completely excluded while Thorotrast of 0·1–0·4 μm in size readily passed into the ducts. The rapid passage of material into the digestive glands confirms a muscular rather than a ciliary mechanism for passing material from the stomach into the digestive glands. The size of particles that do enter the hepatic ducts is so small that it is possible they pass between the cilia. In *Mytilus*, a filter feeding bivalve, the filtering cilia are partially fused together to form cirri but even here the space between the cilia is 0·6 μm (Moore, 1971). Rosén (1952) in a review of his own and earlier work on phagocytosis in the digestive gland of *Helix* showed that biological particles, as large as blood corpuscles (6 μm) could enter the digestive gland. Such results were obtained only after long periods of feeding on these materials alone. It is desirable that this work is repeated using electron microscopy and other modern techniques for if these results are correct, and large biological particles are accepted, but much smaller inert particles are not, then some sensory mechanism at the duct openings appears essential.

Waste material from the digestive glands passes out via the grooves in the walls of the hepatic duct, and then through the channels to the intestinal groove. The faeces contain two easily recognizable components (Pallant, 1970). The first part consists of a membranous sac with a brown fluid at one end and particles at the other, and appears to originate from the digestive gland, i.e. it is the liver string. The second part consists of large particles of material and clearly represents the material that passes directly through the stomach to the intestine. In *Agriolimax* there is therefore powerful extracellular digestion of the food material followed by a slow passage of material into the stomach, where dissolved substrates and fine particles enter the hepatic ducts and larger particles pass to the style sac, and waste material from the digestive gland is kept separate from the other contents of the stomach.

Considering the complex series of events that takes place in the stomach it is not surprising that the arterial and nervous supplies are complex (Carriker, 1946a; Walker, 1969; Bekius, 1972). On the region of the stomach wall where the digestive gland ducts arise, there are several small ganglia and these give rise to a complex anastomosing plexus over the stomach. The neurophysiology of the stomach area has been little studied. Roach (1968) found considerable variation in stomach contractions depending, in part, on whether food material was present and distending it, or if the stomach was empty. Contractions at a rate of 1·5/min were often found but they always stopped within 20 min. In other experiments a regular heart-like beat was found particularly in full stomachs, and contractions sometimes lasted for days. If the preparation slowed or stopped, the beat could be restored by stretching. When under load the stomach contracted at an average rate of 1·5/min, but contractions were in groups at a higher speed separated by pauses of inactivity. The stomach contractions appear to be largely independent of central nervous control. It is to be hoped that future work on stomach neurophysiology will clarify the relation between these findings and stomach function. The blood supply to the stomach arises from the minor hepatic artery which crosses the stomach at the level of the hepatic ducts, and it sends one vessel down each typhlosole and others across the accessory fold.

IX. The digestive gland

The digestive gland is the largest organ in the pulmonate body. Primitively there are two lobes but one may be reduced, or in a few forms completely absent. Each gland opening leads into a large duct which branches to form smaller ducts, then ductules, and finally these end in complex, branched, blind tubules. The whole gland is bound together by connective tissue. Little attention has been paid to the anatomy of the gland, but in *Agriolimax reticulatus* the tubules have a constant arrangement (Fig. 6) which may be of comparative value.

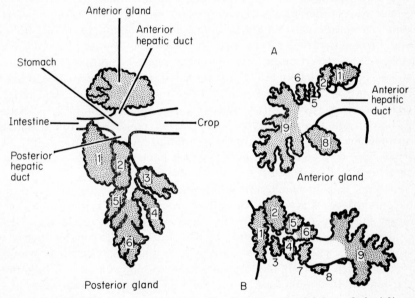

Fig. 6. Arrangement of the lobules in the digestive gland lobes of *Agriolimax reticulatus* (after Walker, 1969).

A large number of workers have studied the cytology of the ducts and tubules but there is still not complete agreement. Some of the difficulties with interpretation are due to fixation – it is difficult to get good fixation of all the cell types in the gland, particularly with normal light microscope methods. Even at the electron microscope level interpretation is still difficult because of changes in ultrastructure associated with phases of activity. Material in the stomach may also get forced into the ducts on fixation.

Nakazima (1956) has published an excellent review of fifteen years' work surveying the structure and function of the digestive gland throughout the Mollusca. This light microscope study has recently been supplemented by extensive accounts of the ultrastructure of the digestive gland cells in *Biomphalaria pfeifferi* (Krauss)(Meuleman, 1972) and *Agriolimax reticulatus* (Walker, 1970a, 1972). The following survey is based largely on these accounts and the original papers should be consulted for discussions of the earlier literature.

The ducts have longitudinal folds in their walls and the epithelium is highly ciliated with some interspersed mucous cells. Striking amounts of lipid and glycogen are present in the ciliated cells, and there are smooth muscle cells in the connective tissue beneath the epithelium. The epithelial cells can be subdivided into four main types: thin cells which are undifferentiated precursors of the other cell types; digestive cells; excretory cells; and (in *Agriolimax*) calcium or (in *Lymnaea* and *Biomphalaria*) secretory cells. Digestive cells predominate and show considerable variation in form obviously related to different functional states. Thus Meuleman (1972) distinguishes the following: 1, a narrow columnar cell with very active organelles involved in protein synthesis, a clear brush border, and a homogeneous apical region; 2, a wider columnar cell showing active endocytosis (pinocytosis and probably also phagocytosis) resulting in the accumulation of small apical granules which fuse to give green granules, which have a granular content, and there are small protolysosomes which also fuse with the green granules; 3, a large cell in which continuing endocytosis has resulted in large accumulations of granules, the green granules have coalesced to form larger and larger yellow granules and eventually a single very large yellow granule; 4, a cell lacking a brush border, in which endocytosis has stopped and where the apical cytoplasm including the large granule together with some smaller granules becomes pinched off and falls into the lumen. It is not known whether one cell can undergo several of these cycles or not; Meuleman thinks most probably not. The colour of the yellow granules was found to vary with the type of food consumed. Walker's findings on the digestive cell of *A. reticulatus* are similar: the same sequence of endocytosis, apical granules, green granules and finally yellow granules is found but the final discharge of the apical cytoplasm together with the large granules was not observed; rarely the tops of the swollen cells lacking microvilli bud off into the lumen. After injection of the animal with histamine, atropine or pilocarpine many cells behaved in this way.

Because of similarities in ultrastructure and histochemistry, coupled with the results of drug treatments, Walker believes that the digestive cell with a well developed large, yellow granule becomes transformed into an excretory cell. This cell type has a brush border, and few cytoplasmic organelles other than the very large granule. Meuleman believes that in *Biomphalaria* the secretory cell gives rise to the excretory cell.

The secretory cell is pyramidal and possesses the well developed rough endoplasmic reticulatum and golgi body characteristic of protein-secreting cells. Large granules accumulate in the apical cytoplasm and are released into the lumen. In *Agriolimax* the morphology of the calcium cell is essentially the same as that of the secretory cell, but the granules contain concentric layers of material and stain strongly for calcium. In this species the apical cytoplasm contains small granules and some protein granules, while the large calcium-containing granules occupy a more basal position. In *Biomphalaria* some secretory cells show signs of degeneration: the protein granules undergo resorption and yellow granules similar to those in the digestive cell appear. Finally the degenerated secretory cell, now termed excretory cell, disintegrates releasing the cell contents to the lumen. The calcium and secretory cells are most frequently found at the ends of the glandular lobules, and Meuleman always found the excretory cells close to normal secretory cells, Neither view of the origin of excretory cells is completely convincing; it may be different in the two animals but it is possible further work will support only one of these views.

Several functions have been ascribed to the digestive gland, namely absorption of food materials at various stages in its digestion, secretion of enzymes, storage of reserve materials and excretion. The evidence for absorption is now firm. Soluble products of digestion and very fine particles enter the hepatic ducts. The muscle cells present around the lobules and ducts give rise to pulsations in the gland, which apparently assist the contractions of the stomach in moving such materials into the glands. What size of particle enters the ducts has been a matter of controversy. Rosén (1952) experimented with many types of particle in *Helix* and found that only the very finest (0·1 μm) carbon particles could enter the gland. He also claimed that pigeon red blood cells and lettuce chloroplasts were taken up by the digestive gland cells, but they were only visible there after the animals had been feeding on these materials for 2–5 days. As Meuleman has shown, the colour of the granules in the digestive gland cells is dependent on the diet and therefore Rosén's work should be repeated using the electron microscope. Using radio-opaque particles and X-raying animals at intervals after the meal, Walker showed that barium sulphate particles (1–3 μm) and chromopaque (1 μm) did not enter the digestive glands. Thorotrast (colloidal thorium dioxide) with a particle size of 0·1–0·4 μm passed rapidly from the stomach into the digestive gland. It therefore seems likely that only the very smallest particles can enter the gland, perhaps because they pass between the cilia around the digestive gland openings. These cilia are present on folds that radiate out from the openings and they all beat away from the openings. In *Agriolimax* these cilia apparently form an effective filter at the entrance to the ducts. The particles that do enter the duct pass very rapidly into the gland and are taken up by the digestive cells by endocytosis. Within the digestive cells the particles appear in sequence in apical granules, green and then yellow granules. No other cell types took up this material and these particles were lost from the gland after about 5 days. Some of the particles left the cells basally, as amoebocytes containing them were found in the underlying connective tissue, but the majority probably passed out apically. No Thorotrast was discovered in excretory cells. Soluble materials labelled with radioactive isotopes (glucose, galactose, glycine and palmitic acid) were taken up by all cell types including the duct cells and they had a diffuse localization. Galactose was however also taken up apically extremely rapidly by cells in the ducts, and to a lesser extent by small areas in the calcium cells. Palmitic acid accumulated in the basal areas of the cells. Calcium was accumulated only in the calcium cells and could still be found after five days in the calcium granules. Fretter (1952, 1953) studied the uptake of radioactive strontium, phosphorus, and iodine by a variety of molluscs. She found active uptake of strontium by the calcium cells in *Arion hortensis* Fér. and in smaller amounts by the digestive cells, but the isotope was retained only in the calcium granules. Strontium from the digestive cells passed into the haemocoel. Phosphorus and iodine were absorbed by both cell types but iodine was more strongly absorbed by the digestive cells. While phosphorus was stored in the calcium granules the majority passed into the haemocoel, and thence to most cells in the body. Iodine was largely excreted by the digestive gland cells and passed out in the faeces; any reaching the haemocoel was excreted by the kidney.

As we have seen above indigestible or unwanted material accumulates in yellow granules and is eventually discharged by the cells back into the lumen of the gland. Such material is conveyed by the well developed ciliary tracts in the ducts of the gland back into the stomach. On the folds in the walls of the duct the cilia beat downwards into the grooves where the cilia beat towards the stomach. A stream of this unwanted material bound together with mucus forms the liver string which is incorporated into the faeces. The liver string is not continuous but is apparently produced periodically, perhaps once a day in *Agriolimax* (Pallant, 1970). It is not known whether this periodic production of a liver string implies that the digestive gland cells discharge the unwanted material only at certain times; whether the material accumulates continuously but passes out of the gland at intervals; or if only part of the discharged material is got rid of in this way. In *Lymnaea* (Carriker, 1946b) the normal stomach pulsations stop while the liver string passes out of the digestive gland openings.

Tissue extracts of the digestive gland reveal the presence of a large number of enzymes (see Table IV). The granules within the digestive cells are shown histochemically to contain such lysosomal enzymes as acid phosphatase and esterase. Taken in conjunction with the electron

Table IV. Enzymes reported in the crop juice of pulmonates. Amylase and some protease activity are probably derived from the saliva.

Enzyme	Substrates tested	Linkages
Amylase[1,2,3]	Starch, glycogen, amylase amylopectin	1-4 and 1-6 α glucose
Cellulase[1,2,4]	Degraded cellulose, cellulose derivatives and native cellulose	1-4 β glucose
α-glucosidase*[1,2,3]	Sucrose, maltose, melezitose methylgalactoside	α-glucoside
β-glucosidase*[1,2]	Salicin, amygdalin, cellobiose, gentiobiose, p-nitrophenyl-β-glucoside	β-glucoside
Trehalase[1]	Trehalose	
α-galactosidase*[1]	Melitose, methyl glucoside raffinose	α-galactoside
β-galactosidae*[1]	Lactose	β-galactoside
α-D-fucosidase*[5]	p-nitrophenyl-α-D-fucoside	
β-D-fucosidase*[5]	p-nitrophenyl-β-D-fucoside	
Xylanase*[4]	Xylan	1-4 β-xylan
Laminarinase[2,6]	Laminarin	1-3 β-glucose
Licheninase[2]	Lichenin	1-3β, 1-4β glucose
Alginase[7]	Alginic acid, sodium alginate	1-4β mannuronic acid
Chitinase*[8]	Degraded chitin	1-4β acetyl glucosamine
Pectinase[1]	Pectin	
Mannanase*[9]	Mannan	1-2, 1-3 α mannose
Glucanase[9]	Glucan	1-2β, 1-3β glucose
β-glucuronidase*[10]		
Arylsulphatase*[10]		

Table IV. continued

Steroid sulphatase[10,11]	3β, 5α, 3β Δ^5, oestrogen and cortico-steroid sulphates	
Cathepsin*[12]	Casein	amide
Protease[13]	Peptone, casein, fibrin, haemoglobin	amide
Gelatinase[3]	Gelatin	amide
Lipase*[3]	Olive oil, Tween 20	ester

* Enzymes with a similar activity have been isolated from lysosomes.
[1]Evans and Jones, 1962a; [2]Stone and Morton, 1958; [3]Walker, 1969; [4]Nielsen, 1962; [5]Marray Got and Jarrilge, 1964; [6]Nielsen, 1963; [7]Franssen and Jeuniaux, 1965; [8]Jeuniaux, 1954; [9]Myers and Northcote, 1958; [10]Leon *et al.*, 1960; [11]Jarrilge, 1963; [12]Evans and Jones, 1962b.

microscope observations of the fate of ingested material in these cells it is clear that material entering the digestive cells is subject to intracellular digestion by a lysosome system (de Duve, 1969). Thus particulate material, or material of large molecular weight enters the cell by endocytosis and becomes localized in the small apical granules, which can be regarded as phagosomes. Small enzyme-rich granules, protolysosomes, produced by the golgi bodies fuse with the phagosomes to form the green granules, which are therefore phagolysosomes. Digestion of material within these green granules results in the release of soluble material and accumulation of resistant material. Granular material present in both the green and yellow granules is shown histochemically to be a lipofuscin, which is the form taken by such resistant or residual material in vertebrate lysosomes. Further fusion of granules, and mixing with freshly ingested material, results in further opportunities for digestion of the resistant material, but eventually it accumulates in the large yellow body, directly comparable with the dense bodies of vertebrate cells. Such dense bodies are either accumulated during the life of the cell or discharged, often at the death of the cell, and all of these methods appear to be used by the digestive gland cells.

The majority of workers are of the view that the crop juice originates from the digestive gland, but how seems rarely to be discussed. A comparison of the enzyme contents of the crop juice and digestive gland extracts reveals many similarities, and it is clear that the salivary glands and crop wall could not possibly contribute more than two or three of the enzymes in the crop juice. The great problems lie in how such a juice passes from the gland, through the stomach, and into the crop while all observations indicate a flow of material only in the reverse direction; and where are the enzymes secreted? If muscular movements of the stomach and digestive glands provide the motive power for the movement of the material into the glands this would entail the presence of some sphincter arrangement at the openings into the crop and intestine. Should this be so, and histological and physiological evidence appears to support this view, then there would seem to be no reason why similar muscular movements could not force material out of the lumen of the gland, and into the crop, if the crop sphincter relaxes. It is not easy to understand how this could occur during the passage of a meal through the gut but, in Stylommatophora at least, feeding activity is largely restricted to the dark hours so that during the period of inactivity in the daytime, when the previous meal had left the crop, it would be possible for such a reversal to take place. The only alternative is that there is an ebbing and flowing of material in and out of the digestive glands, associated with frequent regurgitation of the contents of the digestive gland lumen into the crop.

Digestive cells predominate in the digestive glands and they show little evidence of secretory activity. Calcium cells in *Agriolimax* appear to function mainly as a calcium store (see below) but also absorb soluble material and perhaps could secrete small amounts of a protein secretion. In *Biomphalaria* the secretory cells appear to produce a protein secretion, but there is no voluminous crop juice in these animals. It is possible that *Lymnaea* can feed throughout the whole day and night, and although Carriker (1946b) demonstrated the existence of extracellular enzymes in the stomach lumen the importance of these enzymes in comparison to trituration is not known. There is therefore a paradox: *Agriolimax*, and apparently also other Stylommatophora, where extracellular digestion in the crop is important, appear to lack a specific secretion from the digestive gland. A comparison of the list of enzymes detected in crop juice with those listed as being present in mammalian lysosomes, reveals many similarities (Table IV). Is it possible therefore that the crop juice, at least in part, is derived from the lysosomal enzymes present in the phagolysosomes and residual bodies that are discharged into the digestive gland by the digestive and excretory cells? Should this be true then partially digested food material would be recirculated through the crop. Some evidence for this can be obtained from Walker's observations that barium sulphate particles, which did not enter the digestive gland, had completely cleared the digestive tract in 18 h while Thorotrast, which did enter the digestive gland and was taken up by the digestive cells, was found in the crop for at least three days following a meal. Of the apparently nonlysosomal enzymes, cellulase has been demonstrated by Sumner (1968) to be present in the absorptive cells of the digestive gland tubules. As Sumner points out this cellulase could arise by pinocytosis of bacterial cellulase that has originated in the crop. Although few bacteria are present in the digestive gland this does not of itself prove that the enzyme originated in the digestive gland cells. That the enzyme is localized only in the apical cytoplasm of the cells could indicate a breakdown of enzyme absorbed by pinocytosis from the lumen. Parnas (1961), using antibiotics, found that cellulase disappeared from everywhere but the digestive gland, but this treatment did result in a reduction in the digestive gland cellulase. If these experiments had continued for longer any material pinocytosed should have disappeared within at least five days – this experiment was not apparently carried out. A more detailed comparison, perhaps by electrophoresis of isolated enzymes, is needed to resolve this question. Chitinase although not studied in this way would appear to be a similar problem.

The dark colour of the crop juice is due to two pigments; helicorubin, a haem protein, and an unidentified brown pigment. Similar pigments are responsible for a large part of the coloration of the digestive gland as well (Fox and Vevers, 1960). During periods of starvation or dormancy the amount of pigment in the crop juice increases. From a comparison of their chemical and immunological properties Keilin and Orlans (1969) conclude that helicorubin is derived from cytochromes, and that helicorubin is more closely related to the snail's own cytochromes than to those from bacteria or vertebrates. These authors suggest that helicorubin arises during the breakdown of mitochondria from the animal's own cells, but as plant cytochromes were not tested it can not be ruled out that it arises from the food. Helicorubin disappears from the crop juice after feeding and reappears when the food has been digested. As this pigment does not appear in the intestine or in the faeces, it is apparently absorbed by the digestive gland. Is this soluble material recycled between crop and digestive gland? It is interesting that lysosomes isolated from vertebrate tissues have a haem pigment with an almost identical α-absorption band to helicorubin – 562 compared to 563 μm (Tappel, 1969).

It has been claimed that the digestive gland functions as a major store of reserve food materials such as glycogen. Chemical analyses reveal large amounts of these materials in the gland (Goddard and Martin, 1966) but histochemical examination shows that glycogen is not present in large amounts. Starvation experiments on *Lymnaea stagnalis* (de Jonge-Brink, 1973) indicate that the main storage of glycogen is within vesicular connective tissue cells that are present in large numbers between the digestive gland lobules. Glycogen present in other tissues is lost much more slowly on starvation. Stylommatophora have not been studied in such detail but it would be surprising if similar cells were not also involved in glycogen storage. The role of the calcium cells in storage of minerals is not clear. Burton (1972) records the presence of calcium, magnesium, phosphate and carbonate in these granules and they seem as well developed in slugs as in snails. It has been claimed that these granules are mobilized for shell secretion, particularly for regeneration of the damaged shell; but why then are they present in the slugs which have a very reduced shell and absent from the Basommatophora? Burton was unable to demonstrate any mobilization of calcium from the granules during shell regeneration and from a review of the literature he concludes that there is no firm evidence to support this suggestion. What then is the function of these cells? When zinc is present in the food it is accumulated in the calcium granules (Schoettli and Seiler, 1970) and so also is strontium (Fretter, 1953). It is possible that these granules represent accumulation of material prior to excretion i.e. is it part of a detoxication mechanism? There are suggestions that the digestive gland functions in excretion but there appears to be little evidence for this. The digestive gland contains more uric acid than any other organ except the kidney, and it also has all the enzymes needed for uric acid synthesis (Lee and Campbell, 1965) but it is not known how much is excreted by the digestive gland or if it all passes via the blood to the kidney for excretion. The fate of the calcium granules is not known. It has been suggested that they could function to buffer the digestive juices but there is no clear evidence for this.

X. The intestine and rectum

These regions have been far less studied than the other parts of the digestive system of pulmonates. The intestine is a thin-walled tube of uniform diameter and variously coiled around the other organs. It merges imperceptibly with the rectum which has an even thinner wall. At the junction of the intestine and rectum in *Agriolimax*, and other Limacidae, there is a thin-walled rectal caecum. The anus opens to the exterior to one side of the pneumostome, and in Stylommatophora there seems to be a complex arrangement of folds around this area, but there seems to be no detailed description. As faeces have to leave the animal without contaminating the lung, the anatomy of this region is of some interest.

Walker (1969) has subdivided the intestine of *Agriolimax* into three histologically distinct regions: the pro-, mid-, and post-intestine. The whole of the intestine has folded walls with thin layers of circular and longitudinal muscle, and a lining epithelium of ciliated and goblet mucous cells. The mid-intestine is clearly distinguishable because of its higher epithelium, and the presence of cells containing large numbers of granules – these are termed intestinal cells (Sumner, 1965). Anteriorly the pro-intestine lacks folds in the wall but these develop along its length and all cells including the ciliated cells have a brush border. Towards the posterior end of the mid-intestine the cilia become restricted to the tips of the folds and this continues along the post-intestine. In addition to the absence of intestinal cells, and the epithelium being lower, the post-intestine is distinguished by the presence of a second and large type of mucous cell. In the rectum the folds in the wall are very much reduced in height, except for a single well developed lateral fold. Cilia are well developed on this fold and posteriorly on the tops of the other folds. The other cells in the rectal epithelium include two types of mucous cell. Except for the absence of cilia, the histology of the rectal caecum closely resembles that of the rectum.

Carriker (1946a) also subdivides the intestine of *Lymnaea* into pro-, mid-, and post-intestine but, as has been discussed above, his pro-intestine is probably better interpreted as a style sac. A prominent fold in the wall of the intestine, termed the raphe, stretches its whole length, and could be homologous with the lateral fold in the rectum of *Agriolimax*. Mucous and ciliated cells are present along the intestine, and thin layers of muscle are present beneath the epithelium.

Faeces formation takes place in the style sac along which ciliated tracts convey material posteriorly. In *Lymnaea* the tracts also pass ventrally, so that they converge on the short typhlosole (Fig. 5A). *Agriolimax* (E) has a well developed intestinal groove – the cilia beat into the groove and those in the groove beat posteriorly. A short distance from the end of the intestinal groove the beat of the cilia is reversed, and therefore material accumulates at this point. Particulate material in the style sac accumulates and becomes mixed with mucus. Accumulated material passes into the intestine by peristalsis and at the junction of the style sac with the intestine there is a richly ciliated region of the wall where the cilia beat transversely. Carriker terms this area in *Lymnaea* the pellet compressor, and concludes that it assists in the consolidation of the forming faeces. During passage along the intestine more mucus is secreted around the faeces and they become more consolidated so that fairly solid faeces are defaecated. The intestinal cells seem to play an important role in this encapsulation of the faeces (Walker, personal communication).

Although there are many cilia along the wall of the intestine and rectum it seems likely that peristalsis is responsible for moving the faeces (Carriker, 1946b; Walker, 1969). Roach (1968) in an extensive study of the motility of the intestine in *Arion ater* also describes anti-peristalsis, longitudinal, and pendular contractions, which together must result in a mixing of the contents. Studies of the faeces of *Lymnaea* (Carriker, 1946b) and *Agriolimax* (Pallant, 1970) indicate that the diffe-rent constituents of the faeces remain distinct, and therefore in these animals at least, mixing is not very efficient. Mixing of the faecal mate-rial has been observed in other pulmonates however, e.g. *Oxychilus* (Rigby, 1963).

Passage of material along the intestine was found to be very slow in *Agriolimax*, when radio-opaque markers were used (Walker, 1972). Thus although it took only 20–40 minutes before the first material appeared in the stomach then entered the intestine rapidly, it remained there for 7–$7\frac{1}{2}$ h. Passage through the rectum however appeared to be rapid and faeces were only rarely observed there. It is normally stated that a long intestine is characteristic of herbivores (e.g. Carriker, 1946b). A long intestine, and long sojourn of the faeces there, would imply that impor-tant processes are taking place in this region in addition to the consoli-dation of faeces.

No evidence for secretion of enzymes along the tract has been observed, but strong intracellular enzymes are present in the epithelial cells, e.g. esterase (lipase) and acid phosphatase (Ferreri, 1958; Ferreri and Ducato, 1959; Walker, 1969; and Bowen, 1970, 1971). These are lysosomal enzymes and appear to be localized in apical granules. Uptake of soluble material from the lumen of the intestine has been clearly demonstrated for minerals i.e. calcium, strontium, phosphates, and iodine (Fretter, 1952, 1953) and glycine, galactose, glucose and palmitic acid (Walker, 1972). Palmitic acid was taken up in apical granules while the other material were found to be diffused throughout the cytoplasm of the cells. It thus appears that small molecular weight materials can enter the cells by diffusion, while palmitic acid is taken up by pinocytosis and then dealt with by a lysosomal mechanism. From a comparison of fed and starved animals, Walker concluded that there was an increase in the number of granules in the epithelial cells following feeding.

Some soluble material is certainly present in the faeces when they enter the intestine. It is possible that molecules with a larger molecular weight than those tested, could be taken up by the cells. Besides this soluble material it is possible that further digestion could take place during the stay of the faeces in the gut. Enzymes from the crop juice must be present amongst the faecal material, and bacterial action may also be important. Cellulase and chitinase are notably slow in their action, and it is possible that further digestion by these enzymes is important here. Extensive digestion of cellulose at least, must take place somewhere in the digestive system to account for the 70–80% utilization of plant material. While recirculation of small particles from the digestive gland may help with digestion, a lot of the cellulose is present as large pieces, especially in the Stylommatophora which lack a trituration mechanism, and the Basommatophora apparently lack extensive extracellular digestion. In both groups therefore extended digestion would seem important. Carriker (1946b) implies that in addition to the long stay of material in the intestine, *Lymnaea* often consumes faecal material, so extending the duration of enzyme action even further.

The blood and nervous supplies to the intestine and rectum are well developed (Carriker, 1946a; Walker, 1969; Bekius, 1972; Schmidt, 1916). The posterior aorta and its derivatives give rise to many small branches that supply the intestine directly, or indirectly via the digestive gland. *Agriolimax* appears to lack the separate rectal artery found in *Lymnaea* and *Helix*. From the collection of small ganglia present on the stomach wall, two longitudinal nerves pass along the length of the intestine and give rise to many fine anastomosing branches. A separate nerve supply for the anal region is derived from the right pallial nerve in *Lymnaea* and no supply was demonstrated for the rectum, while in *Agriolimax* the anal nerve is derived from the visceral nerve and supplies the rectum and anus via a complex system of small ganglia and anastomosing nerves. Isolated intestine is capable of independent peristaltic movements (Minker and Koltai, 1961; Roach, 1968) due to the well developed plexus in the wall.

XI. Conclusions

I have attempted in this review to demonstrate that while there is now a large body of information on the structure and function of the pulmonate digestive system, there are still enormous gaps in our knowledge at all levels. Although comparative anatomy appears to be an unfashionable branch of biology at the present time, further work in this field would greatly assist our understanding of the evolution particularly of the crop and stomach in these animals. While Morton has provided much information on the lower Basommatophora, there seems a very great difference between these forms and the higher members of the group such as *Lymnaea*. Our knowledge of crop and stomach structure in Stylommatophora is exceedingly limited. There appear to be several reasons for this neglect. The reproductive system shows very great differences between pulmonate groups and information on this system is very useful for the systematist, while most of these workers give scant attention to the digestive system because superficially differences are small. Most textbooks state that the stomach of the pulmonate is extremely simplified compared to that of other molluscs, but detailed examination reveals they are less simplified than this statement implies. Much interesting work has been published on the adaptations of the molluscan digestive system to diet and habitat, in other gastropods and in lamellibranchs, and it is already apparent that the pulmonate digestive system is equally well adapted. Most of our available information is for the mainly herbivorous pulmonates and it is to be hoped that many more studies will be made on the carnivorous forms.

Compared to our knowledge of comparative anatomy that on function is extremely limited. For most pulmonates we do not know the normal diet, even for otherwise well studied forms such as *Lymnaea stagnalis*. There is neither qualitative nor quantitative information on the majority of the nutritional requirements. Although the radula is well known anatomically, its mode of use, and the involvement of the jaw, during feeding on normal food materials have been investigated only recently. We understand in outline the events taking place during digestion of the ingested material but there remain many details to be clarified, and there is virtually no quantitative information. The effectiveness of trituration, intra- and extra- cellular digestion, the amount of absorption of the various nutrients in the different regions of the digestive system, and the importance of digestion in the intestine, are just three of many areas where there are virtually no facts available.

The studies on *Lymnaea* and *Agriolimax* indicate a very sharp distinction between their digestive mechanisms. Thus *Lymnaea* has an efficient triturating system with a largely unknown contribution from extracellular enzymes originating in the digestive gland. *Agriolimax* has no trituration but a very efficient extracellular digestion. If these mechanisms are so different it would be useful to know their relative effectiveness, by, for example, studying their utilization of the same food material such as lettuce. Our knowledge of digestive mechanisms is restricted to very few pulmonates and again it is to be hoped that studies will be made on less highly evolved members of the group and on those species that feed on animal or fungal material.

An examination of digestion throughout the Mollusca clearly shows a contrast between intracellular digestion in the filter feeding forms and

E

the largely extracellular digestion found in the carnivores and some of
the more highly evolved herbivores. Recent work indicates that these
distinctions are not so clear cut and that, for example in the bivalves
(Reid, 1966), the mode of digestion is adapted to the type of material
ingested. If the extracellular enzymes in the crop juice arise from discharged lysosomal material as suggested above, then this provides a
very simple way of developing this type of digestion. Thus the intracellular digestive mechanism can be converted to the secretion of extracellular enzymes by merely discharging the contents of the phagolysosomes at an earlier stage in their cycle. No special enzyme-secreting
cells need therefore to be evolved.

Co-ordination of the various parts of the digestive system has not been
studied. Roach (1966) studied the time taken by various foods to pass
through the gut of *Arion ater* by simply feeding the animals on different
coloured foods. There were considerable differences in the time taken:
thus carrot took an average of 15 h 20 min and lettuce 11 h 14 min to
pass through the gut. The reasons for these differences are unexplained.
What controls the length of time spent in the various regions of the
digestive system and does it affect the utilization of the food? As we have
seen a study of the faeces indicates that the liver string is produced only
at intervals, thereby implying periodic changes in the physiology of
stomach and possibly also of the digestive gland. Nervous elements are
plentiful in the digestive system but we do not know how the various
parts of the system interact, nor how periodic changes in activity are
brought about.

Lastly I would make a plea for integrated studies on the digestive
system. Ideally anatomical and physiological studies should be combined so that a complete picture of how the digestive system functions
and how it relates to the whole animal, is obtained. Also, physiologists
naturally prefer to work on large convenient animals, and therefore unless
the anatomist does his own physiology it will probably be a long time
before anyone else does.

XII. References

Argaud, L. D. and Bounoure, L. (1910). *J. Anat. Physiol., Paris* **46**, 146–174.
Barry, R. J. C. and Munday, K. A. (1959). *J. mar. biol. Ass., U.K.*, **38**, 81–
    95.
Bekius, R. (1972). *Neth. J. Zool.* **22**, 1–58.
Bierbauer, J. (1957). *Acta Biol., Szeged* **7**, 419–431.
Blain, M. (1957). *Archs Anat. microsc. Morph. exp.* **46**, 489–502.

Boer, H. N., Bonga, S. E. W. and Rooyen, N. van (1967). *Z. Zellforsch. mikrosk. Anat.* **76**, 228–247.

Bovbjerg, R. V. (1968). *Physiol. Zoöl.* **41**, 412–423.

Bowen, I. D. (1970). *Protoplasma* **70**, 247–260.

Bowen, I. D. (1971). *J. Microsc.* **94**, 25–38.

Burton, R. E. (1972). *Comp. Biochem. Physiol.* **43A**, 655–663.

Calow, P. (1970). *Proc. malac. Soc. Lond.* **39**, 203–215.

Calow, P. and Fletcher, C. R. (1972). *Oecologia* **9**, 155–170.

Carriker, M. R. (1943). *Nautilus* **57**, 52–59.

Carriker, M. R. (1946a). *Trans. Wis. Acad. Sci. Arts Lett.* **38**, 1–88.

Carriker, M. R. (1946b). *Biol. Bull. mar. biol. Lab. Woods Hole* **91**, 88–111.

Carriker, M. R. and Bilstad, N. M. (1946). *Trans. Am. microsc. Soc.* **65**, 250–275.

Crampton, D. M. (1973). "The functional anatomy of the buccal mass of one opisthobranch and three pulmonate gastropod molluscs." Ph. D. Thesis, University of Reading.

de Duve, C. (1969). *In* "Lysosomes in Biology and Pathology" (J. T. Dingle, Ed.) Vol. **1**, 3–40. North-Holland, Amsterdam.

Evans, W. A. L. and Jones, E. G. (1962a). *Comp. Biochem. Physiol.* **5**, 149–160.

Evans, W. A. L. and Jones, E. G. (1962b). *Comp. Biochem. Physiol.* **5**, 223–5.

Ferreri, E. (1958). *Z. vergl. Physiol.* **41**, 373–389.

Ferreri, E. and Ducato, L. (1959). *Z. Zellforsch. mikrosk. Anat.* **51**, 65–77.

Fox, H. M. and Vevers, G. (1960). "The Nature of animal Colours." Sidgwick and Jackson Ltd., London.

Franc, A. (1968). *In* "Traité de Zoologie V, Mollusques gastéropodes et scaphopodes" (P.-P. Grassé, Ed.). Masson et Cie., Paris.

Franssen, J. and Jeuniaux, C. (1965). *Cah. Biol. mar.* **6**, 1–21.

Fretter, V. (1952). *Q. Jl microsc. Sci.* **93**, 133–146.

Fretter, V. (1953). *J. mar. biol. Ass. U.K.* **32**, 367–384.

Fretter, V. and Graham, A. (1962). "British Prosobranch Molluscs." Ray Society, London.

Frömming, E. (1954). "Biologie der mitteleuropaïschen Landgastropoden." Duncker & Humblot, Berlin.

Frömming, E. (1956), "Biologie der mitteleuropaïschen Süsswasserschnecken." Duncker & Humblot, Berlin.

Frömming, E. (1957). *Z. angew. Zool.* **44**, 349–357.

Ghose, K. C. (1962). *Proc. R. Soc. Edinb.* **68**B, 186–207.

Ghose, K. C. (1963). *Trans. Am. microsc. Soc.* **82**, 149–167.

Goddard, C. K. and Martin, A. W. (1966). *In* "Physiology of Mollusca" (K. M. Wilbur and C. M. Yonge, Eds) **2**: pp. 275–308. Academic Press, New York and London.

Graetze, E. (1929). *Zool. Jb. allg. Zool. Physiol.* **46**, 375–412.

Graham, A. (1949). *Trans. R. Soc. Edinb.* **61**, 737–778.

Graham, A. (1955). *Proc. malac. Soc. Lond.* **31**, 144–159.

Grime, J. P., MacPherson-Stewart, S. F. and Dearman, R. S. (1968). *J. Ecol.* **56**, 405–420.

Hoffman, H. (1932). *Jena. Z. Naturw.* **67**, 535–550.

Hubendick, B. (1945)., *Ark. Zool.* **36**A, 21, 1–30.

Hubendick, B. (1957). *Ark. Zool.* **10**, 511—521.

Isarankura, K. (1966). "Studies on the replacement of the gastropod radula." Ph.D. thesis, University of Wales.

Isarankura, K. and Runham, N. W. (1968). *Malacologia* **7**, 71–91.

Hyman, L. H. (1967). "The Invertebrates **6** Mollusca 1." McGraw-Hill, New York.

Hunter, P. J. (1968). *Malacologia* **6**, 391–399.

Jager, J. C. (1971). *Neth. J. Zool.* **21**, 1–59.

Jarrige, P. (1963). *Bull. Soc. Chim. biol.* **45**, 761–782.

Jeuniaux, C. (1954). *Mem. Acad. R. Sci. colon. Cl. Sci. nat. med.* 8° **28**, 1–45.

Jeuniaux, C. (1963). "Chitin et Chitinolyse, un Chapitre de la Biologie moleculaire." Masson et Cie., Paris.

de Jonge-Brink M. (1973). *Z. Zellforsch. mikrosk. Anat.* **136**, 229–262.

Kater, S. B. and Fraser-Rowell, C. H. (1973). *J. Neurophysiol.* **36**, 142–155.

Keilin, J. and Orlans, E. (1969). *Nature, Lond.* **223**, 304–306.

Kerth, K. (1971). *Zool. Jb. Anat. Ont. Tiere* **88**, 47–62.

Kerth, K. and Krause, G. (1969). *Wilhelm Roux Arch. EntwMech. Org.* **164**, 48–82.

Kittel, R. (1956). *Zool. Anz.* **157**, 185–195.

Koopmans, J. J. C. (1970). *Neth. J. Zool.* **20**, 445–463.

Lane, N. J. (1964). *Q. Jl microsc. Sci.* **105**, 331–341.

Lee, T. W. and Campbell, J. W. (1965). *Comp. Biochem. Physiol.* **15**, 457–468.

Leon, Y. A., Bulbrook, R. D. and Corner, E. D. S. (1960). *Biochem. J.* **75**, 612–617.

Loisel, G. (1893). *J. Anat. Physiol., Paris* **29**, 466–522.

Malone, C. R. and Nelson, D. J. (1969). *Ecology* **50**, 728–730.

Märkel, K. (1957). *Z. wiss. Zool.* **160**, 213–289.

Marray, A., Got, R. and Jarrige, P. (1964). *Experientia* **20**, 441.

Mason, C. F. (1970a). *Oecologia* **4**, 358–373.

Mason, C. F. (1970b). *Oecologia* **5**, 215–239.

Mead, A. R. (1961). "The giant African snail: a Problem in economic Malacology." University of Chicago Press.

Meuleman, E. A. (1972). *Neth. J. Zool.* **22**, 355–427.

Minker, E. and Koltai, M. (1961). *Acta biol. Szeged* **12**, 199–209.

Moens, R., François, E. and Riga, A. (1966). *Meded. Rijksfac. handb. Gent.* **31**, 1032–1042.

Moore, H. J. (1971). *Mar. Biol.* **11**, 23–27.

Morton, J. E. (1955a). *J. mar. biol. Ass. U.K.* **34**, 113–150.

Morton, J. E. (1955b). *Proc. zool. Soc. Lond.* **125**, 127–168.

Morton, J. E. (1955c). *Phil. Trans. R. Soc. B* **239**, 89–160.

Myers, F. L. and Northcote, D. H. (1958). *J. exp. Biol.* **35**, 639–648.

Nakazima, M. (1956). *Jap. J. Zool.* **11**, 469–566.

Nielsen, C. O. (1962). *Oikos* **13**, 200–215.

Nielsen, C. O. (1963). *Nature, Lond.* **199**, 1001.

Oberholzer, G., Brown, D. S. and van Eeden, J. A. (1970). *Wetenskaplike Bydraes van die P. U. vir C.H.O.*, B. nr **10**.

Owen, G. (1966). *In* "Physiology of Mollusca (K. M. Wilbur and C. M. Yonge, Eds), Vol. 2, pp. 1–51. Academic Press, New York and London.

Owen, G. (1966). *In* "Physiology of Mollusca (K. M. Wilbur and C. M. Yonge, Eds), Vol. 2, pp. 53–96. Academic Press, New York and London.

Pallant, D. (1969). *J. Anim. Ecol.* **38**, 391–8.

Pallant, D. (1970). *J. Conch., Lond.* **27**, 111–113.

Pallant, D. (1972). *J. Anim. Ecol.* **41**, 761—769.

Parnas, I. (1961). *J. cell. comp. Physiol.* **58**, 195–201.

Pruvot-Fol, A. (1926). *Archs Zool. exp. gén.* **65**, 209–343.

Quattrini, D. (1967). *Caryologia* **20**, 191–206.

Reid, R. G. B. (1966). *Comp. Biochem. Physiol.* **17**, 417–433.

Rigby, J. E. (1963). *Proc. zool. Soc. Lond.* **141**, 311–359.

Rigby, J. E. (1965). *Proc. zool. Soc. Lond.* **144**, 445–486.

Roach, D. K. (1966). "Studies of some aspects of the physiology of *Arion ater.*" Ph.D. Thesis, University of Wales.

Roach, D. K. (1968). *Comp. Biochem. Physiol.* **24**, 865–878.

Rodina, A. G. (1957). *Zool. Zh.* **36**, 337–343 (in Russian).

Rosén, B. (1952). *Ark. Zool.* **3**, 33–50.

Runham, N. W. (1962). *Nature, Lond.* **194**, 992–993.

Runham, N. W. (1963a). *Annls Histochim.* **8**, 433–442.

Runham, N. W. (1963b). *Q. Jl microsc. Sci.* **104**, 271–277.

Runham, N. W. (1969). *Malacologia* **9**, 179–185.

Runham, N. W. and Thornton, P. R. (1967). *J. Zool. Lond.* **153**, 445–452.

Schmidt, G. (1916). *Z. wiss. Zool.* **115**, 201–261.

Schoettli, G. and Seiler, H. G. (1970). *Experientia* **26**, 1212–3.

Sminia, T. (1972). *Z. Zellforsch. mikrosk. Anat.* **130**, 497–526.

Sollas, I. B. J. (1907). *Q. Jl microsc. Sci.* **51**, 115–136.

Spek, J. (1921). *Z. wiss. Zool.* **118**, 313–363.

Steen, W. J. van der, Jager, J. C. and Tiemersma, D. (1973). *Proc. K. ned. Akad. Wet.* C, **76**, 47–60.

Sterki, V. (1893). *Proc. Acad. nat. Sci. Philad.* (1893–4), 388–400.

Stern, G. (1970). *Terre Vie* **117**, 403–424.

Stiglingh, I. and van Eeden, J. A. (1970). *Wetenskaplike Bydraes van die P.U. vir C.H.O.* B. nr **22**.

Stone, B. A. and Morton, J. E. (1958). *Proc. malac. Soc. Lond.* **33**, 127–141.

Storey, R. (1970). *J. Conch., Lond.* **27**, 191–195.

Storey, R. (1971). *Proc. malac. Soc. Lond.* **39**, 327–331.

Strasdine, G. A. and Whitaker, D. R. (1963). *Can. J. Biochem. Physiol.* **41**, 1621–1626.

Sumner, A. T. (1965). *Jl R. microsc. Soc.* **84**, 415–421.

Sumner, A. T. (1968). *Histochemie* **13**, 160–168.

Tappel, A. L. (1969). *In* "Lysosomes in Biology and Pathology" (J. T. Dingle and H. B. Fell, Eds), Vol. 2, pp. 209–244 North-Holland Publishing Co., Amsterdam.

Taylor, J. W. (1849–1900). "Monograph of the Land and Freshwater Mollusca of the British Isles. I. Structural and General." Taylor Brothers, Leeds.

Turcek, F. J. (1970). *Biologia, Bratisl.* **25,** 103–8.

Verdcourt, B. (1950). *Microscope* **8,** 225–6.

Vieira, E. C. (1967). *Am. J. trop. med. Hyg.* **16,** 792–6.

Vlieger, T. A. de (1968). *J. Zool. Lond.* **18,** 105–154.

Walker, G. (1969). "Studies on digestion of the slug *Agriolimax reticulatus* (Müller), (Mollusca, Pulmonata, Limacidae)." Ph.D. Thesis, University of Wales.

Walker, G. (1970a). *Protoplasma* **71,** 91–109.

Walker, G. (1970b). *Protoplasma* **71,** 111–126.

Walker, G. (1972). *Proc. malac. Soc. Lond.* **40,** 33–43.

Webb, W. M. (1893). *Zoologist* **17,** 281–9.

Wolda, H., Zweep, A. and Schuitema, K. A. (1971). *Oecologia* **7,** 361–381.

Zylstra, U. (1972). *Z. Zellforsch. mikrosk. Anat.* **130,** 93–134.

Chapter 4

Water relationships

JOHN MACHIN

Department of Zoology, University of Toronto, Canada

I. Introduction

Various aspects of molluscan physiology reviewed recently include osmotic and ionic regulation (Robertson, 1964; Potts and Parry, 1964), environmental physiology (Newell, 1964; Russell Hunter, 1964), nitrogen metabolism and excretion (Florkin, 1966; Martin and Harrison, 1966; Potts, 1967) and carbohydrate metabolism (Martin, 1961; Goddard and Martin, 1966). Chapters containing relevant physiological information also form part of some recent general works (Mead, 1961; Morton, 1967; Hyman, 1967; Runham and Hunter, 1970). All have a bearing on the water relationships of pulmonates. However discussion of this group has usually formed a relatively small component of articles dealing with the whole phylum and some important aspects of pulmonate water balance physiology have been omitted. The most recent volume of "Physiology of Mullusca" (edited by K. M. Wilbur and C. M. Yonge) contains some references up to 1966. Since that time a number of important contributions have been made; about one third of the papers cited below were published after that date.

There are several additional reasons why a more extensive review of water relationships of pulmonates is appropriate at the present time. Of the gastropod groups which have colonized the physiologically demanding environments of land and fresh water, pulmonates have had the greatest success and show the greatest diversity, extending their range even into the most severe desert conditons. In spite of this terrestrial pulmonates are often thought of as rather poorly adapted land animals. However, with information which is now becoming available, this idea is changing. One of the attractions for physiologists in working with pulmonates is that they frequently solve physiological problems in unusual and sometimes unexpected ways.

II. Water content

All soft parts of pulmonates contain appreciable amounts of water. The principal body cavity in molluscs is the haemocoel. However, separated from it by cellular barriers are other fluid-filled cavities which are the remnants of the coelom: the pericardium and the cavities of the kidney and gonad. Pulmonates have a so-called "open" blood system, a network of large and small spaces or sinuses which permit the haemolymph to bathe the tissues directly. There is no separation of interstitial tissue fluid from the circulating blood as there is in phyla with cannular circulatory systems. Haemolymph is circulated from the heart through a branching system of arteries to the various parts of the body and conducted back through the complex system of spaces to the heart by way of the lung.

Quantitative data on the size and concentrations of the fluid compartments are restricted almost entirely to large species. Information about the more primitive pulmonates which inhabit semi-marine environments (Morton, 1954b; 1955a) and of the smaller species living in terrestrial microclimates of high humidity, is completely lacking. The available data (Table I) show that water content is variable, particularly in terrestrial species, where highest values are found during summer in active animals. In view of the limited number of observations made by most authors ranges given are unlikely to represent the maximum for most species. Water seems to be roughly equally distributed between

Table I. Major fluid compartments in adult pulmonates

Species	Total body water content[a] (% wet weight less shell)		Extracellular water (% wet weight less shell)		Authority
	Mean	(Range)	Mean	(Range)	
Basommatophora					
Biomphalaria glabrata	87	—	44·5[e]	—	von Brand et al., 1957
Lymnaea stagnalis (L.)	91·6	—	67·5[e]	—	van Aardt, 1968
Lymnaea stagnalis	—	—	—	—	Greenaway, 1970
Melampus bidentatus Say	47[b]	—	—	—	Hunter and Apley, 1966
Planorbarius corneus (L.)	—	—	58[f]	(40–75)	Borden, 1931
Stylommatophora					
Arion ater	87·5	—	—	—	Pusswald, 1948
Arion ater	86·3	(80·3–91·8)	36·6[d]	(24·0–57·5)	Martin et al., 1958
Achatina fulica Bowdich	86·4	(82·4–92·4)	40·3[d]	(34·8–52·2)	Martin et al., 1958
Cepaea nemoralis	88·2	—	—	—	Trams et al., 1965
Helix aperta Born	—	(78–88)	—	—	Burton, 1966

Table I.—continued

Table I. Major fluid compartments in adult pulmonates

Species	Total body water content[a] (% wet weight less shell)		Extracellular water (% wet weight less shell)		Authority
	Mean	(Range)	Mean	(Range)	
Helix pomatia L.	—	(79·2–91·1)	—	(52·4–81·5)[g]	Burton, 1964
Helix pomatia (active)	83·5	(80·2–88·1)	—	—	von Brand, 1931
Helix pomatia (inactive)	81·2	(79·1–84·6)	—	—	von Brand, 1931
Otala lactea (Müll.)	—	(83–91)	—	—	Burton, 1966
Sphincterochila boissieri (Charp.)	81	—	—	—	Schmidt-Nielsen *et al.*, 1971

[a] Determined by oven dehydration
[b] % wet weight including shell
[c] Determined as inulin space, includes coelomic spaces
[d] Determined as inulin space, on apparently dehydrated animals
[e] Determined as sodium space, includes coelomic spaces and an intracellular component
[f] Determined as haemoglobin space, includes haemocoel only
[g] Determined as copper space, includes haemocoel only; value is over-estimated as wet weight does not include haemolymph solutes and albumen gland

intra- and extra-cellular compartments. This distribution is indicative of the comparatively large body cavity found in animals with open blood systems. Cephalopoda and vertebrates which are structurally and physiologically more advanced, show a departure from this pattern in increasing total cell volume at the expense of open body cavities.

Water is held also in some pulmonates in three locations which are technically outside the body: fluid in the gut, especially the crop, water in the superficial mucus and sometimes the mantle cavity. The total amount of water held in the mucus may be significant. Machin (1964a) found that mucus collected from *Helix aspersa* Müll. contained between 88 and 98% water. Pallial water, found by Blinn (1964) in the polygyrids *Mesodon thyroidus* (Say) and *Allogona profunda* (Say) weighed up to 20% of the total body weight. Although the source of this fluid is not definitely known, it is probably of urinary origin and may serve as an extra water reserve. It appears pallial water occurs in some but not all pulmonates. Blinn failed to demonstrate its presence in some members of the Helicidae, Succineidae and Polygyridae but noted that fluid discharge from the lung was occasionally observed by Künkel (1916) in several species of terrestrial slugs.

III. Osmotic concentration of body fluids

Known body fluid concentrations of pulmonates are summarized in Tables II and III. Data for *Siphonaria pectinata* (L.) (McAlister and Fisher, 1968) and *S. zelandica* (Bedford, 1969) the only genus of marine pulmonates for which haemolymph concentration has been measured, is similar to those for other marine Mollusca in that this fluid is normally iso-osmotic to the surrounding sea water. *Siphonaria*, like other intertidal invertebrates living where restricted volumes of sea water may be diluted by rain or concentrated by evaporation at low tide, is tolerant of a wide range of external concentration.

Freshwater pulmonates, in common with most members of the animal kingdom, maintain much lower body fluid concentrations than their marine relatives. Haemolymph osmotic pressures are not as low as those found in freshwater lamellibranchs, but generally somewhat below those for freshwater crustaceans or vertebrates (Lockwood, 1963). Despite low haemolymph concentrations, freshwater pulmonates are always hyper-osmotic to the outside (Table II). Van Aardt (1968) found, for example, that the haemolymph of *Lymnaea stagnalis* is normally

Table II. Body fluid osmolalities of adult aquatic pulmonates (mOsm/kg H_2O)

	External Medium		Haemolymph		Pericardial Fluid		Urine		Authority
	Mean	(Range)	Mean	(Range)	Mean	(Range)	Mean	(Range)	
Basommatophora	Intertidal								
Siphonaria pectinata	645	(215–2312)	645	(323–2097)	—	—	—	—	McAlister and Fisher, 1968
	Freshwater								
Lymnaea peregra	—	—	134	(78–190)	134	(72–168)	94	(62–137)	Picken, 1937
Lymnaea peregra	—	—	124	—	—	—	—	—	Potts and Parry, 1964
Lymnaea stagnalis	—	—	127	(118–124)	—	—	—	—	Botazzi, 1908
Lymnaea stagnalis		(6—8)	132	(115–140)	—	(118–132)	97	(79–114)	van Aardt, 1968
Lymnaea stagnalis	—	—	132	—	—	—	—	—	Greenaway, 1970
Lymnaea truncatula	—	—	150	(145–166)	—	—	—	—	Pullin, 1971

maintained at about 18 times the concentration of its surroundings, and a tenth the concentration of sea water.

Mean haemolymph osmotic pressures in land pulmonates exeed those of their freshwater relatives, but are not as high as those of terrestrial insects or vertebrates. Haemolymph concentrations are notoriously variable as shown by the ranges given in Table III. The minimum values correspond to the haemolymph concentration found in freshwater species. In snails, where different physiological states are easily discerned, low haemolymph concentrations are found in active animals while the higher concentrations are characteristic of inactive specimens. Burton (1966) considered that haemolymph concentration was related to habitat, being more dilute in the species living in wetter environments. However, this remains to be demonstrated conclusively. A large number of measurements under a variety of experimental conditions by a number of different authors have been accumulated for *Helix pomatia* L. It is likely, therefore, that the observed haemolymph concentration range (128–312 mOsm/kg) is much closer to the maximum tolerated by this species than in others, where data are more limited. On this assumption, comparisons with the other data show that a wider range of haemolymph concentrations (154–430 mOsm/kg) is tolerated by the slug, *Agriolimax reticulatus* (Müll.).

Even in spite of the relatively high salt levels in the haemolymph of *Siphonaria*, water exchange in all pulmonates in air is essentially the same. Since solutes reduce solution vapour pressures very little, pulmonate equilibrium humidities range only from 96·2 to 99·9% r.h. at 20°C (Table IV). It follows therefore, that all land pulmonates are effectively hypo-osmotic to the surrounding air, except at the very highest humidities. The fact that in saturated or slightly subsaturated atmospheres slugs and snails become hyper-osmotic and could condense water vapour from their surroundings is perhaps of some theoretical significance. However, it must be emphasized 0·15°C is all that separates relative humidities of 99 and 100% at 20°C and continuous water absorption by this means is unlikely under natural conditions.

It is possible to compare osmotic concentration differences between haemolymph and external environment in animals living on land and in fresh water. Expressing representative values from Tables II and III, in terms of vapour pressure at 20°C, freshwater snails are exposed to osmotic concentration differences equivalent to 0·04 mmHg. At the same temperature land pulmonates are exposed to the same difference in saturated conditions. However, in subsaturated air, even at levels which are considered quite moderate on land, dehydrating forces are considerably greater.

Table III. Body fluid osmolalities of adult terrestrial pulmonates (mOsm/kg H_2O)

Species	Haemolymph Mean	Haemolymph (Range)	Urine Mean	Urine (Range)	Authority
Stylommatophora					
Agriolimax reticulatus[a]	—	(145–430)	—		Hughes and Kerkut, 1956
Agriolimax reticulatus[a]	349	—	—		Bailey, 1971
Arion ater[a]	151	(69–220)		—	Rouschal, 1940
Arion ater[a]	199	(97–231)		—	Roach, 1963
Arion ater rufus[a]	—	(167–177)		—	Duval, 1930
Achatina achatina[a]	177	—	—		Drilhon and Florence, 1942
Achatina fulica[a]	248	—	153		Martin, 1957
Archachatina ventricosa (Gould)[a]	—	(156–161)	—	(43–89)	Vorwohl, 1961
Helix aperta (active)	215	(170–215)	—		Arivanitaki and Cardot, 1932
Helix aperta (inactive)	215	—	—		Arivanitaki and Cardot, 1932
Helix aperta[a]	—	(150–296)			Burton, 1966
Helix aspersa (active)	—	(160–210)			Arivanitaki and Cardot, 1932
Helix aspersa[a]	261	—			Chiarandini, 1964
Helix aspersa (inactive)	360	(317–419)			Machin, 1966
Helix pomatia (active)	—	(161–215)			Duval, 1930
Helix pomatia (inactive)	—	(199–231)			Duval, 1930
Helix pomatia (active)	—	(128–235)			Kamanda, 1933
Helix pomatia (inactive)	—	(159–271)			Kamanda, 1933

Helix pomatia[a]	—	(135–269)	(34–171)	Vorwohl, 1961
Helix pomatia[a]	—	(140–312)	—	Burton, 1966
Otala lactea[a]	—	(188–274)	—	Burton, 1966
Otala vermiculata (Müll.) (active)	—	(160–200)	—	Arivanitaki and Cardot, 1932
Otala vermiculata (inactive)	220	—	—	Arivanitaki and Cardot, 1932
Strophocheilus oblongus musculus[a]	142	—	—	De Jorge et al., 1965
Theba pisana (Müll.) (active)	—	(200–260)	—	Arivanitaki and Cardot, 1932
Theba pisana (inactive)	240	—	—	Arivanitaki and Cardot, 1932

[a] Physiological state unknown

Table IV. Comparisons of calculated water concentration gradients experienced by representative pulmonates in different environments

Species	Haemolymph osmolality (mOsm/kg) Mean	(Range)	Vapour pressure lowering[a] (P$_0$ – P$_1$) Mean	(Range)	Haemolymph equilibrium humidity (P$_1$/P$_0$ × 100)	Calculated conc. gradient (P$_1$ environment – P$_1$ haemolymph)[b]			
						Freshwater or sat. air	Air 90% r.h.	50% r.h.	0% r.h.
Siphonaria pectinata	645	(323–2097)	0·20	(0·10–0·66)	98·9 (99·4–96·2)	—	–1·5	–8·5	–17·3
Lymnaea stagnalis	127	(115–140)	0·04	(0·04–0·04)	99·8 (99·8–99·8)	+0·04	—	—	—
Agriolimax reticulatus	288	(145–430)	0·09	(0·05–0·14)	99·5 (99·7–99·2)	+0·09	–1·7	–8·7	–17·5
Helix pomatia	220	(128–312)	0·07	(0·04–0·10)	99·6 (99·8–99·4)	+0·07	–1·7	–8·7	–17·5

$$P_0 - P_1 = \frac{\text{milliosmols} \times M \times P_0}{1000 \times 1000}$$

Where P_0 = vapour pressure of pure solvent (17·54 mmHg for water at 20°C)
$P_0 - P_1$ = vapour pressure lowering due to solvent (mmHg)
milliosmoles = milliosmolality/kg H$_2$O
M = molecular weight of solvent (18 for water).

[a] In order to make comparisons between water concentrations in liquid and vapour phases, osmolalities have been recalculated in terms of vapour pressure lowering at 20°C with the following equation (Machin, 1969):

[b] Values for P$_1$ environment are 17·54, 15·79, 8·77 and 0 mmHg for fresh water, 90%, 50%, and 0% r.h., respectively. Positive values indicate water uptake, negative values a loss from the animal

IV. Mechanisms of water exchange

Comparisons between the osmotic concentrations of the body fluids and those of the external environment enable predictions to be made about the direction of water exchange. The magnitude of the exchange, however, not only depends on the concentration differences between compartments, but also on the permeability of the barrier dividing them. In common with those of other soft-bodied invertebrates, most molluscan tissues and organs seem to be highly permeable: water readily flows from one compartment to another when concentration differences exist. Unfortunately, because of their complexity, isolated tissues do not, with a few exceptions, lend themselves to the *in vitro* studies now commonly performed with frog skin and toad bladder.

Using whole animals, McAlister and Fisher (1968) showed that unattached specimens of *Siphonaria pectinata* were unable actively to regulate their haemolymph concentration. Most osmotic exchange following gross alterations in external concentration took place in less than 24 h. Survival rates of 85% and above were obtained following transfer to external concentrations varying between 590 and 1200 mOsm/kg. Some individuals survived concentrations as low as 320 and as high as 2310 mOsm/kg. In common with other intertidal and estuarine molluscs, *Siphonaria* uses its shell as a passive barrier, clamping it to the substratum to prevent temporary osmotic exchange at the extremes of its range. The shell is also important in reducing evaporation in air. *Siphonaria* survives up to 54 h dehydration, completely absorbing lost water in about one hour when re-submerged.

The water balance of freshwater pulmonates has been most extensively studied in *Lymnaea stagnalis*. Van Aardt (1968) has shown that these animals are always hyper-osmotic to the external medium producing a steady flow of urine (Table V) to counteract the osmotic entry of water through the body wall. He provided further evidence that the kidney in *Lymnaea* serves as an efficient volume regulator by demonstrating reduced urine flow in proportion to changes in osmotic gradient across the body wall. In addition, impairment of normal kidney function and the stopping of the heart by nembutal and MS222 anaesthesia rapidly bring about a reversible decrease in urination and a concomitant increase in body weight (5% in eight minutes). The conclusion that water balance in *Lymnaea* is dynamically maintained was further supported by experiments using tritiated water.

It is possible that a number of other processes affect water balance in freshwater pulmonates. *Lymnaea* spp. feed more or less continuously on aquatic vegetation (Carriker, 1946; Calow, 1970) and must therefore ingest some water by this means. Similarly water must be voided by way of the faeces and associated mucus. It is also possible, since the mucus of *Lymnaea truncatula* (Müll.) contains 92% water by weight (Wilson, 1968), significant amounts of water could be eliminated by the secretion of this substance. It is not known, however, what the total rate of mucous secretion is or from what source the mucus becomes hydrated. It is also possible that *Lymnaea stagnalis* (L.) could eliminate water from the body by the direct venting of haemolymph through its haemal pore (Lever and Bekius, 1965). This would presumably occur only under emergency conditions, prompting a sudden contraction of the body. In view of the lack of quantitative information concerning the importance of these alternative processes it can only be concluded at the present time that the kidney is the principal volume regulator in freshwater pulmonates.

It is generally assumed that the initial step in urine formation in the Basommatophora is ultrafiltration, and that this process occurs across the atrial wall of the heart into the pericardial cavity. The evidence for this, first put forward by Picken (1937) was only indirect and walls of the heart of pulmonates do not have the typical podocyte structure seen in mammalian glomeruli (Wendelaar Bonga and Boer, 1969) though they have been identified in some terrestrial prosobranchs (Andrews and Little, 1972). Recent studies, however (Frömter and Diamond, 1972; Newell and Skelding, 1973) have shown that ultrafiltration can occur across simpler epithelia without podocytes. Clearance studies of Van Aardt (1968) are also consistent with the theory of ultrafiltration. Injected inulin, after equilibration, appears in the pericardial fluid in concentration equal to that of the blood, there being no difference between osmotic pressure of the two fluids. Inulin concentration in the final urine is normally slightly increased indicating about 10% water reabsorption (Table V). Higher levels of tubular reabsorption were found in some individuals having inherently low or experimentally impaired urine production rates.

Freshwater pulmonates inhabiting bodies of water which become periodically dried up are known to be able to survive a certain amount of desiccation. Indeed *Lymnaea truncatula* was considered by Boycott (1934, 1936) as a terrestrial species since it spends most of the time out of water. These species seem to occur chiefly among the Lymnaeidae and Planorbidae. However, it is not known whether survival out of water reflects some fundamental structural or physiological characteristic or whether these families have received greatest attention owing to their medical importance as hosts for trematode parasites. There are also

Table V. Rates of urine production in pulmonates

Species	Temp. (°C)	Filtration rate (µl/g body wt/h)		Final urination rate (µl/g body wt/h)		Authority
		Mean	(Range)	Mean	(Range)	
Lymnaea stagnalis	20	—	(60–122)	—	(56–110)	van Aardt[a], 1968
Achatina fulica (active)	25	27	(1·2–120)	—	—	Martin et al.,[a] 1965
Archachatina ventricosa (active)	—	—	—	36	(36–130)	Vorwohl[b], 1961
Helix pomatia (active)	25	280	(18–480)	108	(9–250)	Vorwohl[b], 1961
Helix pomatia (hibernating)	12	1·4	(0–8)	—	—	Vorwohl[b], 1961

[a] Values based in inulin clearance
[b] Values based in direct volume measurement

occasional reports of *Physa* spp. surviving desiccation (Pilsbry, 1926; Mozley, 1939; Ingram, 1940a). Survival studies in the field and laboratory indicate length of survival depends on the rate of water loss from the snail. Survival after loss of the order of 60% of original body water seems possible in the planorbid *Biomphalaria glabrata* (Say) (von Brand *et al.*, 1957; Barbosa and Olivier, 1958) though, due to the metabolism of body reserves, tissue water content remains relatively unchanged. At 27°C, rates of water loss recorded by von Brand for this species during aestivation were 0·25 mg/g body wt/h at 96% r.h. and 2·8 mg/g body wt/h at 15% r.h.

During desiccation snails withdraw deeply into the shell. Many tropical planorbids, species of *Tropicorbis*, *Drepanotrema* and *Biomphalaria*, *glabrata* among others, have various devices for reducing or obstructing the aperture (Richards, 1968). Although ingrowths round the shell opening, aperture lamellae, or dried layers of transparent foamy mucus, are unlikely to reduce the diffusion of water vapour very much they may retard evaporation by eliminating convective transfer. As an alternative means of counteracting water loss in air, Lynch (1966) showed that under certain conditions *Lymnaea tomentosa* (Pfeiffer) in South Australia is able to take up water from drying mud in which it burrows.

Fully terrestrial pulmonates tend to be exposed to a much wider range of conditions than their freshwater relatives. First, much larger variations in absolute water concentration, usually measured in mmHg vapour pressure, are possible in the gas phase. Temperature fluctuations due to the lower specific heat of air have a profound effect on water vapour in the atmosphere and on water exchange in terrestrial animals. The tendency of water vapour to collect in restricted spaces also contributes to great spatial and temporal variation on land. This instability makes it possible for a terrestrial pulmonate to be hyper-osmotic to its environment and then the reverse in a comparatively short time.

Terrestrial pulmonates readily absorb water when immersed in tap water or placed in contact with droplets of dew or rain. Experimental studies of this process are not very satisfactory and results show a great deal of variation (Table VI). Measurement by weighing the animal normally overestimates uptake rate, since water is also absorbed superficially in the mucus and can collect in the lung. Water uptake varies with surface area and the osmotic gradient across the body wall. Neither lends itself to accurate measurement during absorption. Osmotic gradients decrease with haemolymph dilution and uptake values become highly dependent on the duration of the experiment. High rates of

Table VI. Calculated osmotic uptake rates[a] from fresh water in adult pulmonates

Species	Temp. (°C)	Mean uptake rate, partial contact (mg/g body wt/h)		Mean uptake rate, total immersion (mg/g body wt/h)		Authority
		Mean	(Range)	Mean	(Range)	
Lymnaea stagnalis	20	—	—	84[b]	—	van Aardt, 1968
Arianta arbustorum	18–20	460	(315–750)	260	—	Künkel, 1916
Arion ater	18–20	365	—	48	—	Künkel, 1916
Helix aperta	—	50	(25–220)	—	—	Burton, 1966
Helix aspersa	22	—	—	26	—	Machin, 1966
Helix pomatia	25–28	230	(220–295)	250	—	Künkel, 1916
Limax cinereoniger Wolf	18–20	155	—	26	—	Künkel, 1916
Otala lactea	—	50	(10–700)	—	—	Burton, 1966

[a] Uptake rates depend greatly on the duration of water contact.
[b] Determined from inulin measured urine production. The remaining values were obtained by weighing.

water uptake in partly submerged individuals suggest that the hydro-philic properties of superficial mucus (Machin, 1964a) enhance water absorption by promoting surface spreading. Adolph (1932) used this explanation to account for similar uptake rates in fully immersed toads and in individuals in contact only with moist paper.

A further difficulty in accurately establishing rates of osmotic uptake of water lies in the problem of assessing whether or not terrestrial pulmonates drink. Künkel (1916) thought that slugs drank since buccal mass movements were observed as they crawled up a surface collecting water droplets in the region of the mouth as they went. Dainton (1954a) never observed this in her work and considered that water uptake by the gut occurs only by way of the food. Attempts by Machin (1962) to separate osmotic uptake and that due to drinking were never really successful since ligaturing the buccal mass also isolated a significant proportion of the body wall from circulating haemolymph. Comparison between active and anaesthetized animals has not been made.

Terrestrial pulmonates obtain some water from their food. It has been suggested (Arivanitaki and Cardot, 1932) that the movement of water into the haemolymph was delayed until 12–24 h after feeding but this is now considered unlikely (Burton, 1966). Pallant (1970) measured food uptake rate (of carrot) and faecal production in *Agrioli-max reticulatus* in field and laboratory conditions for 24 h. Both food and the faeces were found to have a water content of 85%, however since 78·4% of the carrot was assimilated this resulted in a significant net gain of water to the slug (Table VII). In another study (Mason, 1970), assimilation rates in woodland snails at rather lower temperatures were found to be generally lower for leaf litter and lettuce. Higher rates of consumption and assimilation of earthworms which form part of the natural diet in *Discus rotundatus* (Müll.) and *Oxychilus cellarius* (Müll.) suggest that food of animal origin forms a particularly rich source of water. In spite of this, contribution from the food seems to be rather small in comparison with direct liquid uptake through the skin. Hunter (1968) found that active slugs continued to lose weight in saturated air even though they fed at regular intervals. Further indication that dietary water is only of minor importance to the overall water economy is suggested by the fact that terrestrial pulmonates will not feed normally until they have first absorbed some liquid water (Howes and Wells, 1934b; Burton, 1966; Grime *et al.*, 1968).

Water is lost by evaporation from all exposed surfaces of a terrestrial pulmonate's body, once the ambient humidity falls below the blood equilibrium humidity – about 99·5% r.h. at 20°C (Table IV). Evapora-tion from the surface of active individuals may be described as "vapour

Table VII.   Estimated water uptake rates in the food of terrestrial pulmonates

Species	Type of food	Temp. (C°)	Estimated H_2O gain in food (mg/g body wt/h)	Estimated H_2O loss in faeces (mg/g body wt/h)	Net water gain (mg/g body wt/h)	Authority
Agriolimax reticulatus	carrot root and leaves	18	6·5	1·2	4·3	Pallant, 1970
	woodland litter	—	8·5[b]	1·7	6·8	Pallant, 1970
Discus rotundatus	woodland leaf litter	10	9·4[c]	4·0[d]	5·4	Mason, 1970
	eviscerated earthworm	10	15·1[c]	3·3[d]	11·8	Mason, 1970
Helix aspersa	woodland leaf litter	10	6·0[c]	2·8[d]	3·2	Mason, 1970
	lettuce	10	7·5[c]	2·4[d]	5·1	Mason, 1970
Oxychilus cellarius	lettuce	10	7·9[c]	2·4[d]	5·5	Mason, 1970
	eviscerated earthworm	10	84·6[c]	4·6[d]	80·0	Mason, 1970

[a] Values where temperatures are given were obtained in the laboratory. The remaining value refers to feeding under natural conditions in the field

[b] Estimated using the 0·78 assimilation fraction determined in the laboratory

[c] Calculated from dry weights assuming a 97% water content, as determined for lettuce

[d] Calculated from dry weights using the finding of Pallant that food and faecal water contents are the same

limited" (Beament, 1961) since the rate of loss depends primarily on vapour pressure gradients in the air adjacent to the surface. The maintenance of this state depends however on an equally rapid mechanism of water transport to the surface, in order to keep the superficial epithelium from drying out (Machin, 1964a). Calculations show that the osmotic permeability of the body wall in *Helix aspersa* is too low to permit a sufficiently rapid outward flow in lower humidities. The balance between osmotic flow and evaporation can only be achieved in 50 and 10% r.h. by excessively steep osmotic gradients and superficial solute concentrations reaching 9 and 16 times that of the blood, respectively. Experimental comparisons between living and freshly killed tissues in *Helix aspersa* (Machin, 1964a) have led to the understanding of the important role of mucous secretion in transfer water to the evaporating surface. Mucus extrusion is probably brought about by local increases in haemocoelic pressure, initiated by muscle contraction in active snails (Campion, 1961). The grooves in the skin of larger species, where the mucus seems to collect, apparently act as reservoirs from which it may be conveyed by muscular undulation to areas exposed to excessive desiccation (Machin, 1964a).

Evaporation from animals with moist skins is a notoriously variable phenomenon, because the factors which determine water vapour transfer in the air are so complex. Practically any environmental parameter—humidity, temperature, wind speed, wind direction—affects evaporative water loss. Attempts have long been made by biologists to relate evaporation with strictly environmental parameters such as relative humidity or saturation deficiency. Since these parameters are in effect comparisons between existing water vapour concentrations and those at saturation, they are approximately equivalent to the real driving force for evaporation which is the difference in vapour pressure at the surface and that in the air some distance away. In most situations failure to take proper account of surface vapour pressure does not lead to serious error. However, in large temperature gradients or in high humidities, solute vapour pressure lowering becomes significant and analysis of evaporation in terms of ambient humidity alone can lead to serious error (Thornthwaite, 1940).

Empirical analysis of evaporation in still air is relatively simple and rates of loss can be shown to depend, at a given temperature, on surface area and the difference in vapour pressure at the surface and that of the ambient air. Measurements using specimens of *Helix aspersa* at 20°C (Machin, 1962), indicate that surfaces covered with freshly secreted mucus lose water by evaporation at a rate of about $2 \cdot 5 \, \text{mg/cm}^2/\text{h/mmHg}$ (Table VIII). The coefficient of diffusion of water vapour itself, of

Table VIII. Water transfer coefficients for terrestrial snails under different conditions at 20°C

Species	Experimental conditions	Barrier to which coeff. principally applies	Transfer coefficient (mg/cm^2/h/mmHg)	Authority
Helix aspersa	aqueous media	dorsal body wall	47	Machin, 1966
Helix aspersa	dorsal body wall in still air withdrawn snail	still air around snail	2·5	Machin, 1962
Helix aspersa	secreting mucus	air in shell opening	2·6	Machin, 1965
Helix aspersa	shell with distilled water withdrawn snail	still air in shell opening	2·3	Machin, 1965
Helix aspersa	regulating evaporation withdrawn snail	mantle	0·039	Machin, 1966
Otala lactea	regulating evaporation	mantle	0·016	Machin, 1972

which the above value is an effective measure, may be expected to increase with absolute temperature (Mason and Monchick, 1965). Small differences in observed evaporation rates between extended animals and those withdrawn inside the shell are most readily explained by the presence of convection currents in apparently "still air". The more exposed the surface the greater effect of convection in speeding up evaporative loss.

Analysis of evaporation is more complex in air moving at measurable speed. Under such conditions subtle factors such as the animal's size, orientation and shape, including those of non-evaporating areas, also begin to influence evaporation (Machin, 1964b, c). As a general rule evaporation per unit area is greater in small animals than in larger ones. Loss from terrestrial pulmonates and other small cylindrical surfaces increases with the power 0·66 of the wind speed. For flat evaporating surfaces the rate of increase is slightly less, the wind speed function being 0·5. Even though terrestrial pulmonates may limit their activity to wind conditions below one or two miles per hour evaporation in these conditions is about five times more rapid than in still air.

Dainton (1954a) and Machin (1962) have both attempted to separate water lost during locomotion from that due to evaporation. In one study with *Helix aspersa*, Machin found losses were about equal at 20°C and 55% RH. However pedal losses show a great deal of variability and are frequently very much greater than evaporation (Table IX). The amount of mucus lost during continuous progression markedly decreased with time and the degree of hydration of the animal. Since pedal loss and to a certain extent, evaporation, depend on the degree of extension and activity, all terrestrial pulmonates are able to limit dehydration by becoming inactive, at least intermittently. Although terrestrial slugs are able to reduce surface area slightly by acquiring more spherical proportions, they often lose water very rapidly and completely desiccate in a few days (Künkel, 1916; Hogbin and Kirk, 1944; Howes and Wells, 1934b). Similar rates of desiccation were also observed in the marsh species *Succinea putris* (L.) whose widely open shell prevents full withdrawal of the animal (Künkel, 1916).

The reduction of water loss following withdrawal is most marked in fully terrestrial snails (Table X) and rates of desiccation are considerably lower than in the Basommatophora or in slugs. The shell itself is obviously important in controlling evaporation, being comprised of an organic matrix and crystalline calcium carbonate (Wilbur, 1964; Saleuddin and Chan, 1969) which together are almost completely impermeable to water at normal temperatures. At increased temperatures, Machin (1967) found no significant differences between the

Table IX. Rates of non-urinary water loss by continuously active terrestrial pulmonates

Species	Mode of water loss	Temp. (°C)	r.h. (%)	Rate of water loss[a] (mg/g body wt/h)		Authority
				Mean	(Range)	
Agriolimax reticulatus	locomotion	—	100	—	(30–250)	Dainton, 1954a
Slugs in general	evaporation	—	45	—	(30–50)	Dainton, 1954a
Helix aspersa	locomotion	20	—	40	(18–110)	Machin, 1962
Helix aspersa	evaporation	20	55	45	—	Machin, 1962
During intermittant activity						
Agriolimax agrestis (L.)	total	18–20	—	8·3	—	Künkel, 1916
Arianta arbustorum	total	28	65	5·8	(3·6–9·8)	Cameron, 1970a
Arion ater	total	15–19	40–80	10	—	Howes and Wells, 1934b
Arion ater	total	18–20	—	2·6	—	Künkel, 1916
Cepaea hortensis (Müll.)	total	28	65	3·7	(1·7–6·2)	Cameron, 1970a
Cepaea nemoralis	total	28	65	3·9	(1·9–8·2)	Cameron, 1970a
Lehmannia marginata (Müll.)	total	18–20	—	17·9	—	Künkel, 1916
Limax maximus L.	total	18–20	—	2·4	—	Künkel, 1916
Limax tenellus Müll.	total	18–20	—	8·7	—	Künkel, 1916

[a] All values apply to still air and are expressed in terms of shell-less body weight

Table X. Rates of water loss by terrestrial snails withdrawn into the shell

Species	Temp. (°C)	r.h. (%)	Water loss without epiphragm (mg/g body wt/h)	Water loss with epiphragm[a] (mg/g body wt/h)	Authority
Arianta arbustorum	—	—	—	0·20	Künkel, 1916
Arianta arbustorum	24	35	—	0·185	Gebhardt-Dunkel, 1953
Arianta arbustorum	28	65	0·61	—	Cameron, 1970a
Buliminus detritus (Müll.)	24	35	—	0·033	Gebhardt-Dunkel, 1953
Cepaea hortensis	24	35	—	0·052	Gebhardt-Dunkel, 1953
Cepaea hortensis	28	65	0·38	—	Cameron, 1970a
Cepaea nemoralis	28	65	0·33	—	Cameron, 1970a
Cepaea nemoralis	—	—	—	0·10	Künkel, 1916
Eulota fruticum (Müll.)	24	35	—	0·093	Gebhardt-Dunkel, 1953
Helicella virgata	30	5	—	0·039	Andrewartha, 1964
Helicella virgata	24	<40	—	0·02	Pomeroy, 1968a
Helicigona lapicida (L.)	—	—	—	0·032	Künkel, 1916
Helix aspersa	20	<20	0·17	0·10	Machin, 1967
Helix pomatia	—	—	—	0·041	Künkel, 1916
Otala lactea	20	<20	0·066	0·042	Machin, 1967
Sphincterochila boissieri	20	<20	0·014	0·0096	Machin, 1967
Sphincterochila boissieri	—	—	—	0·038	Schmidt-Nielsen et al., 1971
Succinea putris	—	—	3·75	—	Künkel, 1916

a Animals for some of these values may have been sealed to the substratum

Table XI. Loss of water through the shell in terrestrial pulmonate snails[a]

Species	Temp (°C)	r.h. (%)	Shell loss (mg/g body wt/h)	% of total inactive loss	Authority
Arianta arbustorum	26	45	0·0139	2·3	Cameron, 1970a
Cepaea hortensis	26	45	0·0084	2·2	Cameron, 1970a
Cepaea nemoralis	26	45	0·0059	1·8	Cameron, 1970a
Helix aspersa	22	70	0·0079	4·6	Machin, 1967
Otala lactea	22	70	0·017	25·8	Machin, 1967
Sinumelon remissum Iredale	20	dry air	0·014	70	Warburg, 1965
Sphincterochila boissieri	22	70	0·0086	61·5	Machin, 1967

[a]Based on weight loss of distilled water through empty shells with aperture sealed

permeability coefficient of the shell material from three species. However differences in shell thickness and surface area gave rise to widely
different total losses from the whole shell (Table XI). Cameron (1970a)
found similar specific differences related to the environmental availability of water. His data show that mean rates of water loss through the
shell are less than 3% of the total loss by inactive snails, though much
higher proportions obtained by Machin (1967) and Warburg (1965)
suggest perhaps that some of their methods for sealing the shell aperture
may have been inadequate. Since the aperture is the principal site of
water loss from inactive snails, adaptations which reduce evaporation
from this area are of great value. Among the most obvious of these are
apertural modifications such as ribs, teeth and the clausilium. Gebhardt-
Dunkel (1953) and Machin (1967) investigating different series of terrestrial snails, found that the relative size of the aperture decreases in species living in dryer habitats. Adaptations which promote a closer
sealing of the aperture to the substratum such as the development of a
strengthened or reflected apertural lip or changes in the plane of growth
(Mazek-Fialla, 1934; Machin, 1967) are also found better developed in
xerophilous species. In addition to structural improvements in shell
design, the secretion of an epiphragm, either to seal the peristome to
the substratum or to close the aperture itself temporarily, is related to
the reduction of water loss. Although broad generalizations cannot be
made, epiphragm thickness, and degree of calcification, together with
the tendency to produce multiple epiphragms (Gammon, 1943) seem
to increase with the aridity of the habitat in some series of species
(Machin, 1967). Cameron (1970a) noted a similar correlation between
habitat and the readiness with which a given species would form an
ephiphragm before withdrawal.

The epiphragm, though recognized as being permeable to water to
some extent (Kühn, 1914; Fischer, 1950) has traditionally been regarded
as the principal barrier to water loss for the aperture. Wide acceptance
of this view has been demonstrated as recently as 1965 by Warburg,
when he described the air space behind the epiphragm as saturated with
water vapour, acting as an "insulation" between the animal and the
outside environment. Contrary to this, there has been a growing
realisation that the principal barrier to water loss from the aperture lies
with the mantle tissues of the snail itself. The most persuasive evidence
for this is direct measurements of epiphragm permeability from several
species of snail (Machin, 1967) and the failure of these structures in
experimental models, even the thick hibernation epiphragm of *Helix
pomatia*, to reduce evaporative loss to the low levels found in intact,
inactive snails. In support of this, various reports are now appearing in

the literature describing evaporation rates from the mantle of helicid snails (Table VIII) in the absence of an epiphragm, which are much below the levels expected from a moist surface coated with mucus (Machin, 1965, 1966, 1967, 1972; Pomeroy, 1966; Andrewartha, 1964; Gebhardt-Dunkel, 1953). Mantle surface temperatures, in inactive *Helix pomatia* measured by Hogbin and Kirk (1944), showed evaporation from the mantle was very slow. However the lack of surface cooling was thought to be due to the presence of a thick hibernation epiphragm. Unfortunately they failed to observe the effects of removing it.

Weight loss from inactive terrestrial snails (Table X) is at least three orders of magnitude less than rates observed in active animals. The data also suggest that losses are environmentally correlated, lowest values occurring in species from arid habitats. This conclusion should be viewed with some caution, however, since all measurements were not performed under identical conditions. Pomeroy (1966) has shown that rates of loss depend on temperature, humidity and the size of the snail. Observed rates of loss also, to some degree, depend on the experimental procedure used to measure weight loss. Inactive snails are highly sensitive to any form of mechanical disturbance and respond by secreting mucus, resulting in increased water loss (Machin, 1965). This observation was confirmed by Schmidt-Nielsen *et al.*, (1971) who found that rates of loss from *Sphincterochila boissieri* decreased with the frequency of weighing.

The mechanism responsible for regulation of evaporation from the mantle is largely unknown. Machin (1965, 1966, 1972) has shown that reduced evaporation and mucus extrusion are incompatible and regulating snails drastically reduce glandular activity. Andrewartha (1964) and Pomeroy (1966) have speculated about the presence of a water pump in the mantle tissues. However, at the present time a passive impermeable barrier seems most likely. This barrier does not lie in the mucus but apparently is located beneath a superficial hygroscopic compartment which is in concentration equilibrium with the ambient air (Machin, 1972). Changes in ambient humidity induced a measurable gain or loss of water to bring the appropriate concentration change about. It must be emphasized however that these changes are relatively small and do not represent significant gains or losses to the animal's internal reserves.

There are some indications that the epiphragm may reduce the dispersal of humid pulmonary air, thereby conserving water. Continuous weight recordings of inactive *Otala lactea* (Müll.) with an epiphragm (Machin, unpublished observations) show occasional losses up to 1 mg followed by a gradual increase in weight (Fig. 1). The most reasonable

explanation of these events, since they occur in weight but not surface
temperature traces, is that they represent the release of humid air from
the lung. The subsequent gain of weight is due to the absorption of water
by the dehydrated mantle surface.

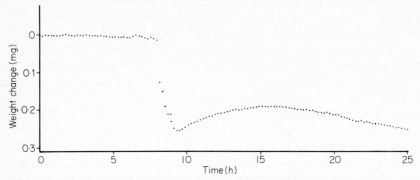

Fig. 1. Weight record of inactive *Otala lactea* with an epiphragm showing presumed
pulmonary water loss followed by partial water uptake by the mantle. The snails
were maintained at 25°C and 60% r.h.

Urine formation in terrestrial pulmonates differs slightly from that
found in freshwater species. The discovery by Rolle (1907) that the
reno-pericardial duct in terrestrial slugs and snails was a tenth the
diameter found in *Lymnaea stagnalis*, has always introduced some doubt
as to the site of initial urine formation. Vorwohl (1961) however was
able to demonstrate by measuring flow from the various kidney compartments, that urine in *Helix pomatia* and *Archachatina ventricosa*
Gould is first formed by ultrafiltration across the wall of the kidney sac.
Although podocytes are present at the site of ultrafiltration in some
terrestrial prosobranch molluscs (Andrews and Little, 1971), ultrafiltration in terrestrial pulmonates appears to take place through
relatively unspecified epithelia in the kidney sac wall (Skelding, 1973;
Newell and Skelding, 1973). Results of other experimental studies are
consistent with the ultrafiltration theory of urine formation. The primary urine is similar in composition to that of the haemolymph with
respect to osmotic pressure and the concentrations of electrolytes, urea
and inulin (Vorwohl, 1961; Martin *et al.*, 1965 Skelding, 1973);
Martin *et al.* also showed that back pressure on the kidney decreases
urine flow. Urine formation is prevented by pressures exceeding 12 cm
of water. Systolic pressure in the ventricle of *Helix pomatia* is about
24 cm H_2O (Jones, 1971).

Filtration rates in active hydrated snails are very variable but of the
same general level found in freshwater *Lymnaea stagnalis* (Table V).

During its formation in the secondary ureter, final urine becomes hypotonic to the blood (Vorwohl, 1961) by electrolyte reabsorption (Skelding, 1972). Inulin concentrations observed by Martin *et al.* (1965) indicate very little tubular reabsorption of water at intermediate to rapid urine flow rates, though there is some reabsorption over the slower ranges. Inulin concentrations measured by Skelding (1972) indicate about 33% water reabsorption in final urine formation in *Achatina achatina* (L.). Urine production rate is greatly reduced following experimental dehydration by a decrease in filtration (Vorwohl, 1961) and increased tubular reabsorption to 70% (Skelding, 1972). Hibernating individuals show further reductions in filtration rate (Vorwohl, 1961), some individuals showing no measurable urine production at all. The kidney obviously contributes in an important way to fluid elimination in hydrated animals. Vorwohl showed that urine elimination contributed significantly to observed weight losses in active specimens of *Archachatina ventricosa*. Blinn (1964) showed that in some polygyrids, fluid, probably originating from the kidney, is not eliminated immediately but retained for some time in the mantle cavity.

V. Water exchange in different environments

There is very little physiological information concerning primitive pulmonate families such as the Ellobiidae, which are found in marine intertidal and supratidal habitats (Morton, 1954b; 1955a). Microclimates at low tide under rocks, among plant roots and in crevices more closely resemble subtidal conditions than they do truly terrestrial habitats. Larger intertidal species are unable to take advantage of such environmental protection and must survive a wider range of temperatures (Newell, 1969) as well as greater fluctuations in the availability of water from diluted sea water to quite low atmospheric humidities. In common with other intertidal gastropods *Siphonaria pectinata* tolerates a wide range of body water contents and osmotic pressures and on a short term basis, is able to restrict osmotic exchange at the extremes of its range by clamping the shell to the substratum (McAlister and Fisher, 1968).

In contrast with the intertidal zone, freshwater environments provide comparatively constant conditions and in addition, much lower levels of dissolved salts. Theoretically, temperature should directly affect osmotic exchange, but not very much. Indirect temperature effects on the activity of an aquatic animal are probably of greater physiological significance. In colder climates some aquatic snails pass the winter half

F

grown in a partially inactive state buried in mud (Boycott, 1936). Withdrawal into the shell during this hibernation may serve to reduce exposed surface area and the osmotic entry of water. However, some hardy species remain fully active and extended even when the temperature is just above freezing point. At the other extreme some species remain active in temperatures exceeding 30°C in hotter climates during the summer.

There is no physiological reason, at least from the point of view of water balance, why freshwater species should not survive well in water with a higher salt content. As long as the body fluids remain iso-osmotic or hyper-osmotic to their surroundings, no fundamental change in water balance physiology is required. Under experimental conditions van Aardt (1968) has shown that freshwater *Lymnaea stagnalis* readily survive saline solutions and respond to them by decreasing urine production in proportion to the reduced osmotic gradient across the body wall. There are also a number of recorded instances of freshwater pulmonates inhabiting natural brackish water environments. Abdel Malek (1958) reported that *Bulinus contortus* (Mich.) and *Biomphalaria glabrata* tolerate up to 2·1 and 3·6 parts per thousand NaCl, respectively, in various aquatic situations in North Africa. Since both these levels are well below haemolymph concentrations normally found in freshwater species (Table II), these snails are probably hyper-osmotic to their surroundings. On the other hand maximum salinities of 11 parts per thousand (320 mOsm/kg) in which *Lymnaea stagnalis*, *L. peregra* (Müll.), *L. palustris* (Müll.), *Physa fontinalis* (L.) and *Planorbis vortex* (L.) are found living in the Baltic (Jaeckel, 1951) are considerably higher than salt levels found in normal freshwater individuals. Presumably since hypo-osmotic regulation is unknown in the Mollusca, Robertson (1964) thought it likely that these species were iso-osmotic to their surroundings, having acquired a tolerance for higher haemolymph salt levels. Hubendick (1948) found that the kidney of Baltic *Lymnaea limosa* (L.) was reduced in size compared with freshwater individuals, presumably because of the decreased osmotic entry of water.

Though freshwater pulmonates do not show the extreme resistance to desiccation found in terrestrial species, some are able to survive fairly prolonged periods without water. Some bodies of water inhabited by snails, dry up on a regular seasonal basis, usually during the hottest months of the summer. For some snails living in irrigation canals of Egypt enforced desiccation regularly comes during the winter months (Barlow, 1933, 1935) when the canals are drained. Individuals of some species have been known to survive several hundred days out of water (Table XII). However, these values seem to be atypical since high mortality rates are usually observed after much shorter periods of desiccation (Abdel Malek, 1958). In general the survival of two to four months without water seems fairly common in species inhabiting temporary bodies of water which is usually enough to maintain the snail population to the following wet season. There is some correlation however, between the ability to survive desiccation and the permanence of the aquatic habit to which the snail is acclimated. In two Brazilian pulmonates, Olivier (1956) found lower mortalities and higher maximum survival times in populations from temporary bodies of water compared with individuals from habitats which did not dry out.

Aestivation in sheltered situations under vegetation or in deep crevices in mud increases the likelihood of survival (Abdel Malek, 1958; Barbosa and Olivier, 1958). Locations in which humidities are higher and temperatures lower reduce the rate of evaporative loss and increase survival time (von Brand *et al.*, 1957). In hot climates animals which are protected from direct exposure to the sun may also avoid high, potentially lethal day-time temperatures. Some species are better able to resist high temperature when kept out of water. High thermal death point in *Biomphalaria glabrata* in water was found to be 42°C whereas members of the same species were able to survive exposure to 45°C after a period of desiccation (Barbosa and Olivier, 1958). In most instances basommatophorans probably arrive in suitable locations for aestivation largely by chance. *Lymnaea tomentosa* in South Australia, however appears to be an exception. Lynch (1966) showed that this species at the beginning of a period of aestivation burrows in response to a fall in soil water. Water content is the critical factor, since experimental immersion of *Biomphalaria glabrata* in mud saturated with water and presumably poorly oxygenated resulted in premature death (Barbosa and Olivier, 1958).

The presence of environmental water is a major factor in determining the distribution and behaviour of terrestrial pulmonates. Boycott (1934) has pointed out that each type of habitat is usually occupied by its own characteristic group of species. It is unusual to find species from wet environments in dry locations and vice versa. Even when similar species occur in the same locality, differences in habitat selection relating to environmental humidity are found. *Arianta arbustorum* (L.) which laboratory studies established as being less well adapted to dry conditions than *Cepaea* spp. (Cameron, 1970c) preferred areas of greater plant cover (Cameron, 1970a).

Except in wet, marshy habitats, terrestrial pulmonates survive only by regulating their activity according to the availability of water (Arivanitaki and Cardot, 1932; Pomeroy, 1968b). The rapidity with

Table XII. Recorded survival times in air, without food or water, in aquatic and terrestrial pulmonates

Species	Conditions	Temp (°C)	r.h. (%)	Duration of dessication period	Survival (%)	Authority
Basommatophora						
Siphonaria pectinata	laboratory	24	—	54 h	—	McAlister and Fisher, 1968
Biomphalaria glabrata (from permanent pools)	laboratory	27	96	64 days	50	von Brand *et al*., 1957
(from permanent pools)	laboratory	25–30	70–80	113 days	5	Olivier, 1956
(from temporary pools)	laboratory	25–30	90–95	113 days	66	Olivier, 1956
Biomphalaria glabata	dried mud, field	—	—	180 days	—	Barlow, 1933, 1935
Biomphalaria glabata	dried mud, lab	—	—	300 days	—	Barlow, 1933, 1935
Bulinus truncatus (Audovin)	dried mud, field	—	—	120 days	—	Humphreys, 1932
Bulinus truncatus (Audovin)	dried mud, lab	—	—	360 days	2	Barbosa and Olivier, 1958
Lymnaea truncatula	dried mud, lab	—	—	190 days	18	Kendal, 1949
Physa gyrina Say	dried stream bed	—	—	60 days	—	Ingram, 1940[a]
Stagnicola bulimoides techella (Hald.)	dried mud, field	40	—	158 days	—	Olsen, 1944

Tropicorbis centrimetralis (Lutz) (from permanent lake)	laboratory	25–30	90–95	99 days	1	Olivier, 1956
(from temporary pools)	laboratory	25–30	90–95	99 days	11	Olivier, 1956
Stylommatophora						
Allogona profunda	refrigerator	—	—	390 days	—	Blinn, 1963
Arion ater	laboratory	15–19	80	3–4 days	—	Howes and Wells, 1934b
Eremina desertorum (Forskål)	laboratory	—	—	4 year	—	Baird, 1850
Helicella neglecta (Drap.)	laboratory	30	6	134 days	50	Pomeroy, 1966
Helicella virgata	laboratory	30	5	234 days	—	Andrewartha, 1964
Helicella virgata	laboratory	25	50–54	350 days	50	Pomeroy, 1966
Helix aspersa	laboratory	—	—	390 days	—	Ward, 1897
Helix pomatia	laboratory	—	—	300 days	—	Künkel, 1916
Otala lactea	laboratory	—	—	2 year	—	Tryon, 1882
Otala lactea	laboratory	—	—	3·5 year	—	Cook *et al.*, 1895
Otala lactea	laboratory	—	—	4 year	—	Gaskoin, 1850
Sphincterochila boissieri	laboratory	15	20	3 year	—	Machin, 1967
Micrarionta vietchi (Newcomb)	laboratory	—	—	6 year	—	Stearns, 1877

which water is either gained or lost from active, extended terrestrial pulmonates apparently makes control of the water content at a constant level impossible. Terrestrial pulmonates apparently keep internal fluctuations within physiologically tolerable limits, behaving in a way that generally encourages water uptake and reduces water loss.

Probably all species take advantage of cooler night-time temperatures and the presence of dew by being active at night especially between sunset and midnight when feeding frequently occurs (Barns and Weil, 1954a; South, 1965). Many species in fact undergo a semi-regular diurnal rhythm in which nocturnal activity is followed by a period of quiescence during the day (for snails see Tercafs, 1961; Blinn, 1963; Pomeroy, 1968a; Cameron, 1970b and for slugs Dainton, 1954a; Lewis, 1969a). This cycle in larger snails such as *Helix pomatia* (Howes and Wells, 1934a) and *Archachatina ventricosa* (Vorwohl, 1961) may occur instead at intervals of several days. The nocturnal rhythm is not as rigidly adhered to as commonly supposed: snails (Ingram, 1940b; Blinn, 1964) and slugs (Barnes and Weil, 1945a) frequently become active during or following day-time rain. Night-time activity may be also restricted by excessively high winds or extremes of temperature (Barnes and Weil, 1954a; Cameron, 1970b). The above observations support the principle that environmental water plays the dominant role in regulating activity in terrestrial pulmonates. However, moist-skinned animals are poorly equipped to detect ambient humidity directly since superficial hydrated mucus ensures that the body surface is always close to saturation (Dainton, 1954a). Laboratory experiments with *Agriolimax reticulatus* showed that the animal is activated below 20°C instead by falling temperatures and remains unaffected by changing light intensities after a period of acclimation. Lewis (1969a), however, could find no such effect of temperature in *Arion ater* (L.) and confirmed the results of Karlin (1961), Getz (1963), Newell (1968) and Cameron (1970b) which showed activity to be regulated by diurnal light changes.

Once active, both slugs and snails show preferences for the higher humidities, including saturated air (Tercafs, 1961; Lewis, 1969a) when presented with a choice. In constant humidity Cameron (1970a) found specific differences in overall water loss from intermittently active snails, lower values occurring in *Cepaea* species which are known to inhabit drier, more exposed situations than *Arianta arbustorum* (Cameron, 1969, 1970c). Differences in the peak activity of these snails in relation to temperature were also observed. *Arianta*, which extends into cooler more northern localities than *Cepaea* was relatively more active at lower temperatures (Cameron, 1970b). Terrestrial pulmonates also show preferences for fairly narrow temperature ranges presumably since they represent a compromise between tolerable levels of evaporation and a suitably high rate of progression. Perhaps for the same reason several slug species show marked seasonal differences in peak activity, which were found to occur during the summer months for *Agriolimax reticulatus* and the winter for *Arion hortensis* Fér. (Barnes and Weil, 1945b). Stimulation of activity above the preferred range by temperature increases (Dainton, 1954a) possibly respresents an emergency response to remove the animals from conditions in which rapid dehydration would occur. The tendency for terrestrial pulmonates to avoid or to move away from air currents (Shelford, 1913; Dainton, 1954b; Tercafs, 1961) would similarly lead the animal to more favourable conditions.

There is some evidence that the activity cycles of terrestrial pulmonates persist in constant conditions favouring continuous activity (Howes and Wells, 1934a, 1934b; Dainton, 1954b; Lewis, 1969b). *Oxychilus cellarius* maintains a diurnal rhythm (Tercafs, 1961) even when isolated from daily fluctuations in subterranean caves. Newell (1968) has suggested that an inherent rhythm, modulated by environmental factors such as light, might form the basis of the activity cycle of slugs.

Generally terrestrial pulmonates remain inactive and avoid exposure to heavy rain (Boycott, 1943; Barnes and Weil, 1945b; Blinn, 1963). One explanation of this behaviour is that prolonged contact with liquid water is potentially dangerous to terrestrial pulmonates and can lead to overhydration and eventual death. Occasionally, slugs and snails are trapped in pools of water apparently because excessive swelling mechanically impairs efficient crawling. Small species such as *Carychium tridentatum* (Risso) seem well adapted to live in saturated air or complete immersion but are hampered by the surface tension of water films (Morton, 1954a). It is possible that instances of death in water are atypical. A large number of hydrophilic species (Boycott, 1934) successfully inhabit saturated conditions where water is freely available for much of the time. Even in the larger helicids the rate of urine production is similar in magnitude to observed rates of water uptake and there seems no reason why the kidney cannot regulate the body volume in wet environments. The rates of urine production in *Lymnaea stagnalis* and in some of the larger terrestrial species are given in Table V.

Terrestrial pulmonates survive unfavourable climatic conditions by a period of inactivity, called "hibernation", during winter months when low temperatures are unsuitable for activity and feeding, and "aestivation" during the hottest and driest part of the summer (Howes and Wells, 1934a). Hibernation, it seems, has some of the characteristics of the mammalian phenomenon involving seasonally determined metabolic preparations (von Brand, 1931), which cannot be prevented by artificially maintained conditions favourable for activity (Künkel, 1916). Though snails can be roused from hibernation and aestivation by high humidities and temperatures, Kühn (1914) and Howes and Wells (1934a) found it takes hibernating *Helix pomatia* much longer to become fully active. There are other external differences between aestivation and hibernation. The epiphragm secreted at the onset of hibernation is thicker and more heavily calicified than during aestivation in *Helix pomatia* (Howes and Wells, 1934a) and *Mesodon thyroidus* (Blinn, 1963). The latter species shows other behavioural differences in burying itself during winter but climbing trees before summer inactivity.

In the dormant state in subsaturated conditions animals steadily lose water by evaporation. Among those adaptations which keep losses to a minimum, thus increasing the likelihood of survival, is the ability to select suitable sites for dormancy. In doing so many species, particularly those from temperate regions, adopt their normal behavioural responses in selecting areas of high humidity, moderate temperatures and low light levels. Some slugs and snails show a form of homing behaviour in repeatedly returning to the same sheltered location (*Helix aspersa*, Taylor 1894–1921); *Agriolimax reticulatus*, South, 1965), where they often aggregate in large numbers. Water lost by snails in this condition is reduced by sealing the shell aperture directly to the substratum (Machin, 1967) or to the shell of another animal. However some species, *Mesodon thyroidus* and *Allogona profunda* (Blinn, 1963) and *Helix pomatia* (Taylor, 1894–1921), consistently hibernate partially buried in soil with the aperture uppermost, sealed entirely by the epiphragm. In *Allogona ptychophora* (Brown), individuals hibernating with the aperture facing downwards showed higher winter mortality (Carney, 1966). Many slugs also spend periods of inactivity buried in moist soil. The distribution of *Arion fasciatus* (Nilsson) and *Agriolimax agrestis* (L.) for example is closely correlated with soil water content; however the environmental requirements of others appear not nearly so exacting (Boycott, 1934; Carrick, 1939; Getz, 1959; South, 1965). *Agriolimax reticulatus* and *Arion intermedius* Normand seem to be more resistant to desiccation and show no such correlation with soil moisture. There is some evidence that slugs also compensate for their poor control over evaporation by being able to tolerate greater body water losses than snails (Table XIII).

In some of the larger snails living in hotter climates, dormant periods are spent in exposed rather than sheltered locations. Many species of land pulmonate including slugs (Crozier and Pilz, 1924) show a marked

Table XIII. Survived water loss in pulmonates

Species	Survived water loss (% body wt)	Authority
Basommatophora		
Biomphalaria glabrata	70	von Brand *et al.*, 1957
Stylommatophora		
Agriolimax agrestis	68	Künkel, 1916[a]
Arion ater	65	Künkel, 1916
Arion ater	66	Howes and Wells, 1934b
Cochlicella acuta	16	Pomeroy, 1966
Cochlicella ventrosa	24	Pomeroy, 1966
Helicella neglecta	20	Pomeroy, 1966
Helicella virgata	33	Pomeroy, 1966
Helix aspersa	29	Pomeroy, 1966
Helix pomatia	58	Pomeroy, 1966
Lehmannia marginata	78	Künkel, 1916
Limax cinereoniger	60	Künkel, 1916
Limax maximus	76	Künkel, 1916
Limax tenellus	80	Künkel, 1916
Theba pisana	25	Pomeroy, 1966

[a] Künkel's data are based on animals with initially high water contents

negative geotaxy. Presumably this response in some species leads snails to become inactive and sealed with the epiphragm at the top of any available vegetation, including trees (Gammon, 1943; Yom-Tov, 1971b) or any other raised objects such as fence posts or buildings (Pomeroy, 1968a). There is some indication that differences in site selection behaviour occur in the same species depending on the climate. *Helix aspersa* select exposed sites on plants in South Australia (Pomeroy, 1968a) and so do *Cepaea* species in Spain (Machin, personal observation); these snails normally select a more protected site for dormancy in Britain.

Snails inhabiting hotter, drier climates have no hibernation period, but winter months are usually the only time of the year when regular activity is possible. Pomeroy (1968b) showed that winter activity of *Helicella virgata* (da Costa) in South Australia was regulated by environmental conditions in much the same way as spring and autumn activity in North European species. Winter activity is even more restricted for snails in truly desert conditions. For *Trochoidea seetzeni*

(Pfeiffer) and *Sphincterochila boissieri* (Charp.) which live in the Judean
and Negev deserts of Israel this is related to an average of 15–25 rainy
days per year (Yom-Tov, 1970). It is not precisely known whether the
accumulation of metabolic toxins, the effects of starvation or dehydration
or the prolonged exposure to high temperature eventually cause death
in inactive terrestrial snails. There is some indication however, that body
water at least partially determines the length of a period of inactivity.
In the most extensive study of the phenomenon to date, Pomeroy (1966)
showed that conditions which increase evaporative water loss such as
high temperatures at low humidities decrease survival time. Both he
and Gebhardt-Dunkel (1953) have shown that there is an abrupt
increase in weight loss following a prolonged period of reduced evaporation. Pomeroy also showed that larger individuals of the same species
survived longer. Whatever the cause, death following a protracted
period of desiccation appears to occur fairly rapidly perhaps when some
critical blood concentration is reached. Machin (1966) found, in dead
snails, that mechanisms by which evaporation from the mantle is regulated break down at death.

Unfortunately records of survival times under artificial conditions
without food and water (Table XII) only occasionally include the
temperatures and humidities in which the animals were kept. However,
since survival times are generally longer in snails from more arid
environments, and all values exceed likely environmental periods of
drought, survival of seasonal extremes seems likely, at least in some
individuals. A considerable margin of safety is particularly desirable in
desert-dwelling species. Since winter rain is infrequent and may not fall
at all in a given year, snails may have to survive inactive periods which
are two or three times longer than normal. Observed survival times of
several years as in *Sphincterochila boissieri* together with low rates of
energy utilization (Schmidt-Nielsen *et al.*, 1971) suggest that these species are well adapted to survive even the most severe terrestrial
conditions.

Very arid environments impose a further stress on animals living in
them since they are also very hot. Owing to the lack of cloud cover,
prolonged periods of direct solar radiation, particularly in summer
months, can lead to soil temperatures in the Israeli deserts for example
as high as 70°C (Yom-Tov, 1971b). Snails survive hot dry conditions by
water conservation not by evaporative cooling. Machin (1972) has
shown that low rates of water loss observed in inactive *Otala lactea*
reduce mantle temperates only by a maximum of 0·07°C. Snails living
in hot climates have a number of ways of avoiding maximum temperatures which occur at ground level. The habit of aestivating in exposed

locations, above the surface of the ground is one of them. Yom-Tov (1971b) has shown that the desert snail *Troichoidea seetzeni* is as much as 14°C less than ground temperature when attached to a bush, 10 cm above the ground. Pomeroy (1968a) showed that temperatures experienced by dormant snails decrease with their height from the ground. There is at least one desert snail, *Sphincterochila boissieri*, that apparently avoids higher plants, and remains dormant on the desert floor (Schmidt-Nielsen *et al.*, 1971) or buried just beneath it (Yom-Tov, 1971b). Withdrawal of the animal within the shell so that an insulating air space in the empty body whorl is formed between the living animal and the ground is apparently critical. Temperature differences of as much as 10°C exist between the base of the shell in contact with the ground and its apex which is occupied in the snail.

The colour and reflecting properties of the shell become important when aestivation occurs in locations exposed to the sun. Desert and semi-desert species are either white as in *Sphincterochila* and *Helicella* or very lightly banded as in *Trochoidea*. Yom-Tov (1971b) and Schmidt-Nielsen *et al.* (1971) found 90–95% reflectance of visible and infrared wave lengths in *Sphincterochila* and slightly lower values for *Trochoidea*. The high reflectance of the shell is undoubtedly important in reducing shell temperatures. *Trochoidea* aestivating on plants were only 1°C higher than the surrounding air whereas somewhat larger differences were recorded for *Sphincterochila* on the ground and *Helicella virgata* (da Costa) on fence posts. Selection of most favourable site appears to be less important for *Helicella virgata* which was introduced to South Australia from Europe. Most specimens aestivate about 2–3m above the ground whereas they would be cooler if they were higher up. These species also apparently retain behaviour patterns which are more appropriate to the northern hemisphere in selecting the sun-exposed north-west aspects of fence posts. By contrast site selection by truly desert species must be important since the lethal temperature of *Sphincterochila* and *Trochoidea*, 55°C (Yom-Tov, 1971b; Schmidt-Nielsen, 1971), is well below the maximum environmental temperatures.

VI. Physiological consequences of water exchange

A. Electrolytes

Rapid water exchange in an environment of low salt concentration presents some problems in electrolyte balance. At the mean rate of urine production it takes *Lymnaea stagnalis* only 5·4 h to eliminate a volume equivalent to that of the haemolymph. Since a reduction of only 24% in the osmotic pressure of the final urine almost certainly means that substantial amounts of electrolyte remain in it even after tubular reabsorption, it is likely that the kidney is a serious source of salt loss to the snail. Salts are probably also lost through the body wall. Greenaway (1970) has demonstrated that *Lymnaea stagnalis* loses significant proportions of its blood sodium following exposure to a sodium-free medium. Although the various sites of sodium exchange were not identified by Greenaway, he was able to show that normal snails can maintain balance in media containing as little as 0·025 mM/l, and attributed this to active sodium transport, presumably through the body wall. A potential difference of about 16 mV with haemolymph negative to the outside, was demonstrated.

Plant foods eaten by terrestrial snails have a higher proportion of potassium to sodium than found in the haemolymph (De Jorge and Haeser, 1968). Burton (1968a) observed a temporary rise in blood potassium and calcium following feeding but no significant change in sodium and magnesium. The kidney is important in maintaining normal haemolymph levels of electrolytes since a major proportion of calcium, sodium and chloride is actively reabsorbed from the ultrafiltrate in *Achatina achatina* (L.), while there is a small net secretion of potassium (Skelding, 1973). In spite of reabsorption significant amounts of chloride were shown by Martin *et al.* (1965) to remain in the final urine of terrestrial pulmonates which they thought often suffered from salt deficiencies.

Significant amounts of salt may also be lost as a result of mucous secretion. This conclusion is based on direct measurements which show the electrolyte content of freshly secreted mucus to be only slightly lower than that of the blood (Machin, 1962; Burton, 1965b). Differences in the proportions of ions between mucus and blood have led Burton to propose two phases of mucus formation. First ions which have higher intracellular concentration such as potassium and magnesium become associated with macromolecules of mucus or their precursors, presumably when these substances are being synthesized and stored as droplets within the glands (Wondrak, 1967; 1968b). Secondly, perhaps at the time of extrusion, when swelling probably occurs, a watery component, rich in sodium resembling an ultrafiltrate of the blood, is added.

Unfortunately there is no information concerning possible changes in electrolyte composition on the animal's surface since mucus cannot be collected without initiating further extrusion. Water-rich mucus in terrestrial pulmonates serves the same purpose as an aquatic environment in preventing superficial desiccation. It is tempting to extend this idea a little further by postulating an electrolyte-reabsorbing mechanism in the integument. The free epithelial surface, apparently over most of the body, including those parts covered by the shell, is composed of a layer of microvilli (Hubendick, 1958; Schwalbach and Lickfield, 1962; Lane, 1963; Wondrak, 1968a; Lloyd, 1969; Saleuddin, 1970; Rogers, 1971). The presence of microvilli, membrane invaginations and mitochondria is commonly associated with active electrolyte and water transport, for example, in the kidney (Wendelaar Bonga and Boer, 1969). Small vacuoles at the bases of the microvilli on the external epithelium, together with electron-dense bodies resembling lysosomes suggest that some of the organic molecules in the mucus might also be recycled.

The tissues of all terrestrial pulmonates are subject to some fluctuations in water content arising from rapid water exchange with environment. Burton (1964) found that the greatest changes in the water content of *Helix pomatia* occurred in the haemolymph and gave rise to corresponding fluctuations in haemocyanin, sodium and chloride concentration. It was also reported that electrolyte composition may also be affected by urine and mucus production, feeding and exchange with the other tissues. The organs of terrestrial pulmonates also decrease in water content as the animal becomes dehydrated (Pusswald, 1948; Gebhardt-Dunkel, 1953; Burton, 1965c). In general, muscular organs change less than the more glandular tissues, presumably reflecting consistent differences in the size of solvent and non-solvent compartments of each. Water content in all but one of the tissues taken from *Helix pomatia* was inversely proportional to haemolymph sodium concentration (Burton, 1965c) suggesting osmotic exchange (Dick, 1959). No determinations of extracellular space, which could be large in animals with open blood systems, have been made. Unfortunately without this information full analyses of solvent and solute exchange between intracellular and extracellular compartments in terrestrial pulmonates cannot be made. It is well known that the magnitude of the volume of cells is usually somewhat less than predicted. This volume regulation has been variously attributed to the presence of a non-solvent compartment within the cell (Lucké and McCutcheon, 1932), or the exchange of electrolytes and nitrogenous solutes across the membrane (Potts, 1958; Little, 1965), or differences in charge and Donnan equilibrium which may vary with intracellular protein concentration (Gary-Bobo and Solomon, 1968). Only one of these mechanisms has been investigated in pulmonates. Bedford (1969) demonstrated changes in levels of free amino acids in the foot and digestive gland of *Siphonaria zelandica* (Quoy and Gaimard) following external osmotic change, though their role in cell volume regulation was not conclusively demonstrated.

Fluctuations of fluid volumes and tissue water contents pose further problems of interest to physiologists. Varying water content accounted for dehydrated and hydrated *Helix pomatia* (Postma, 1936). In the same snail excessive loss of water could lead to impared circulation and gas exchange. Although observed haemocyanin concentrations appear well below the point at which the haemolymph viscosity becomes too high, Burton (1965a) admits that his analysis is somewhat simplified, and possible haemodynamic complexities may make some of his conclusions invalid. Burton (1969) has also shown that increased haemocyanin concentration in dehydrated *Helix pomatia* leads to an increased buffering capacity of the haemolymph. Other phenomena such as changes with concentration in colloid osmotic pressure and Donnan equilibrium remain to be investigated.

Since excitable tissues depend so much on electrolytes, the functioning of nerves and muscles in the low or fluctuating ion concentrations typical of pulmonate haemolymph is of considerable physiological interest. Nerve cells have a number of special properties which are perhaps related to the unusual features of haemolymph composition and its inherent instability in terrestrial species. Neurones from terrestrial snails continue to produce action potentials long after the removal of sodium from the bathing medium. This behaviour is best explained by the presence of an extracellular reservoir of sodium ions close to the nerve membrane (Moreton, 1968b) of the type already identified in *Anodonta* (Glupta *et al.*, 1969). A compartment which resists the dissipation of sodium ions would tend to reduce the effects of rapid concentration fluctuations in the haemolymph. Experiments with small neurones also indicate that the depolarization phase of the action potential is less dependent on sodium and rather more on calcium than is usual in other animals (Chamberlain and Kerkut, 1967; 1969). This increased importance of calcium also extends to freshwater pulmonates (Jerelova *et al.*, 1971) and possibly reflects lower haemolymph sodium-calcium ratios (Prosser and Brown, 1961; Burton, 1968b) compared with other animals.

It has been known for some time that the electrical activity of slug pedal ganglia varies with the osmotic pressures of the bathing medium (Hughes and Kerkut, 1956; Kerkut and Taylor, 1956). Rates of electrical discharge increase as the bathing fluid becomes more dilute and have a stimulatory effect on locomotion (Kerkut, 1959). More recently it has been shown that peripherally applied osmotic stimulants have the same effect on the cerebral and pedal ganglia of both slugs and snails (Rózsa, 1963). The resting potentials of certain nerve cells from active and inactive snails differ (Kerkut and Walker, 1961) although Moreton (1968a) failed to show that this was widespread in the nervous system.

B. Metabolism

Most pulmonates from temperate regions consume oxygen at rates between 50 and 300µl/g body wt/h, at temperatures around 20°C, presumably depending on the degree of activity (Liebsch, 1928; von Brand et al., 1957; Kienle and Ludwig, 1956; Wesemeier, 1960). In both aquatic and terrestrial species, withdrawal brings about a fairly rapid initial drop in metabolic rate, followed by a plateau or slow decline. Values between 10 and 80 µl/g body wt/h at 20°C are typical of oxygen consumption in inactive animals (Muller, 1943; Wells, 1944; Blaẑka, 1954). Starvation also leads to a decline in metabolic rate. Lack of food however is not the only reason for metabolic adjustment, reduction of oxygen consumption in aestivating Biomphalaria glabrata depending also on the rate of dehydration (von Brand et al., 1957). In the terrestrial snail Arianta arbustorum, oxygen consumption during aestivation and hibernation is about half that of individuals which have simply been denied food (Nopp, 1965). A sudden rise in environmental temperature produces a temporarily elevated metabolic rate (Blaẑka, 1954; Rising and Armitage, 1969; Schmidt-Nielsen et al., 1971). Semi-regular periods of increased oxygen consumption in Sphincterochila boissieri kept in apparently constant conditions (Schmidt-Nielsen et al., 1971) may indicate some form of internal rhythm.

Oxygen consumption in aquatic snails increases markedly with temperature giving high Q_{10} values of about 2 or greater (von Brand et al., 1948; Berg and Ockelmann, 1958; Grainger, 1969). There is conflicting information about the effect of temperature in terrestrial pulmonates. High Q_{10} values are indicated by the data of Blaẑka (1954) and Schmidt-Nielsen et al. (1971) for Helix pomatia and Sphincterochila boissieri, respectively. On the other hand Nopp (1965) reports that oxygen consumption in aestivating Helicella obvia and Cernuella variabilis remains the same between 25 and 40°C. Low Q_{10} values over wide temperature ranges have also been demonstrated in a number of in vitro preparations from land snails – mitochondria

(Newell, 1966) and the digestive gland and gonad (Nopp and Farahat, 1967). Regulation of metabolic rate in different temperatures particularly during inactive periods is an attractive theory for terrestrial snails which restrict evaporative cooling, but much more work needs to be done to fully establish it.

There is some information on the nature of metabolic processes during prolonged inactivity and starvation. *Biomphalaria glabrata* during aestivation out of water consumes mainly polysaccharide but also some lipid (von Brand *et al.*, 1957). Lactic acid also decreased suggesting there was no switch to anaerobiosis. Von Brand *et al.* (1948) interpreted a decline in respiratory quotient during starvation as increased fat or protein consumption, although they were aware of the difficulties in interpreting r.q. values in animals with calcareous shells. Perhaps for the same reason there are conflicting values reported for terrestrial species. Fischer and Duval (1931) give r.q. values for active *Helix pomatia* and *H. aspersa* around 0·8 and about 1·0 for hibernating individuals. Values around 1·0 were also reported by Kienle and Ludwig (1956) for three other species. By contrast Liebsch (1928) found very much lower r.q. values in three inactive terrestrial pulmonates, including *Helix pomatia*.

Terrestrial snails show the same dependence on carbohydrate metabolism as seen in *Biomphalaria Helix pomatia* accumulates polysaccharide as galactogen in the albumen gland and glycogen in other tissues (May, 1932) during early summer and autumn (von Brand, 1931). During hibernation polysaccharide is consumed at a mean rate of 0·1 mg/g body wt/day. At this rate roughly 60% of the total had been consumed to the point of arousal in one hundred and eighty days. At least on the basis of remaining carbohydrate reserves, hibernation might be extended for a further 120 days. Unaccounted for weight loss in von Brand's data (1931) together with increases in kidney nitrogen during hibernation (Jezewska *et al.*, 1963; Speeg and Campbell, 1968a) indicate some protein consumption during hibernation; lipids however remained constant. The qualitative relationship between polysaccharide and oxygen consumption and water production is as follows:

$$C_6H_{10}O_5 + 6O_2 = 6CO_2 + 5H_2O$$
$$162g \qquad 134\ 1 \qquad\qquad 90g$$

i.e. 0·1 mg polysaccharide/g/day oxidized by 83 µl O_2/g/day yields 0·056 mg H_2O/g/day.

At 0°C, mean hibernation temperature estimated from von Brand (1931), winter oxygen consumption for *Helix pomatia* is about 60 µl/g body wt/day. The fairly close agreement between the two estimates of oxygen consumption confirms that breakdown of carbohydrate is the principal energy source during hibernation. Since *Helix pomatia* loses about 1 mg H_2O per day (Table X) metabolic contributions to the animal's water reserves are probably negligible. On the other hand, desert snails lose water during aestivation much more slowly. In an essentially similar study to that of von Brand, Schmidt-Nielsen *et al.* (1971) showed that the body components of *Sphincterochila boissieri* were essentially the same as in *Helix*. However they failed to show any significant seasonal trends. It is interesting to find a temperature shift in the metabolism of this desert snail, oxygen consumption per g body weight per day at 25°C being about the same as the rate for *Helix pomatia* at 0°C. At 25°C *Sphincterochila* loses 0·25 mg/g body wt/day which is only five times the metabolic production of water from carbohydrate oxidation.

Nitrogen metabolism and excretion have been recently reviewed by Florkin (1966), Martin and Harrison (1966) and by Potts (1967). Since that time a number of new papers have appeared. It is generally agreed that freshwater pulmonates eliminate nitrogen principally in the form of ammonia. This compound is the simplest nitrogenous excretory product used by animals and its elimination avoids the loss of potentially valuable biochemical energy when more complex molecules are excreted. Ammonia, however is highly soluble in aqueous media (890 mg/ml H_2O at 20°C) and toxic at concentrations of the order of 0·5 mg/ml (Standon, 1955; Wilson *et al.*, 1968). Freshwater pulmonates are particularly well suited to keeping ammonia levels below lethal levels. A comparatively permeable body wall together with a high rate of urine production make rapid elimination of ammonia possible. Bayne and Friedl (1968) found that *Lymnaea stagnalis* produces ammonia at a rate of 0·0019 mg/g body wt/h whereas the same species produces urine at rates exceeding 56 µl/g body wt/h (Table V). Because of its high solubility significant amounts of ammonia probably diffuse out through the body wall. Wilson (1968) suggested this possibility after finding high ammonia concentrations in the mucus of immersed *Lymnaea truncatula*. The same species was observed by Pullin (1971) to have 0·0053 mg/ml ammonia in the haemolymph.

A variety of other nitrogenous compounds are excreted by freshwater pulmonates. The most important of these, next to ammonia, is urea. Bayne and Friedl (1968) found that urea is eliminated from *Lymnaea stagnalis* about one quarter as rapidly as ammonia. Urea is even more soluble than ammonia (1193 mg/ml H_2O at 20°C). Its elimination results in the loss of one carbon atom for every two nitrogens.

The great advantage of urea is that it is much less toxic. The terrestrial snail *Bulimulus dealbatus* (Say) after 10 months aestivation was found by Horne (1971) to tolerate urea concentrations as high as 23 mg/g tissue wet weight.

The excreta and kidneys of freshwater pulmonates also contain various purines, amino acids and perhaps protein (Haggag and Fouad, 1968; Bouillon and Delhaye, 1970). Most of these occur in too low concentrations to be significantly related to water balance. In a recent survey of the uric acid contents of a series of gastropods Duerr (1967) has confirmed an early finding that *Lymnaea* and *Planorbis* spp., amongst others, tend to accumulate suprisingly large amounts of uric acid in the kidney. This phenomenon, Duerr points out, cannot be of great adaptive significance, though its value to aquatic snails exposed to periods of desiccation was not considered.

Substances having very low solubility are in many ways ideal excretory products for animals which undergo prolonged periods of inactivity. These compounds can be allowed to accumulate to high concentrations, usually as crystalline solids, without being physiologically harmful. In the undissolved state substances are neither chemically toxic nor do they contribute significantly to the osmotic pressure of the surrounding fluid. Should the substance be formed from soluble precursors, it can be readily transported through cellular and multicellular membranes in this form. If precipitation occurs in the lumen of the kidney, for example, the resulting decline in osmotic pressure can lead to passive water reabsorption and simultaneous concentration of the excretory substance with the minimum use of energy.

It has been known for some time that inactive terrestrial snails are uricotelic. Recent work, using improved extraction and detection techniques have shown that a large number of different terrestrial pulmonates, including slugs, rely more on purines such as xanthine and guanine than was previously suspected. Estimates based on the nitrogen contents of the kidneys or excreta of inactive terrestrial pulmonates indicate that purines account for about 80–90% of the total nitrogen in snails and about 70–80% in slugs (Jezewska *et al.*, 1963; Speeg and Campbell, 1968a; Jezewska, 1969; Badman, 1971). Although in many species uric acid is still the principal nitrogenous excretory substance, xanthine is the most abundant purine in inactive *Helix pomatia* and guanine in *Limax maximus* (Jezewska, 1969). In view of this terrestrial species may be more correctly termed "purinotelic" (Jezewska *et al.*, 1963; Badman, 1971). Of the three purines described guanine would appear to be most suitable for nitrogen elimination and water conservation. It is the most insoluble (0·005 mg/ml

H_2O at 20°C, Jezewska *et al.*, 1963) and contains five nitrogen atoms to the same number of carbons, whereas xanthine and uric acid contain one less nitrogen. Xanthine is a good deal more soluble (0·5 mg/ml H_2O) than either guanine or uric acid (0·025 mg/ml H_2O). During aestivation in *Otala lactea* Speeg and Campbell (1968a) showed that all three purines steadily accumulate in the kidney at a total rate of about 0·019 mg/g/body wt/day. Both Speeg and Campbell and Jezewska *et al.* (1963) observed an increase in the ratio of guanine to xanthine as inactivity progressed, possibly reflecting differences in their suitability as storage materials. Presumably as a means of increasing the storage capacity of the kidney and reducing the likelihood of obstruction, large amounts of crystalline deposits accumulate in the cellular lining of the kidney sac (Bouillon, 1960; Bouillon and Delhaye, 1970). Marked decreases in purine concentration and a loss of cytoplasmic inclusions in the kidney are observed once activity and the elimination of excretion is resumed. Guanine, xanthine and uric acid together account for 89% of total kidney excretion in active *Helix pomatia* compared with 91% during hibernation (Jezewska *et al.*, 1963). In active *Otala lactea* the kidney contains 83% of total nitrogen compared with 98% during aestivation (Speeg and Campbell, 1968a).

Some terrestrial snails such as *Helix aspersa* and *Otala lactea* (Speeg and Campbell, 1968b), but not all (Badman, 1971), also eliminate nitrogenous waste extrarenally as gaseous ammonia. This method of elimination has obvious advantages since it does not require the loss of water to the animal. Ammonia accounts for 30% of the total nitrogen excreted in aestivating *Otala lactea* and about 5% of the total in active snails (Speeg and Campbell, 1968a). It is of interest to note in the inactive state that ammonia permeates the shell and epiphragm two orders of magnitude more slowly than water, presumably because of smaller concentration gradients. The importance of ammonia as an excretory end product and the correspondingly low concentrations of urea in some terrestrial pulmonates are due to the widespread presence of the enzyme urease (Razet and Dagobert, 1968; Horne and Boonkoom, 1970). Absence of urea-splitting enzymes may account for high urea levels in *Strophocheilus oblongus musculus* Becquaert (De Jorge and Petersen, 1970).

VII. Some aspects of activity regulation

In recent years molluscan physiologists have become increasingly aware of the importance of neurosecretory substances in water and electrolyte balance and many other activities of aquatic and terrestrial pulmonates. Physiological changes associated with varying seasonal activity in land species for example are very likely under hormonal control. However day to day activity seems to depend on external stimuli for its initiation and on direct osmotic stimulation of the pedal ganglion for its continuation (Kerkut, 1959).

More recent observations made by the author and his associates show that snails also respond to the same external stimuli described earlier for slugs and appear unresponsive, at constant temperature, to frequent changes in humidity ranging between saturated and 1·5% r.h. During water loss experiments with *Otala lactea* (Machin, 1972), several withdrawn animals sometimes became active after turning off the laboratory light, even though mantle tissue covering the eyes in this species is darkly pigmented. Foh (1932) found that the mantle of *Helix pomatia* is more sensitive to light than the eyes. *Otala lactea* is also readily activated by a decrease in ambient temperature. Results of laboratory experiments, summarized in Table XIV, show that the magnitude of

Table XIV. Effect of ambient humidity on temperature activation in *Otala lactea*

Duration of temp. decrease, 26 to 10°C (min)	r.h. at 26°C (%)	Calculated r.h. at 10°C [a] (%)	Activation[b] (%)
Control: (no temp. decrease)	32	—	4
65	4	11	4·5
180	4	11	2·2
80	32	88	58
205	32	88	45
55	100	100	89
245	100	100	60

[a] Calculated assuming no further addition of water from snail
[b] Values based on 48 animals per experiment

the response depended on the humidity and the rate of cooling. Howes and Wells (1934a) also reported that *Helix aspersa* is activated by mechanical disturbance. In a more detailed study of this phenomenon using *Otala lactea*, Ross (1967) found that gentle striking, shaking and rotation of the animal produced responses characteristic of the initial stages of emergence. The mantle of *Otala lactea* first shows surface pitting and rippling which becomes progressively more extensive until the entire structure is involved in muscular movement. Following this the pneumostome usually opens and the posterior tip of the foot is protruded. However these experiments were performed during winter in the laboratory at about 20°C and in less than 40% r.h. and only one of the large number of animals experimentally stimulated became fully active. The proportions of snails responding to the different mechanical stimuli by pneumostome opening or foot protrusion are shown in Fig. 2.

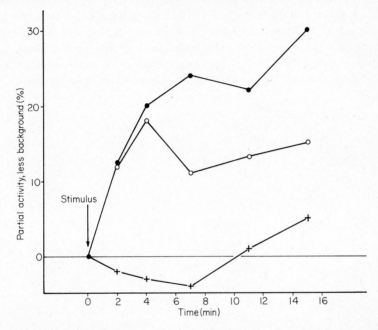

Fig. 2. Percentages of *Otala lactea* showing partial emergence following different types of gentle mechanical stimulation: percussion (●) shaking (○) and rotation (+). Values were obtained by subtracting mean "background" activity before stimulation from the data of Ross (1967). Each point represents the mean of at least 10 separate experiments using groups of 36 snails.

These observations suggest that full emergence in snails involves two distinct phases of response: the first arising from external stimuli and the second in which full activity occurs only when high ambient humidity is sensed. A possible reason for the initial response to rather unspecific stimuli is that they are the only indications of environmental change detectable within the shell. In the unstimulated snail evaporative cooling is very slight and probably unsuited to the accurate monitoring of the outside humidity. However decreasing light and temperature, or mechanical disturbances could indicate the coming of night or of a storm and as such represent a slightly increased chance of suitable environmental conditions for activity. Occasional response, even to a somewhat ambiguous stimulus, presumably results in greater water economy than would be the case with humidity monitoring based on continuously elevated evaporation.

As far as can be observed all types of stimulus result in renewed mucous secretion by the mantle (Machin, 1965). The association of mucous secretion with the demonstrated ability to discriminate between high and low humidities after initial stimulation suggests a psychrometric method of humidity detection. To test if this was possible, shell and mantle temperatures were measured in withdrawn *Otala lactea* during exposure to a 16°C drop in temperature. Results summarized in Fig. 3 show that snails in humid air experienced little or no evaporative

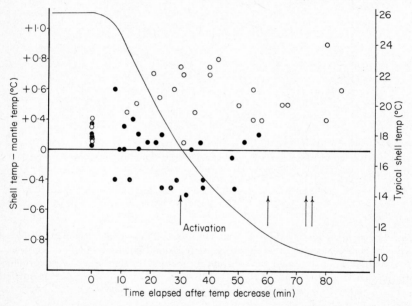

Fig. 3. Graph showing humidity dependent differences in evaporative cooling in *Otala lactea* following a 16°C decrease in temperature. All four snails in humid air (●) became fully active (arrows); those in dry air (○) showed evidence of mucous secretion by increased cooling but failed to emerge from their shells.

cooling and all became fully active. Animals in dry air failed to emerge but showed considerable mantle cooling, indicating that they had at least secreted mucus. More experimental work needs to be done before the mechanism of activation is fully understood and its neurophysiological basis should be worked out. Kerkut and Ridge (1962) have already shown that certain neurones in land snails are sensitive to temperature changes.

VIII. Water relationships of developing stages

Eggs of aquatic and terrestrial pulmonates appear to be rather less hardy than adults. Many freshwater species live for several years, surviving adverse conditions in winter or during drought as adults rather than as eggs or juveniles (Boycott, 1936). The eggs of freshwater Basommatophora develop within a fluid-filled perivitelline capsule, protected externally by layers of jelly and mucus. Capsular or perivitelline fluid is thought to have a nutritive as well as protective function containing 3–6% of the polysaccharide galactogen and 6–8% protein in *Lymnaea stagnalis* (Horstmann, 1956; Bayne, 1968b). Beadle (1969a) found that ^{22}Na distributes itself roughly equally between the capsular fluid and the outside, suggesting passive ion exchange. Isolated perivitelline capsules show temporary shrinkage in solutions of substances of low molecular weight. Pores in the capsular membrane permit the entry of molecular weights up to 342 in *Biomphalaria sudanica* Martens and 504 in *Lymnaea stagnalis*. Membranes with these properties retain macromolecules of nutritive value and permit the entry of electrolytes such as sodium which are actively absorbed by the developing embryo (Beadle, 1969b; Beadle and Beadle, 1969). Excretory products would also be small enough to escape through the membrane. Comparatively large pore size reduces the magnitude of osmotic exchange. The fluid within the capsule has an internal colloid osmotic pressure of 1·5 mM/l (25 mmHg) in *Biomphalaria* and 4–5 mM/l (75 mmHg) in *Lymnaea* (Beadle, 1969a). The mechanism by which its volume is regulated is unknown, but it could be by hydrostatic back pressure. By contrast egg cells of *Lymnaea stagnalis* between oviposition and first cleavage undergo considerable swelling, having an effective intercellular osmotic pressure of 93 mOsm/kg (Raven and Klomp, 1946). Electrolyte-coupled water transport is responsible for volume regulation in the prenephridial embryo (Beadle and Beadle, 1969b). Short-term water exchange in the developing embryo is somewhat buffered by the capsular fluid (Raven and Klomp, 1946) and surrounding mucus and jelly layers. Apart from increasing the total bulk of the egg mass which contains numerous capsules, these layers of water-rich mucopolysaccharide provide no protection against drying up (Bayne, 1968a).

Even in *Biomphalaria glabrata*, which is seasonally exposed to drought conditions, desiccation of the egg masses is rapid and the embryos show very little tolerance of dehydration (Chernin and Adler, 1967).

Eggs in the Stylommatophora are laid in clutches, each egg being a separate entity and nearly spherical. As in freshwater species the developing embryo is protected within a capsule of nutritive perivitelline fluid surrounded externally by several layers, principally of mucopolysaccharide. All layers are of jelly-like consistency in *Agriolimax* spp. In *Helix aspersa*, *Cepaea nemoralis* (L.) and *Arion ater* (L.) they contain crystalline deposits of calcium carbonate, concentrated enough to form a semi-rigid shell in the last two (Bayne, 1966; 1968b). The outer layer of *Arion ater* eggs, but not the others, is also impregnated with lipid (Smith, 1965).

Rates of water loss from eggs of terrestrial pulmonates in sub-saturated atmospheres do not greatly differ from those of freshwater species. Bayne (1968a) found that the eggs of *Agriolimax reticulatus* are completely dehydrated in 50% r.h. and 25°C in about an hour. Variations in the permeability in the egg's outer layers were rather small, there being a factor of two between the most (*Agriolimax*) and the least permeable (*Helix*). Yom-Tov (1971a) found that the osmotic pressure of the perivitelline fluid was 54 mOsm/kg in *Sphincterochila boissieri* and 150 mOsm/kg in *Trochoidea seetzeni*. Carrick (1939) reported that permanent contact with a moist surface is necessary for the maintenance of the egg's full volume and turgidity. Although this has not been established, the perivitelline membrane is probably permeable to ions and other solutes of small molecular weight. Bayne (1966) observed that healthy and presumably well hydrated vitelline sacs from the eggs of *Agriolimax reticulatus* swelled only slightly in dilute solutions and hence have some mechanism for preventing excessive water uptake and bursting. To ensure as far as possible that the optimum water content of the egg is maintained during development terrestrial pulmonates bury their eggs in holes and crevices where ground water and humidity are high. Some snails excavate their own subterranean nests (Herzberg, 1965; Yom-Tov, 1971a). Both snails and slugs avoid egg-laying in unsuitably dry conditions (Carrick, 1942; Arias and Crowell, 1963; Wolda, 1965).

There seems to be variability in the resistance of embryo of slugs and snails to desiccation. Carmichael and Rivers (1932) found that eggs of *Limax flavus* L. survived a weight loss of 80–85% and embryos just before hatching of 70–75% loss. Bayne (1969) found that newly laid eggs and advanced embryos of *Agriolimax reticulatus* survived a temporary weight loss of about 70–75%, except at 90% r.h., when hatchability of newly laid eggs was much reduced. Recent work suggests that resistance to desiccation, even in desert snails, is variable. Yom-Tov (1971a) reported that the eggs of *Trochoidea seetzeni* develop normally after a temporary loss in weight of 71%. All eggs of *Sphincterochila boissieri* failed to hatch after this treatment. Further varied reports of the tolerance or otherwise of snail eggs to desiccation are given by Gugler (1963) and Wolda (1965) though quantitative information is lacking.

What little is known of juvenile pulmonates suggests that they are under greater physiological stress and less able to survive adverse conditions than adults. The larger surface to volume ratios in smaller individuals result in more rapid rates of water and electrolyte turnover in freshwater *Lymnaea stagnalis* (van Aardt, 1968) and more rapid depletion of body water in terrestrial species. The danger of desiccation is compounded by the fact that the immature shell is poorly adapted to prevent water loss. Shells of juvenile snails are thinner and consequently more permeable to water. They also lack the special apertural modifications of the adult shell which either reduce its size or assist in its closing during inactivity. Chatfield (1968) found that many juvenile specimens of *Monacha cantiana* (Montagu) secrete an apertural rib in preparation for hibernation. In spite of precautions like these, mortality is high in all but the larger juveniles and most temperate species do not survive the winter (Boycott, 1934). Rapid growth rate therefore has considerable survival value in terrestrial snails. Perhaps the need to feed at every available opportunity explains the observations of Blinn (1963) and Pomeroy (1968b) which show that the frequency of activity is greater in juvenile snails than in adults.

IX. Evolutionary considerations

Two aspects of water balance physiology may be related to the previous evolutionary history of the animal. These are nitrogen metabolism and body fluid concentration. Needham (1938) reported high uric acid concentrations in the kidneys of *Lymnaea stagnalis* and *Planorbarius corneus* but much lower concentrations in *Lymnaea peregra* and *Ancylus fluviatilis* Müll. He interpreted this as an indication of a previous terrestrial history of advanced Basommatophora. More recent studies of nitrogen metabolism in this group have confirmed significant levels of uric acid synthesis and accumulation in various tissues in *Lymnaea stagnalis*, *L. palustris*, *Planorbarius corneus* and *Physella parkei* (Currier) (Duerr, 1966; 1967; Conway *et al.*, 1969). It must be emphasized however, with increasing biochemical knowledge, that the factors governing the use of a particular nitrogenous substance of excretion are more complex than was previously thought. The work of Jezewska *et al.* (1963) and Jezewska (1969) has led to a shift in the stress on uric acid excretion in terrestrial species in favour of purines as a whole and the significance of uric acid excretion as an indication of previous terrestrial existence has perhaps been overestimated. All *Lymnaea* species do not show high levels of uric acid in the kidney. The concentrations of uric acid, even when very high, give an exaggerated impression of their importance in nitrogen metabolism as a whole. Spitzer (1937) calculated that uric acid represents only 5% of the total nitrogen excreted by *Lymnaea stagnalis*.

The variety of intertidal, supratidal and inland semi-terrestrial and terrestrial habitats in which primitive pulmonates are found, particularly in the family Ellobiidae (Morton, 1954b, 1955a,b) suggests early environmental experimentation. It is particularly unfortunate that information about body fluid concentration in these pulmonates is lacking since marked differences in haemolymph salt levels in terrestrial members of other phyla, the Arthropoda and Chordata, can be related either to direct colonization of land from the sea or to indirect invasion from fresh water. Observed blood concentration ranges from freshwater pulmonates are very low indicating full adaptation to a low salt environment. Haemolymph osmotic pressures in these animals contrast with those of intertidal *Siphonaria* (Table II). Even though *Siphonaria* shows a very wide tolerance, low lethal salt concentrations are 2 or 3 times the normal haemolymph levels in freshwater lymneids. Concentrations in terrestrial slugs and snails, though somewhat higher, are much more characteristic of fresh water than of marine environments. This suggests an ancestral migration, perhaps of an ellobiid-like animal through estuaries to fresh water before the invasion of fully terrestrial habitats. Again physiological evidence for previous evolutionary history must be treated with caution. It can be argued that low salt levels in terrestrial pulmonates could have arisen in response to low dietary salt intake and the need to survive prolonged periods of dehydration and increased haemolymph concentration.

X. References

Abdel-Malek, E. T. (1958). *Bull. Wld Hlth. Org.* **18**, 785–818.
Adolph, E. F. (1932). *Biol. Bull. mar. biol. Lab., Woods Hole* **62**, 112–125.
Andrewartha, H. G. (1964). *Proc. Linn. Soc. N. S. W.* **89**, 287–294.
Andrews, E. and Little, C. (1971). *Nature, Lond.* **234**, 411–412.

Andrews, E. and Little, C. (1972). *J. Zool. Lond.* **168**, 395–422.

Arivanitaki, A. and Cardot, H. (1932). *J. Physiol. Path. gén.* **30**, 577–592.

Arias, R. O. and Crowell, H. H. (1963). *Bull. Sth Calif. Acad. Sci.* **62**, 83–97.

Badman, D. G. (1971). *Comp. Biochem. Physiol.* **38A**, 663–673.

Bailey, T. G. (1971). *Comp. Biochem. Physiol.* **40A**, 83–88.

Baird, W. (1850). *Ann. Mag. nat. Hist.* (2) **6**, 68–69.

Barbosa, F. S. and Olivier, L. (1958). *Bull. Wld Hlth. Org.* **18**, 895–908.

Barlow, C. H. (1933). *Am. J. Hyg.* **17**, 724–742.

Barlow, C. H. (1935). *Am. J. Hyg.* **22**, 367–391.

Barnes, H. F. and Weil, J. W. (1945a). *J. Anim. Ecol.* **13**, 140–175.

Barnes, H. F. and Weil, J. W. (1945b). *J. Anim. Ecol.* **14**, 71–105.

Bayne, C. J. (1966). *Comp. Biochem. Physiol.* **19**, 317–338.

Bayne, C. J. (1968a). *J. Zool., Lond.* **155**, 401–411.

Bayne, C. J. (1968b). *Proc. malac. Soc. Lond.* **38**, 199–212.

Bayne, C. J. (1969). *Malacologia* **9**, 391–401.

Bayne, R. A. and Friedl, F. E. (1968). *Comp. Biochem. Physiol.* **25**, 711–717.

Beadle, L. C. (1969a). *J. exp. Biol.* **50**, 473–479.

Beadle, L. C. (1969b). *J. exp. Biol.* **50**, 491–499.

Beadle, L. C. and Beadle, S. F. (1969). *J. exp. Biol.* **50**, 481–489.

Beament, J. W. L. (1961). *Biol. Rev.* **36**, 281–320.

Bedford, J. J. (1969). *Comp. Biochem. Physiol.* **29**, 1005–1014.

Berg, K. and Ockelmann, K. W. (1958). *J. exp. Biol.* **36**, 690–708.

Blažka, P. (1954). *Zool. Jb. (Zool. Physiol.)* **65**, 430–438.

Blinn, W. C. (1963). *Ecology* **44**, 498–505.

Blinn, W. C. (1964). *Physiol. Zool.* **37**, 329–337.

Borden, M. A. (1931). *J. mar. biol. Ass. U.K.* **17**, 709–738.

Botazzi, F. (1908). *Ergebn. Physiol.* **7**, 161–402.

Bouillon, J. (1960). *Annls Sci. nat. (Zool.)* (12) **2**, 719–749.

Bouillon, J. and Delhaye, W. (1970). *Annls Sci. nat. (Zool.)* (12) **12**, 1–26.

Boycott, A. E. (1934). *J. Ecol.* **22**, 1–38.

Boycott, A. E. (1936). *J. Anim. Ecol.* **5**, 116–186.

Burton, R. F. (1964). *Can. J. Zool.* **42**, 1085–1097.

Burton, R. F. (1965a). *Can. J. Zool.* **43**, 433–438.

Burton, R. F. (1965b). *Comp. Biochem. Physiol.* **15**, 339–345.

Burton, R. F. (1965c). *Can. J. Zool.* **43**, 771–779.

Burton, R. F. (1966). *Comp. Biochem. Physiol.* **17**, 1007–1018.

Burton, R. F. (1968a). *Comp. Biochem. Physiol.* **25**, 501–508.

Burton, R. F. (1968b). *Comp. Biochem. Physiol.* **25**, 509–516.

Burton, R. F. (1969). *Comp. Biochem. Physiol.* **29**, 919–930.

Calow, P. (1970). *Proc. malac. Soc. Lond.* **39**, 203–215.

Cameron, R. A. D. (1969). *J. Linn. Soc. (Zool.)* **48**, 83–111.

Cameron, R. A. D. (1970a). *J. Zool., Lond.* **160**, 143–157.

Cameron, R. A. D. (1970b). *J. Zool., Lond.* **162**, 303–315.

Cameron, R. A. D. (1970c). *Proc. R. Soc.* B **176**, 131–159.

Campion, M. M. (1961). *Q. Jl microsc. Sci.* **102**, 195–216.

Carmichael, E. B. and Rivers, T. D. (1932). *Ecology* **13**, 375–380.

Carney, W. P. (1966). *Nautilus* **79,** 134–136.

Carrick, R. (1939). *Trans. R. Soc. Edinb.* **59,** 563–597.

Carrick, R. (1942). *Ann. appl. Biol.* **29,** 43–55.

Carriker, M. R. (1946). *Biol. Bull. mar. biol. Lab., Woods Hole* **91,** 88–111.

Chamberlain, S. G. and Kerkut, G. A. (1967). *Nature, Lond.* **216,** 89.

Chamberlain, S. G. and Kerkut, G. A. (1969). *Comp. Biochem. Physiol.* **28,** 787–801.

Chatfield, J. E. (1968). *Proc. malac. Soc. Lond.* **38,** 233–245.

Chernin, E. and Adler, V. L. (1967). *Ann. trop. Med. Parasit.* **61,** 11–14.

Chiarandini, D. J. (1964). *Life Sci.* **3,** 1513–1518.

Conway, A. F., Black, R. E. and Morrill, J. B. (1969). *Comp. Biochem. Physiol.* **30,** 793–802.

Cooke, A. H., Shipley, A. E. and Reed, F. R. C. (1895). *In* "Cambridge Natural History" (S. F. Harmer and A. E. Shipley, Eds) Vol. 3, pp. 1–535. Macmillan, London.

Crozier, W. J. and Pilz, G. F. (1924). *J. gen. Physiol.* **6,** 711–721.

Dainton, B. H. (1954a). *J. exp. Biol.* **31,** 165–187.

Dainton, B. H. (1954b). *J. exp. Biol.* **31,** 188–197.

De Jorge, F. B. and Haeser, P. E. (1968). *Comp. Biochem. Physiol.* **26,** 627–637.

De Jorge, F. B. and Petersen, J. A. (1970). *Comp. Biochem. Physiol.* **35,** 211–219.

De Jorge, F. B., Ulhôa Cintra, A. B., Haeser, P. E. and Sawaya, P. (1965). *Comp. Biochem. Physiol.* **14,** 35–42.

Dick, D. A. T. (1959). *Int. Rev. Cytol.* **8,** 387–448.

Drilhon, A. and Florence, G. (1942). *Bull. Soc. Chim. biol.* **24,** 96–103.

Duerr, F. G. (1966). *Physiologist* **9,** 172.

Duerr, F. G. (1967). *Comp. Biochem. Physiol.* **22,** 333–340.

Duval, M. (1930). *Annls Physiol.* **6,** 346–364.

Fischer, P. H. (1950). "Vie et Moeurs des Mollusques". Payot, Paris.

Fischer, P. H. and Duval, M. (1931). *Annls Physiol. Physicochim. biol.* **7,** 88–93.

Florkin, M. (1966). *In* "Physiology of Mollusca" (K. M. Wilbur and C. M. Yonge, Eds) Vol 2, pp. 309–351. Academic Press, New York and London.

Foh, H. (1932). *Zool. Jb. (Zool. Physiol.)* **52,** 1–78.

Frömter, E. and Diamond, J. (1972). *Nature, Lond.* **235,** 9–13.

Gammon, E. T. (1943). *Bull. Calif. Dep. Agric.* **32,** 173–187.

Gary-Bobo, C. M. and Solomon, A. K. (1968). *J. gen. Physiol.* **52,** 825–853.

Gaskoin, J. S. (1850). *Proc. zool. Soc. Lond.* **18,** 243–244.

Gebhardt-Dunkel, E. (1953). *Zool. Jb. (Zool. Physiol.)* **64,** 233–266.

Getz, L. L. (1959). *Am. midl. Nat.* **61,** 485–498.

Getz, L. L. (1963). *Ecology* **44,** 612–613.

Gupta, B. L., Mellon, D., Jr. and Treherne, J. E. (1969) *Tissue and Cell* **1,** 1–30.

Goddard, C. K. and Martin, A. W. (1966). *In* "Physiology of Mollusca" (K. M. Wilbur and C. M. Yonge, Eds) Vol 2, pp. 275–308. Academic Press, New York and London.

Grainger, J. N. R. (1969). *Comp. Biochem. Physiol.* **29**, 671–687.

Greenaway, P. (1970). *J. exp. Biol.* **53**, 147–163.

Grime, J. P., MacPherson-Stewart, S. F. and Dearman, R. S. (1968). *J. Ecol.* **56**, 405–420.

Gugler, C. W. (1963). *Trans. Kans. Acad. Sci.* **66**, 195–201.

Haggag, G. and Fouad, Y. (1968). *Z. vergl. Physiol.* **57**, 428–431.

Herzberg, F. (1965). *Veliger* **7**, 234–235.

Hogben, L. and Kirk, R. L. (1944). *Proc. R. Soc.* B **132**, 239–252.

Horne, F. R. (1971). *Comp. Biochem. Physiol.* **38A**, 565–570.

Horne, F. R. and Boonkoom, V. (1970). *Comp. Biochem. Physiol.* **32**, 141–153.

Hortsmann, H. J. (1956). *Biochem. Z.* **328**, 342–347.

Howes, N. H. and Wells, G. P. (1934a). *J. exp. Biol.* **11**, 328–343.

Howes, N. H. and Wells, G. P. (1934b). *J. exp. Biol.* **11**, 344–351.

Hubendick, B. (1948). *J. Conch.*, *Paris* **88**, 1–10.

Hubendick, B. (1958). *Ark. Zool.* (2) **11**, 31–36.

Hughes, G. M. and Kerkut, G. A. (1956). *J. exp. Biol.* **33**, 282–294.

Humphreys, R. M. (1932). *Trans. R. Soc. trop. Med. Hyg.* **26**, 241.

Hunter, P. J. (1968). *Malacologia* **6**, 391–399.

Hunter, W. R. (1964). *In* "Physiology of Mollusca" (K. M. Wilbur and C. M. Yonge, Eds) Vol 1, pp. 83–126. Academic Press, New York and London.

Hunter, W. R. and Apley, M. L. (1966). *Biol. Bull. mar. biol. Lab., Woods Hole* **131**, 392–393.

Hyman, L. H. (1967). "The Invertebrates Vol. 6 Mollusca I". McGraw-Hill, New York.

Ingram, W. M. (1940a). *Nautilus* **54**, 84–87.

Ingram, W. M. (1940b). *Nautilus* **54**, 87–91.

Jaeckel, S. (1951). *Arch. Hydrobiol.* **44**, 214–270.

Jerelova, O. M., Krasts, I. V. and Viprintsev, B. N. (1971). *Comp. Biochem. Physiol.* **40A**, 281–293.

Jezewska, M. M. (1969). *Acta biochim. pol.* **16**, 313–320.

Jezewska, M. M., Gorzkowski, B. and Heller, J. (1963). *Acta biochim. pol.* **10**, 55–65.

Jones, H. D. (1971). *Comp. Biochem. Physiol.* **39A**, 289–295.

Kamanda, T. (1933). *J. exp. Biol.* **10**, 75–78.

Karlin, E. J. (1961). *Nautilus* **74**, 125–130.

Kendal, S. B. (1949). *J. Helminth.* **23**, 57–68.

Kerkut, G. A. (1959). *Proc. 15th. Intern. Congr. Zool., London*, **1958** 845–848.

Kerkut, G. A. and Ridge, R. M. A. P. (1962). *Comp. Biochem. Physiol.* **5**, 283–295.

Kerkut, G. A. and Taylor, B. J. R. (1956). *J. exp. Biol.* **33**, 493–501.

Kerkut, G. A. and Walker, R. J. (1961). *Comp. Biochem. Physiol.* **2**, 76–79.

Kienle, M. and Ludwig, W. (1956). *Z. vergl. Physiol.* **39,** 102–118.

Kühn, W. (1914). *Z. wiss. Zool.* **109,** 128–184.

Künkel, K. (1916). "Zur Biologie der Lungenschnecken". Carl Winters, Heidelberg.

Lane, N. J. (1963). *Q. Jl microsc. Sci.* **104,** 495–504.

Lever, J. and Bekius, R. (1965). *Experientia* **21,** 1–4.

Lewis, R. D. (1969a). *Malacologia* **7,** 295–306.

Lewis, R. D. (1969b). *Malacologia* **7,** 307–312.

Liebsch, W. (1928). *Zool. Jb.* (*Zool. Physiol.*) **46,** 161–208.

Little, C. (1965). *J. exp. Biol.* **43,** 23–37.

Lloyd, D. C. (1969). *Protoplasma* **68,** 327–339.

Lockwood, A. P. M. (1963). "Animal Body Fluids and their Regulation". Heinemann, London.

Lucké, B. and McCutcheon, M. (1932). *Physiol. Rev.* **12,** 68–139.

Lynch, J. J. (1966). *Aust. J. Zool.* **14,** 65–71.

Machin, J. (1962). "The water relations of snail integument". Ph.D. Thesis, University of London.

Machin, J. (1964a). *J. exp. Biol.* **41,** 759–769.

Machin, J. (1964b). *J. exp. Biol.* **41,** 771–781.

Machin, J. (1964c). *J. exp. Biol.* **41,** 783–792.

Machin, J. (1965). *Naturwissenschaften* **52,** 18.

Machin, J. (1966). *J. exp. Biol.* **45,** 269–278.

Machin, J. (1967). *J. Zool., Lond.* **152,** 55–65.

Machin, J. (1972). *J. exp. Biol.* **57,** 103–111.

Martin, A. W. (1957). *In* "Recent Advances in Invertebrate Physiology" (B. T. Scheer, Ed.) pp. 247–276. University of Oregon Publications, Eugene.

Martin, A. W. (1961). *In* "Comparative Physiology of Carbohydrate Metabolism in Heterothermic Animals" (A. W. Martin, Ed.) pp. 35–64. University of Washington Press, Seattle.

Martin, A. W. and Harrison, F. M. (1966). *In* "Physiology of Mollusca" (K. M. Wilbur and C. M. Yonge, Eds) Vol 2, pp. 353–386. Academic Press, New York and London.

Martin, A. W., Harrison, F. M., Huston, M. J. and Stewart, D. M. (1958). *J. exp. Biol.* **35,** 260–279.

Martin, A. W., Stewart, D. M. and Harrison, F. M. (1965). *J. exp. Biol.* **42,** 99–123.

Mason, C. F. (1970). *Oecologia* **4,** 358–373.

Mason, E. A. and Monchick, L. (1965). *In* "Humidity and Moisture" (A. Wexler and W. A. Wildhack, Eds) Vol 3, pp. 257–272. Reinhold, New York.

May, F. (1932). *Z. Biol.* **92,** 319–324.

Mazek-Fialla, K. (1934). *Z. Morph. Ökol. Tiere* **28,** 445–468.

McAlister, R. O. and Fisher, F. M. (1968). *Biol. Bull. mar. biol. Lab., Woods Hole* **134,** 96–117.

Mead, A. R. (19 61). "The giant African Snail". University of Chicago Press, Chicago.

Moreton, R. B. (1968a). *J. exp. Biol.* **48,** 611–623.

Moreton, R. B. (1968b). *Nature, Lond.* **219,** 70–71.

Morton, J. E. (1954a). *Proc. malac. Soc. Lond.* **31,** 30–46.

Morton, J. E. (1954b). *J. mar. biol. Ass. U.K.* **33,** 187–224.

Morton, J. E. (1955a). *Proc. zool. Soc. Lond.* **125,** 127–168.

Morton, J. E. (1955b). *Phil. Trans. R. Soc.* B **239,** 89–160.

Morton, J. E. (1967). "Molluscs". 4th. edition. Hutchinson, London.

Mozley, A. (1939). *Int. Revue ges. Hydrobiol. Hydrogr.* **38,** 243–249.

Muller, I. (1943). *Riv. Biol. (Perugia)* **35,** 48–95.

Needham, J. (1938). *Biol. Rev.* **13,** 225–251.

Newell, G. E. (1964). *In* "Physiology of Mollusca" (K. M. Wilbur and C. M. Yonge, Eds) Vol 1, pp. 59–81. Academic Press, New York and London.

Newell, P. F. (1968). *In* "The Measurement of Environment Factors in Terrestrial Ecology" (R. M. Wadsworth, Ed.) pp. 141–146. Blackwell, Oxford.

Newell, P. F. and Skelding, J. M. (1973). *Malacologia* **14,** 89–91.

Newell, R. C. (1966). *Nature, Lond.* **212,** 426–428.

Newell, R. C. (1969). *Am. Zool.* **9,** 293–307.

Nopp, H. (1965). *Z. vergl. Physiol.* **50,** 641–659.

Nopp, H. and Farahat, A. Z. (1967). *Z. vergl. Physiol.* **55,** 103–118.

Olivier, L. (1956). *J. Parasit.* **42,** 137–146.

Olsen, W. O. (1944). *J. agric. Res.* **69,** 389–403.

Pallant, D. (1970). *Proc. malac. Soc. Lond.* **39,** 83–87.

Picken, L. E. R. (1937). *J. exp. Biol.* **14,** 20–34.

Pilsbry, H. A. (1926). *Proc. Acad. nat. Sci. Philad.* **77,** 325–328.

Pomeroy, D. E. (1966). "The ecology of *Helicella virgata* in South Australia". Ph.D. Thesis, University of Adelaide.

Pomeroy, D. E. (1968a). *Aust. J. Zool.* **16,** 857–869.

Pomeroy, D. E. (1968b). *Aust. J. Zool.* **17,** 495–514.

Postma, N. (1936). *Proc. K. ned. Akad. Wet.* **39,** 891–896.

Potts, W. T. W. (1958). *J. exp. Biol.* **35,** 749–764.

Potts, W. T. W. (1967). *Biol. Rev.* **42,** 1–41.

Potts, W. T. W. and Parry, G. (1964). "Osmotic and Ionic Regulation in Animals". Pergamon Press, Oxford.

Prosser, C. L. and Brown, F. A., Jr. (1961). "Comparative Animal Physiology" 2nd edition. Saunders, Philadelphia.

Pullin, R. S. V. (1971). *Comp. Biochem. Physiol.* **40A,** 617–626.

Pusswald, A. W. (1948). *Z. vergl. Physiol.* **31,** 227–248.

Raven, C. P. and Klomp, H. (1946). *Proc. K. ned. Akad. Wet.* **49,** 101–109.

Razet, P. and Dagobert, C. (1968). *Archs Sci. physiol.* **22,** 172–181.

Richards, C. S. (1968). *Am. J. trop. Med. Hyg.* **12,** 254–263.

Rising, T. L. and Armitage, K. B. (1969). *Comp. Biochem. Physiol.* **30,** 1091–1114.

Roach, D. K. (1963). *J. exp. Biol.* **40,** 613–623.

Robertson, J. D. (1964). *In* "Physiology of Mollusca" (K. M. Wilbur and C. M. Yonge, Eds) Vol 1, pp. 283–311. Academic Press, New York and London.

Rogers, D. G. (1971). *Z. Zellforsch. microsk. Anat.* **114,** 106–116.

Rolle, G. (1907). *Jena. Z. Med. Naturw.* **43,** 373–416.

Ross, R. J. (1967). "The effect of mechanical stimulation on snails differing in water content." M.Sc. Thesis, University of Toronto.

Rouschal, W. (1940). *Z. wiss. Zool.* **153,** 196–218.

Rózsa, K. S. (1963). *Acta biol. hung. Suppl.* **5,** 44.

Runham, N. W. and Hunter, P. J. (1970). "Terrestrial Slugs." Hutchinson, London.

Saleuddin, A. S. M. (1970). *Can. J. Zool.* **48,** 409–416.

Saleuddin, A. S. M. and Chan, W. (1969). *Can. J. Zool.* **47,** 1107–1111.

Schmidt-Nielsen, K., Taylor, C. R. and Shkolnik, A. (1971). *J. exp. Biol.* **55,** 385–398.

Schwalbach, G. and Lickfeld, K. G. (1962). *Z. Zellforsch. microsk. Anat.* **58,** 277–288.

Shelford, V. E. (1913). *Biol. Bull. mar. biol. Lab., Woods Hole* **25,** 79–120.

Skelding, J. M. (1973). *Malacologia* **14,** 93–96.

Smith, B. J. (1965). *Ann. N.Y. Acad. Sci.* **118,** 997–1014.

South, A. (1965). *J. Anim. Ecol.* **34,** 403–417.

Speeg, K. V., Jr. and Campbell, J. W. (1968a). *Comp. Biochem. Physiol.* **26,** 579–595.

Speeg, K. V., Jr. and Campbell, J. W. (1968b). *Am. J. Physiol.* **214,** 1392–1402.

Spitzer, J. M. (1937). *Zool. Jb. (Zool.* Physiol.) **57,** 457–496.

Standon, B. W. (1955). *J. exp. Biol.* **32,** 84–94.

Stearns, R. E. C. (1877). *Am. Nat.* **11,** 100–102.

Taylor, J. W. (1894–1921). "Monograph of the Land and Freshwater Mollusca of the British Isles". Taylor Brothers, Leeds.

Tercafs, R. R. (1961). *Annls Soc. r. zool. Belg.* **91,** 85–116.

Thornthwaite, C. W. (1940). *Ecology* **21,** 17–28.

Trams, E. G., Lauter, C. J., Bourke, R. S. and Tower, D. B. (1965). *Comp. Biochem. Physiol.* **14,** 399–404.

Tryon, W. W. (1882). "Structural and Systematic Conchology". Philadelphia Acad. Nat. Sci.

Van Aardt, W. J. (1968). *Neth. J. Zool.* **18,** 253–213.

Von Brand, T. (1931). *Z. vergl. Physiol.* **14,** 200–264.

Von Brand, T., McMahon, P. and Nolan, M. O. (1957). *Biol. Bull. mar. biol. Lab., Woods Hole* **113,** 89–102.

Von Brand, T.. Nolan, M. O. and Mann, E. R. (1948). *Biol. Bull. mar. biol. Lab., Woods Hole* **95,** 199–213.

Vorwohl, G. (1961). *Z. vergl. Physiol.* **45,** 12–49.

Warburg, M. R. (1965). *Proc. malac. Soc. Lond.* **36,** 297–305.

Ward, J. (1897). *Nature, Lond.* **20,** 363.

Wells, G. P. (1944). *J. exp. Biol.* **20,** 79–87.

Wendelaar Bonga, S. E. and Boer, H. H. (1969). *Z. Zellforsch. microsk. Anat.* **94,** 513–529.

Wesemeier, H. (1960). *Z. vergl. Physiol.* **43,** 1—28.

Wilbur, K. M. (1964). *In* "Physiology of Mollusca" (K. M. Wilbur and C. M. Yonge, Eds) Vol 1, pp. 243–282. Academic Press, New York and London.

Wilson, R. H. (1968). *Comp. Biochem. Physiol.* **24,** 629–633.

Wilson, R. P., Muhrer, M. E. and Bloomfield, R. A. (1968). *Comp. Biochem. Physiol.* **25,** 295–301.

Wolda, H. (1965). *Archs néerl. Zool.* **16,** 387–399.

Wondrak, G. (1967). *Z. Zellforsch. microsk. Anat.* **76,** 287–294.

Wondrak, G. (1968a) *Protoplasma* **66,** 151–171.

Wondrak, G. (1968b). *Z. Zellforsch. microsk. Anat.* **80,** 17–40.

Yom-Tov, Y. (1970). *Ecology* **51,** 907–911.

Yom-Tov, Y. (1971a). *Israel J. Zool.* **20,** 231–248.

Yom-Tov, Y. (1971b). *Proc. malac. Soc. Lond.* **39,** 319–326.

Chapter 5

Nervous system, eye and statocyst

G. A. KERKUT and R. J. WALKER

Department of Physiology and Biochemistry, University of Southampton, England

I. General account of the brain and peripheral nerves

This section will deal mainly with the nervous system of the land snail *Helix aspersa*, Müll., though examples will also be given of the nervous systems of *Planorbarius*, *Lymnaea*, *Archachatina*, *Agriolimax* and *Onchidium*.

The brain consists of a condensed ring of circumoesophageal ganglia (Fig. 1). Fig. 1 shows a series of diagrams referring to four pulmonate molluscs, *Helix*, *Stenogyra*, *Lymnaea* and *Chilina*. The ganglia are most separated in *Chilina* and *Lymnaea* and most condensed in *Helix*. In *Chilina* the supra- and sub-intestinal (=oesophageal) ganglia are still separated from the pleural ganglia by long connectives. A parietal ganglion is located on the left connective. The connective is still visible in *Lymnaea*, less so in *Stenogyra* and not at all in *Helix* (until one examines sections through the ganglia). The main sense organs that will be considered on pp. 210–237 are the eyes at the tip of the anterior tentacles and the statocysts next to each of the pedal ganglia.

The brain (circumoesophageal ganglionic ring) of *Helix aspersa* is made up from 9 ganglia and their connectives. These are numbered in Fig. 1. There are two cerebral ganglia dorsal to the gut (numbered 1 and 2) and each of these can be subdivided into a procerebrum, meso-cerebrum and metacerebrum (Figs 2, 3A, 16). Hanström (1925)

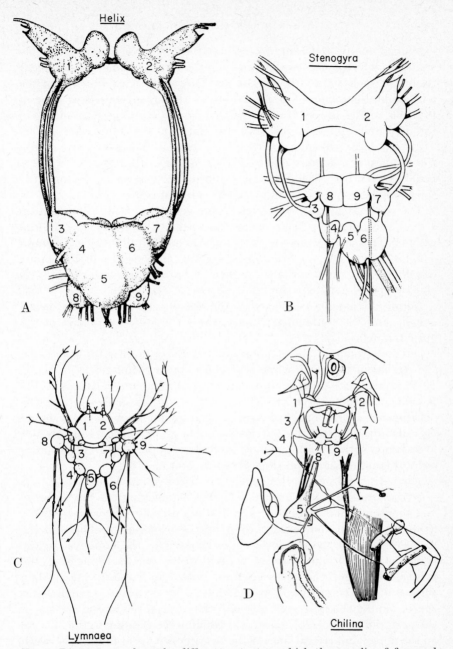

Fig. 1. Diagrams to show the different extent to which the ganglia of four pulmonate molluscs are concentrated to form the circumoesophageal mass. A. *Helix*. B. *Stenogyra*. C. *Lymnaea*, pedal ganglia widely separated and their commissure not shown. D. *Chilina* (after Plate). The condensation is greatest in *Helix*, less in *Stenogyra* and *Lymnaea* and least in *Chilina*. The numbers refer to the ganglia. 1. Left cerebral; 2, Right cerebral; 3, Left pleural; 4, Left parietal; 5, Visceral (abdominal); 6, Right parietal; 7, Right pleural; 8, Left pedal; 9, Right pedal.

examined many *Helix* brains prepared by the Golgi method which stains up selected nerve cell bodies and their axones. The diagrammatic results of such a staining method are shown in Fig. 2 which is of the left cerebral ganglion of *Helix*. 2A shows the arrangements of some of the cells in the procerebrum with their axons forming connections in the neuropil and sending branches out via the tentacular nerve. There is a large cell shown in themesocerebrum. This sends axones into the cerebral commissure and as shown in Fig. 2B into the cerebro-pedal connective. The posterior cerebrum cells send axones into the cerebropedal connectives and into the cerebral commissure.

The cerebral ganglia are linked to the ganglia ventral to the gut by a double connective. These connectives lead to the pedal and pleural ganglia. The suboesophageal ganglionic mass contains the left pleural ganglia (3, Fig. 1A), a left parietal ganglion (4), a visceral or abdominal ganglion (5), a right parietal ganglion (6), a right pleural ganglion (7) and two pedal ganglia (8 and 9).

Additional ganglia connected to the oesophageal ring are the buccal ganglia and the tentacular ganglia; the latter are at the base of the optic tentacles.

Our knowledge of the structure of the *Helix* brain still depends on the studies of the early workers (Nabias, 1898; Schultze, 1879; Bang, 1917; Kunze, 1917, 1921; Hanström, 1925.) Some indication as to the possible functional role of each ganglion can be deduced from the distribution of the nerve trunks from that ganglion. However it should be remembered that the cell bodies for a given nerve axone are not necessarily within that ganglion, and may in fact be some distance away in another ganglion (Figs 9 and 15 and p. 176).

The peripheral nerves from the *Helix* brain run as follows—

Cerebral ganglia: 1, a tentacular nerve, connecting the procerebrum and the tentacular ganglion, and leading to the posterior tentacle; 2, an optic nerve connecting the eye on the posterior tentacle and the postcerebrum; 3, the statocyst nerve running to the statocyst in the pedal ganglia; 4, an internal peritentacular nerve, connecting the procerebrum and the mesocerebrum with the skin on the median surface of the large (posterior) tentacle; 5, an external peritentacular nerve, joining the postcerebrum with the skin on the lateral surface of the posterior tentacle; 6, an internal labial nerve running to the muscles on the dorsal and lateral sides of the buccal cavity and the skin between the tentacles; 7, an external labial nerve running to the muscles on the ventral and lateral sides of the buccal cavity; 8, the medial labial nerve supplying the oral lobe and the small tentacle: the small tentacular branch may have an accessory ganglion; 9, the penis nerve, whose cell bodies lie in the pedal ganglion and the axones run through the pedal-cerebral connective and emerge from the cerebral ganglion, and 10 a cerebral artery nerve.

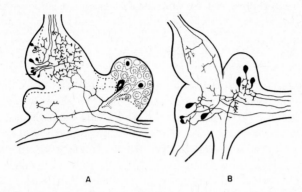

A   B

Fig. 2. Golgi-stained neurones in the left cerebral ganglion of *Helix*. A. Diagram showing several nerve cells in the procerebrum and their axones running to the tentacle and neuropile. B. Several neurones in the mesocerebrum and postcerebrum with their axones running towards the connectives. (After Hanström, 1925.)

Buccal ganglia. These are paired and each contains a large nerve cell which can be up to 200μm in diameter. The nerve trunks arising from them are as follows—

two salivary gland nerves (one large and one small); an anterior gastric nerve running to the oesophagus; a posterior gastric nerve running to stomach and liver, and four pharyngeal nerves running to the muscles of the pharynx.

Tentacular ganglion. This is almost an extension of the cerebral ganglion, lying at the base of the optic tentacle and has a very dense neuropil (Lane, 1962). The main nerve trunk runs to the eye.

Pleural ganglia. These lie on the connective between the cerebral ganglia and the parietal ganglia, giving origin to the pharyngeal retractor nerve, and the columellar muscle nerve.

The *parietal ganglia* supply the nerves to the skin and muscles of the body wall and the mantle skirt.

The *abdominal ganglion* (also called visceral ganglion) has two branches, the intestinal nerve passing to the kidney, liver, heart, albumen gland, gonad, intestine and skin, and the anal nerve to the rectum.

The *pedal ganglia* have nerve trunks branching as follows—
three cutaneous nerves supplying the skin of the head area from the large tentacle to the region of the penis; one pair of anterior pedal

muscle nerves; ten pairs of posterior pedal muscle nerves, and two pedal artery nerves (anterior and posterior).

The nerve trunks are generally paired, one nerve trunk from each of the paired ganglia except for the visceral ganglion and the penis nerve.

II. General histology of the ganglia

The ganglia are covered by an epineural sheath. This is in two layers, a thick outer layer and a thin inner layer. The thick outer layer has a complex structure the histology of which has been described by Fernandez (1966), Newman *et al.* (1968), Benjamin and Peat (1968), Sanchis and Zebrano (1969a, b) and Rogers (1969). In *Helix* brain the epineural sheath has a large number of globular cells that contain glycogen; the glycogen content of the cells increases during the winter. In addition there are muscle cells, fibroblasts, nerve axones and collagen in the layer. The inner layer of connective tissue is much thinner and has a lamellate structure.

The blood system penetrates the epineural sheath. The blood supply to the ganglia has been described by Pentreath and Cottrell (1970). Each paired ganglion is supplied by symmetrically arranged branches from the anterior aorta. Capillaries from these branches open into a blood space which is adjacent to and continuous over the surface of the nervous tissue. Blood passes through the epineural sheath to the body-cavity sinus. Three tissue layers separate the blood space from the neurones: a, the luminal endothelium; b, a fibrous connective tissue layer mainly of collagen; c, glial cells. Both the luminal endothelium and the connective tissues are freely permeable to uncharged particles of 10 nm diameter or less.

Within the inner layer of the connective tissue, the nerve cells are arranged with their cell bodies (perikarya) peripherally and their axones running centrally into the neuropil. The axones form a dense network of interconnections within the neuropile (Figs 2, 3, 4).

Figure 3 shows three sections through the *Helix* ganglia based on drawings from Kunze (1921). Figure 3A is a diagram through the cerebral ganglia. The larger cell bodies are arranged peripherally in the lobes of the ganglia, i.e. mesocerebrum. There is also a mass of more dense nuclear material placed laterally in the ganglia, forming the globuli cells. The pedal ganglia similarly have the cell bodies placed peripherally and the section in 3B is taken through the posterior end of the ganglia so that the cells appear to be located centrally in the ganglionic mass. The tracts leading from the cells are also indicated. Figure 3C shows a

Fig. 3. Diagrammatic transverse sections through the brain of *Helix*. A, cerebral ganglia; B, pedal ganglia; C, parietal, pleural and visceral ganglia. (After Kunze, 1921.)

section through the pleural, parietal and visceral ganglia. The large cells are in the parietal and visceral ganglia. The ganglia have a complex three dimensional shape and the section shows some of the tracts linking between the ganglia.

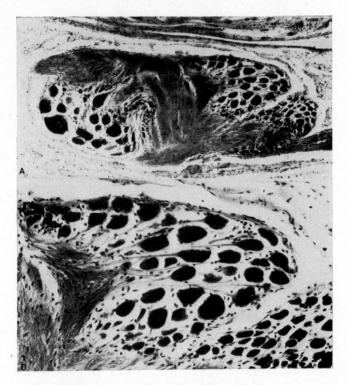

Fig. 4. Photomicrographs of sections through the A, parietal and B, visceral ganglia of *Helix*. Note the peripheral arrangement of the cell bodies, and the axones running centrally.

Figure 4 is a photomicrograph of sections through the ganglia showing the position of the nerve cell bodies and the axones. There has been a shrinkage of the cell bodies during the fixation, otherwise the cell bodies would have occupied the vacant space around them. The cell bodies vary in size between 20μm and 200μm as shown here. The thinner outer connective tissue layer is immediately above the cell bodies, and there is a space between this and the thicker outer connective tissue layer. Note in Fig. 4B the axones running inwards towards the neuropile.

The superficial location of the cell bodies in the ganglia makes it relatively easy to dissect away the connective tissue and reveal the cells in the living brain. Figure 13 is a photograph of such a brain lightly stained with methylene blue. With practice it is possible to recognise specific nerve cell bodies in the *Helix* brain, and we have identified

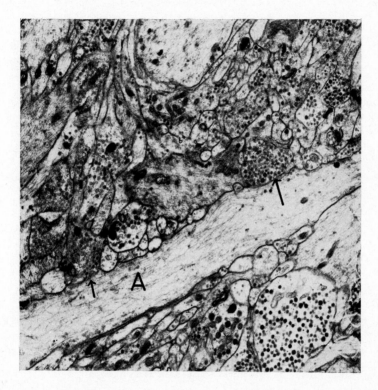

Fig. 5. Electron micrograph of the brain of *Archachatina*. Axons and synapses in the neuropile of the pedal ganglion. × 10 500. (Nisbet and Plummer, unpublished.)

about 150 cells on the bases of their position in the brain, electrical activity, connections and pharmacological activity (Kerkut *et al.* 1974).

Amoroso *et al.* (1964) and Baxter and Nisbet (1963) have published details of the fine structure of the neurones and glial cells of *Archachatina marginata* (Swainson). There are three types of cells in addition to the neurones within the ganglia: 1, sheath cells which surround and may penetrate the neurones; the sheath cells contain a large amount of glycogen and may have a nutritive function in relation to the nerve cells; 2, glial cells which lie between the sheath cells and are also present in the walls of the blood vessels. They have a better developed endoplasmic reticulum and Golgi complex and more mitochondria than the sheath cells; 3, supporting cells which may penetrate the larger axones in a manner similar to the sheath cells (Schlote, 1957), and differ from the sheath cells in that they contain bundles of electron-dense

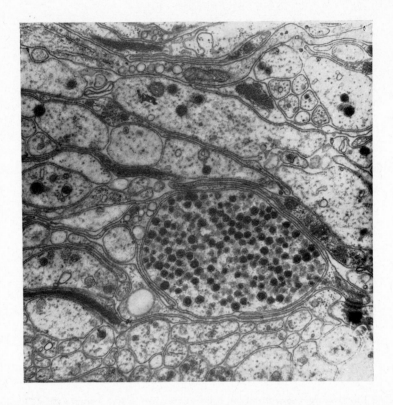

Fig. 6. Electronmicrograph of the brain of *Archachatina*. Electron-dense granules and axones from the neuropile of the cerebral ganglion × 20 000. (From Amoroso *et al.*, 1964.)

material. They also have well marked desmosomes. The supporting cells function to provide a firm and resiliant framework around the nerve fibres.

Amoroso *et al.* (1964), Baxter and Nisbet (1963) and Nisbet and Plummer (1968) have made a general study of the electron microscopic structure of the pulmonate brain (Figs 5, 6). Figure 5 is a photograph of the axones and synapses in the neuropile of the pedal ganglion of

Archachatina. It has a main axone running across the picture (A) and a synaptic vesicle coming into contact with it (marked with an arrow). Figure 6 is a photograph at a higher magnification of the neuropile in the cerebral ganglion. The centre of the picture shows an ending containing many electron-dense vesicles, and also shows clearly the many thin

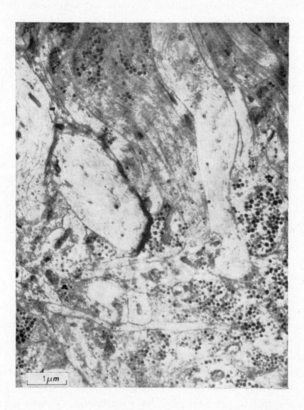

Fig. 7. Electron micrograph of the neuropile of the visceral ganglion of *Helix* stained to show the localization of choline esterase. The esterase is localized at some synapses and is shown as a black deposit in the region of the synapse. (From Newman *et al.*, 1968.)

processes of the supporting cells around the axones. These help insulate the axons from each other and provide a framework and continuity between the neurones.

The fine structure of the synapses has been studied by Gerschenfeld (1963) in the brains of *Vaginula solea* d'Orbigny, *Helix aspersa* Müll. and *Helix pomatia* L. The synaptic junctions are always of the axo-axonal type. The perikarya never show any synaptic contacts. The post-synaptic membranes are mainly located on small finger-like axonal branches, and the presynaptic endings can be either *en passant* or the terminal type. Gerschenfeld classified the synaptic endings into three main types depending on their vesicular content (clear or dense) and the size of the vesicles. Newman *et al.* (1968) localized the enzyme choline esterase by histochemical means at synapses in the *Helix* neuropile (Fig. 7) and showed that some of the dense vesicles in the synapses lost their density after pretreatment with reserpine. This suggests that the vesicles could contain a catecholamine such as dopamine (Curtis and Kerkut, 1969). We already have the evidence that dopamine is present in *Helix* neurones (see p. 195 and Fig. 10).

A. Nerve–muscle junctions

These have been described for the pulmonate nervous system by Benjamin and Peat (1968), Rogers (1968) and Barrantes (1970). Figure 8 is a diagram reconstructed from electron micrographs of an axone branching and making two simple junctions with snail muscle fibres (Kerkut *et al.* 1966).

B. Procion Yellow marking of axonal pathways

Using the technique devised by Stretton and Kravitz (1968) it is possible to place a microelectrode filled with the dye Procion Yellow into a selected neurone in the brain of *Helix aspersa* and to pulse the dye into the cell body. If the preparation is left for several hours, the dye is carried along the branches of the axone of that neurone. The technique is slightly unpredictable in that many of the fine branches of an axone may fail to fill, but positive results clearly indicate the anatomical pattern of the branching axons.

Ten specific nerve cells were repeatedly studied by this method and their axonal pathways determined (Kerkut *et al.*, 1970). Figure 9 is a simplified diagram adapted from their paper and the axonal pathways of three of the cells are shown. Note the complex pathway taken by some of the axons through the ganglia. It is probable that complex pathways are due to embryological and functional development of the nervous system. It is possible to check on the results of the Procion Yellow technique by electrophysiological mapping, i.e. seeing which axons on stimulation lead to antidromic action potentials in known nerve cells.

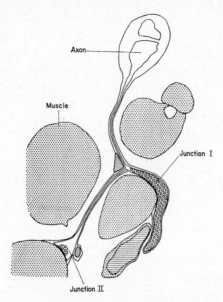

Fig. 8. Diagram of a nerve–muscle junction in *Helix aspersa*. The nerve muscle junction is a simple finger-like process that presses against the muscle fibre. (From Kerkut *et al.*, 1966.)

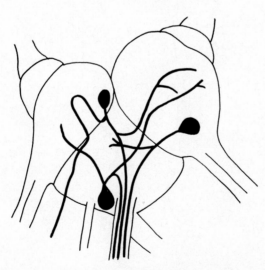

Fig. 9. Procion Yellow location of axons. Diagram of the left parietal, abdominal and right parietal ganglia of *Helix*. The axonal pathway of three nerve cells is shown here as elucidated from Procion Yellow marking of the nerve cells. (Modified from Kerkut *et al.*, 1970.)

C. Nerve cells containing catecholamines and 5-HT

If snail brains are freeze dried and exposed to paraformaldehyde vapour under conditions of controlled humidity (Falck, 1962; Kerkut et al., 1967; Sedden et al., 1968; Marsden and Kerkut, 1969) those cells that contain catecholamines such as dopamine fluoresce a green-blue colour when illuminated with u.v. light, while cells that contain 5-HT (5-hydroxytryptamine) fluoresce a yellow colour. If the animals are pretreated by injecting DOPA (dihydroxyphenylalanine) or 5-hydroxytryptophan into their body cavity two hours before removing the brain, the drug becomes decarboxylated in the body and the brain takes up the amine. This increases the amount of amine in the nerve cells and the specific neurones fluoresce much more strongly and brightly.

It is possible to map out which neurones contain 5-HT and dopamine in the brain of *Helix aspersa* (Sedden et al., 1968; Loker, 1973). There is a large fluorescent yellow cell body in each of the cerebral ganglia. The right cerebropleural connective contains a chain of small nerve cells containing 5-HT along its peripheral boundary. There is a group of large neurones containing 5-HT in the right parietal ganglion, the visceral ganglion and in the pedal ganglia. The neurones containing

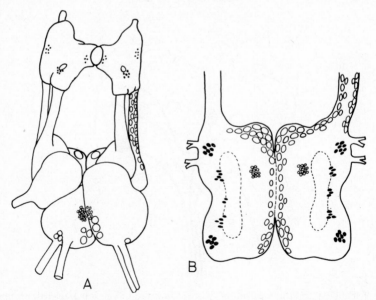

Fig. 10. Localization of the cells in central nervous system of *Helix* that contain 5-HT (○) and dopamine (●). A. Dorsal view of the brain B. T.S. through pedal ganglia. (Fig. provided by Janet Loker.)

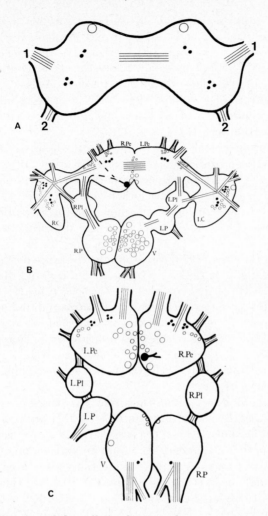

Fig. 11. Localization of the cells in the central nervous system of *Planorbarius corneus* that contain 5-HT (○) and dopamine (●). A. Dorsal surface of the cerebral ganglia = fluorescent fibres. 1. Cerebral nerve trunk. 2. Cerebro-pedal connective. B. Ventral view of the circumoesophageal ganglia. C. Dorsal view of the suboesophageal ganglia. L.C, left cerebral; L.P, left parietal; L.Pe, left pedal; L.Pl, left pleural; R.C, right cerebral; R.P, right parietal; R.Pe, right pedal; R.Pl, right pleural; V, visceral, (From Marsden and Kerkut, 1969.)

dopamine are in general much smaller than those containing 5-HT (Fig. 10).

Loker (1973) has studied the relationship between the pharmacological activity of specific nerve cells that contain 5-HT and dopamine in the *Helix* brain. She did this by marking specific nerve cells with Procion Yellow after she had tested a group of adjacent cells' pharmacological responses to iontophoretic application of drugs. The tissue was then freeze dried and the yellow fluorescent cells (containing 5-HT) corresponded in position with the Procion Yellow cells (these fluoresce a more orange colour). The 5-HT cells were all excited by iontophoretic application of acetylcholine and also excited by iontophoretic application of 5-HT.

The distribution of the monoamines in the brain of *Planorbarius corneus* (L.) has been studied by Marsden and Kerkut (1970) using fluorescence microscopy and microspectrofluorimetry. Certain neurones in the brain contain dopamine whilst others contain 5-HT (Fig. 11). One giant neurone in the right pedal ganglion contains dopamine, which is in contrast with the situation in *Helix* where dopamine is confined to small neurones.

Cottrell and Osborne (1970) dissected out the large nerve cells in the metacerebral ganglia of *Limax maximus* L. and isolated the 5-HT from the cells, the 5-HT being assayed on the *Helix* heart. From seven assays, the 5-HT content of the giant cell was estimated as 0.7ng/cell. The total volume of the giant cell was about 1·3 nl but one fifth of this is occupied by the nucleus which does not fluoresce. Thus the concentration of 5-HT in the cytoplasm of the cell soma was 4×10^{-3}M, and a similar concentration of 5-HT was found in the giant cells in the cerebral ganglia of *Helix*.

Sakharov and Zs-Nagy (1968) localized two large neurones in the cerebral ganglia of *Lymnaea* that contained 5-HT (see p. 198).

Cottrell (1971) showed that the two large cerebral cells of *Helix pomatia* contain 5-HT and that they send an axone along the cerebro–buccal connective to make synaptic connections with three neurones in each buccal ganglion. It is suggested that the EPSP in the buccal cells can be driven by the large cerebral cells and that 5-HT is the probable excitatory transmitter.

III. Synaptic potentials

Many *Helix* nerve cells, on penetration by a microelectrode, show spontaneous postsynaptic potentials. It is also possible to stimulate a peripheral nerve trunk and drive a post-synaptic potential in the impaled nerve cell body.

EPSP. Many cells show excitatory postsynaptic potentials (EPSP). These potentials tend to depolarize the membrane and can lead to action potentials. Figure 12A shows a series of spontaneous EPSP in a nerve cell. They range in size from 3mV to 10mV depolarization. Figure 12B

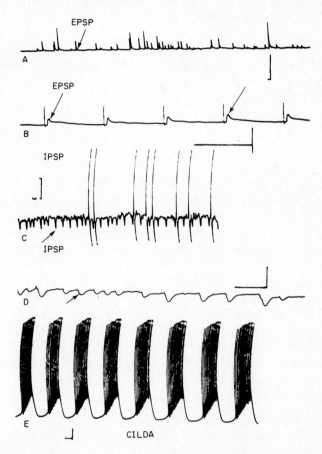

Fig. 12. Post-synaptic potentials in *Helix* neurones. (Fig. provided by John Lambert.) Calibration: vertical lines 5mV, horizontal lines 1 s.

shows the EPSP on a faster time scale. They are driven by stimulation of a peripheral nerve trunk and the fast spike just before each EPSP is due to the stimulus artifact. An evoked EPSP is marked with an arrow.

IPSP. Figure 12C shows spontaneous inhibitory postsynaptic potentials in a nerve cell. Note that they are downward (hyperpolarizing) potentials. Figure 12D shows spontaneous IPSP on a faster time scale and the different sizes of the IPSP can be seen. This difference in size is probably due to the IPSP being formed at synapses at different sites from the cell body and the microelectrode only sees a fraction of the evoked potential from each synapse, depending on the value of the space constant of the membrane.

CILDA. Figure 12E shows CILDA (cells with an inhibition of long duration). The cell shows a series of action potentials followed by a hyperpolarization that lasts for two seconds after which there is another series of action potentials.

A. Two excitatory transmitters onto one nerve cell

Gerschenfeld and Stefani (1966, 1968) found that some neurones in the snail *Helix aspersa* could be excited by iontophoretic application of either 5-HT or acetylcholine, ACh. They could also drive two different types of EPSP, (1) a fast EPSP (duration 200–500 ms) by stimulation of the anal nerve, this EPSP being blocked by hexamethonium; (2) there was also a slow EPSP (duration 800–2000 ms) driven by the intestinal nerve and this was blocked by LSD. It is likely that the fast EPSP has acetylcholine as the transmitter and the slow EPSP has 5-HT as the transmitter.

A similar system was investigated by Kerkut *et al.* (1970). Possible advantages of having two different transmitters are as follows:

1. It would allow for the depletion of synthesis of one transmitter at the presynaptic ending.

2. It would reduce the competition for presynaptic uptake of the released transmitter.

3. It would allow for any adaptation or desensitization of the post-synaptic receptor to one transmitter.

4. It would allow a fast and slow control system similar to that seen in the control of crustacean muscle systems.

5. It would allow for a background discharge setting the level of the membrane potential and for a quick on-and-off effect.

6. If one of the transmitter systems or the receptors was selectively altered by hormonal action, this would allow a change or selection of the neuronal circuits in the central nervous system according to the level of the hormone present.

B. Complex synaptic activity

It is often possible to find more complex synaptic activity such as biphasic potential (excitatory/inhibitory) fast and slow compound synaptic potentials, etc. These can involve two or more ionic mechanisms at the one synapse. It is mainly through work on the molluscan synaptic systems that such complex transmission systems have been recognized and analysed (Kehoe, 1969, 1973).

IV. Mapping of nerve cells in *Helix* brain

Kerkut and Walker (1961) found that many of the nerve cells in the *Helix* brain responded to application of drugs, but the responses appeared variable. It was later discovered that repeatable results could be obtained provided that one worked on the same cell on each occasion. With practice it became possible to work consistently on specific nerve cells in the *Helix* right parietal, abdominal, left parietal and cerebral ganglia (Kerkut and Walker, 1962; Kerkut and Meech, 1966; Walker 1967; Glaizner, 1968a; Walker and Hedges, 1968; Kerkut and York, 1969; Kerkut *et al.*, 1969).

It was also possible to map out the axonal branching of ten neurones in the *Helix* brain using the Procion Yellow injection technique (Kerkut *et al.*, 1970), and Fig. 9 shows the branching pattern of the axones of three of these cells.

Figure 13 is a photograph of the dorsal surface of *Helix* brain after the outer connective tissue sheath has been removed. The cells have been lightly stained with methylene blue to increase the photographic contrast, and with practice it is possible to recognize about 100 of the large nerve cell bodies and to penetrate these with microelectrodes. Figure. 14 is a model of the snail brain showing positions of some of the large cells that have been identified and whose synaptic inputs and axonal connections are known (Kerkut *et al.*, 1974). When the peripheral nerve trunks are taken up into suction electrodes and stimulated it is possible to drive EPSP, IPSP or antidromic action potentials in given cell bodies. Figure 15 shows excitatory inputs, inhibitory inputs and the peripheral axonal connections to nerve trunks of approximately 100 nerve cells. When a given nerve trunk is stimulated it is marked in black. Thus Fig. 15A has the left parietal nerve trunk stimulated. The nerve cells that show EPSP when the left parietal nerve trunk is stimulated are shown in black on the diagram. Similarly

Fig. 15E shows which nerve cells have inhibitory post-synaptic potentials when the specific nerve trunks are stimulated; Fig. 15 I–L show which cell bodies have their peripheral axones emerging in a given nerve trunk (i.e. give an antidromic action potential in that cell body when the nerve trunk is stimulated). We have also tested the pharmacological responses of many of these cells, the work being an extension of the study by Glaizner (1968b) Kerkut *et al.*, (1974).

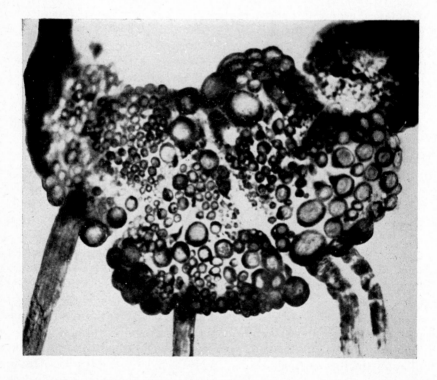

Fig. 13. Photograph of living *Helix* brain with the outer connective tissue removed to expose the nerve cells. It is possible to record the electrical activity from specified cells. (Photograph provided by Janet Loker.)

Fig. 14. Model of *Helix* brain showing the location of nerve cell bodies whose electrical, axonal and pharmacological properties are known. (Photograph provided by Janet Loker.)

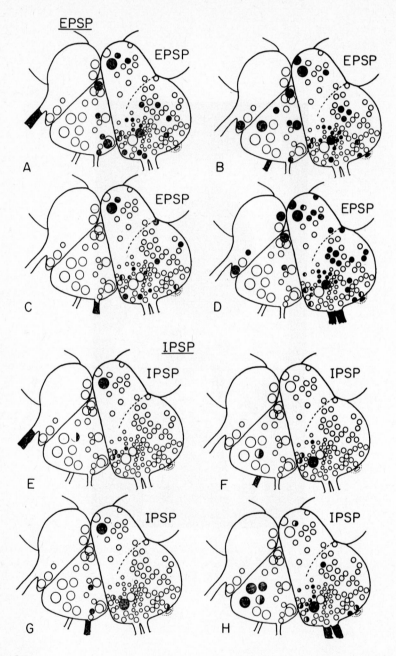

Fig. 15.

ANTIDROMIC

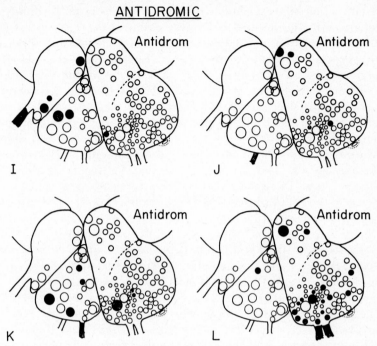

Fig. 15. The peripheral connections of known nerve cells in *Helix* brain. Axones in the peripheral nerves marked in black are stimulated through a suction electrode. Those cells that show an EPSP, IPSP or an antidromic action potential are shown in black.

V. Special applications of the pulmonate CNS

The pulmonate central nervous system has been used to examine and extend the classical studies of neurophysiology based on the squid axon and the frog sciatic nerve trunk. In general much of the information obtained from the pulmonate nervous system is similar to that obtained from other nerves and Tauc (1966) gives a good general account of gastropod neurones based mainly on studies on the opisthobranch *Aplysia*. Pulmonates such as the snail have indicated previously unsuspected features of neurophysiology and some of these are briefly listed here.

A. High and low chloride neurones

It is normally assumed that all nerve cells have the same ionic composition. Oomura (1963) suggested that one explanation of the different action of acetylcholine on snail neurones could be in terms of a difference in the internal chloride concentration of the nerve cells. Some nerve cells are depolarized (D) by acetylcholine and other cells are hyperpolarized (H) by acetylcholine (Tauc and Gerschenfeld, 1962). Oomura suggested that the H cells could have a low internal chloride concentration whilst the D cells had a higher chloride concentration.

Kerkut and Meech (1966) developed an electrode with a silver chloride tip and this could be inserted into a nerve cell and measure a potential in the cell that was proportional to the internal chloride concentration They found that the concentration in the H neurones of *Helix aspersa* was $11 \cdot 2 \pm 0 \cdot 6$ mM and that for the D neurones was $24 \cdot 7 \pm 0 \cdot 8$ mM. Increasing the chloride concentration changed the response of the H cell to that of a D cell. Oomura *et al.* (1968) measured the chloride concentration in *Onchidium* neurones by means of a similar electrode and found that there was a difference between the chloride concentrations of the H and D cell. In *Aplysia*, Chiarandini *et al.* (1967) and Chiarandini and Gerschenfeld (1967) calculated that the Cl_i was 43mM for D cells and 12mM for H cells. An ion exchange electrode has been used to measure the chloride concentration in *Aplysia* neurones.

The chloride level in the large cells was $27 \cdot 7 \pm 1 \cdot 8$mM with a range from 21–31mM. The E_{Cl} value was from 57–70mV and the recorded membrane potential was 56mV. The smaller nerve cells had a chloride level of $40 \cdot 7 \pm 1 \cdot 5$mM (range 37–47). The E_{Cl} was 53mV while the membrane potential was 58mV (Walker and Brown, 1971).

B. Sodium and potassium levels

Kostyuk *et al.* (1969) found that the internal K^+ concentration in *Helix* neurones was $0 \cdot 073$M, the value for Na^+ being $0 \cdot 013$M. These values were less than those determined by chemical analysis ($K^+ = 0 \cdot 093$M and $Na^+ = 0 \cdot 031$M). They suggested that the high levels of electrolytes within the nerve cell might affect the activity of the sodium and potassium within the cell. Kostyuk and his colleagues (Kostyuk, 1967; Sorkina and Zelenskaya, 1967) have studied the variations in the ionic concentrations of the haemolymph of *Helix* and *Planorbarius* throughout the year. In *Planorbarius* living in ponds, the concentrations of $K^+Cl^-Na^+$ and Cl, are maximal during the summer and autumn; in the spring they reach minimal levels. In *Helix* the seasonal haemolymph variations are greater. There are two maxima: one at the end of summer and the other at winter–early spring.

There are also seasonal variations in the concentrations of ions within the nerve cells. The sodium concentration is minimal in summer and increases in the winter-spring. The lowest-highest values in $mM/kg/$ ww for *Helix* are $K^+70\cdot46$–$82\cdot36$; $Na^+20\cdot38$–$38\cdot76$. For *Planorbarius* the levels were $K^+50\cdot48$–$71\cdot64$; $Na^+10\cdot94$–$39\cdot74$. There were also variations in the membrane potential of the *Helix* neurones through a yearly cycle.

Using insulin, sucrose, sulphate etc. Zelenskaya *et al.* (1968) found that the extracellular space for the CNS of *Helix pomatia* was $37\cdot37\%$ and that for *Planorbarius corneus* was $34\cdot14\%$.

C. Calcium action potentials

In squid axon the sodium ions carry most of the inward current during the action potential though there is a small inward current carried by calcium ions (Hodgkin and Huxley, 1945; Baker *et al.* 1971). In gastropod nerve cells the situation is more complex. Nerve cells can have the inward current carried by sodium ions (Chamberlain and Kerkut, 1967, 1969; Moreton, 1968). Some nerve cells have the inward current carried by a mixture of sodium and calcium ions (Kerkut and Gardener, 1967; Geduldig and Junge, 1968), while other nerve cells have their inward current carried by calcium ions (Oomura *et al.*, 1962 Meves, 1966, 1968; Gerasimov *et al.*, 1965; Gerasimov, 1964; Gerasimov and Maisky, 1963; Magura 1969; Krishtal and Magura, 1970). In nerve cells where calcium is the main ion carrying the inward current the relationship between the height of the overpotential and the external calcium concentration is such that the regression line for a tenfold change in external calcium concentration is 29mV (i.e. for a divalent ion it would be expected to be $58/2=29$). Krishtal and Magura (1970) using the voltage clamp technique indicated that for some neurones of *Planorbarius corneus* (L.), *Lymnaea stagnalis* (L.) and *Helix pomatia* part of the inward current was carried by the calcium ions.

Wald (1972), working on *Helix aspersa*, found that there was a difference between the somatic and the axonal action potentials. The somatic action potentials showed an increase of 10mV in their overpotential for a 10-fold increase in external calcium ion concentration and the action potentials decreased when the external calcium concentration was reduced. They could be eliminated with 2mM EGTA. The axonal action potential appeared to be mainly sodium-dependent whilst the somatic action potential appeared to be calcium-dependent. The sodium-dependent action potentials were resistant to tetrodotoxin ($5 \times 10^{-6}g/ml$) (Chamberlain and Kerkut, 1967, 1969; Wald, 1972).

D. *Alteration of solution around the synapse and nerve cells*

In the vertebrates the nerve cells and synapses of the CNS lie deep within the nervous system and it is not easy to change the concentration of the ionic media immediately bathing the nerve cells and synapses. In invertebrates such as *Helix* it is easy to change the ionic composition around the nerve cells. During the IPSP there is a change in the permeability of the post-synaptic membrane so that ions of about 0·3nm diameter penetrate the membrane (Kerkut and Thomas, 1964). This was investigated by putting solutions containing ions of different size around the cells and seeing how this affected the IPSP.

Though one can change the bulk of the solution it is important to consider the actual concentration at the cell surface. There is some suggestion that one can not necessarily wash all the sodium ions away from the nerve cell surface in *Helix* (Chamberlain and Kerkut, 1969; Moreton, 1968) and also that microenvironments can exist within the neuropil and these may lead to local increases or differences in ionic concentration that can markedly affect the responses of the nerve cell (Kerkut *et al.*, 1973).

E. *Ionic permeability*

The post-synaptic potentials are due to an increase in the membrane permeability to one or more ions. The effect of a transmitter is to increase the membrane permeability and this is specific to the cell membrane under investigation, i.e. cell membranes differ in their properties. Thus ACh can act on some nerve cells and increase their membrane permeability to Na^+. In other nerve cells it increases the permeability to Cl^- and in other cells to K^+. Whether the membrane is depolarized or hyperpolarized depends on the relationship between the membrane potential of the cell and the equilibrium potential for the ions. The pulmonate CNS offers special advantages in that one can map the individual nerve cells in the brain, determine the pharmacological and ionic properties of the membranes, and check this by iontophoretic application of the drug on to one specific part of the cell under controlled ionic and membrane potential conditions. In this way one can obtain a clearer identification of the changes in ionic permeability of the cell, and parts of the cell.

F. *Axoplasmic transport*

Invertebrate preparations based on *Helix aspersa* were one of the first systems to show fast transport of labelled material from the CNS to the periphery. It is possible to isolate the brain, pharyngeal nerve trunks and pharyngeal retractor muscles and arrange these with lanolin barriers so that the brain is in one compartment, the nerve trunks in a second compartment and the muscle in a third compartment. Radioactive material bathing the CNS passes into the nerve cells and along the nerve trunks to the muscles. The rate of transport can be as high as 1 cm in 20 min (Kerkut et al., 1967). This is much faster than the rate of 2mm/24 h that was the previously accepted "standard" rate of transport along axones. Since then fast rates have been discovered for most vertebrate and invertebrate nerve trunks and rates up to 1cm/h and faster are quite common (reviewed in Barondes, 1967). The snail also shows transport of material from the muscle to the CNS. Similar centripetal transport systems have also been found for the vertebrate nervous system (Kristensson and Olsson, 1971).

The invertebrates offer a special experimental advantage in that their low pressure blood system and their poikilothermy allows the nervous system to survive under conditions that would rapidly destroy mammalian central neurones.

G. Giant cells

Though a considerable amount of work has been carried out on the giant neurones of the abdominal ganglia of *Aplysia* by Tauc and his colleagues (Tauc, 1966; 1967) and Kandel and his colleagues (Kandel *et al.*, 1967; Kandel and Kupfermann, 1970) they have also investigated some of the properties of the larger cells in *Helix*.

Kandel and Tauc (1966a, b) studied the properties of the symmetrical giant cells in the metacerebrum of *Helix aspersa* and *Helix pomatia* (Fig. 16). These cells lie on the ventral surface of the ganglia and are 140μm in diameter. The input and output organization of the two cells is completely symmetrical. Simultaneous recordings from both cells failed to reveal interconnections but did show that the cells share in common, the output of at least two interneurones. The ipsilateral input to a given cell produced EPSPs that were consistently more effective and of a shorter latency than the contralateral input. Both cells were depolarized by acetylcholine. The cells showed identical biophysical properties; both showed anomalous rectification especially either side of the resting level of the membrane potential. This rectification could be important as a post-synaptic determinant of synaptic efficiency.

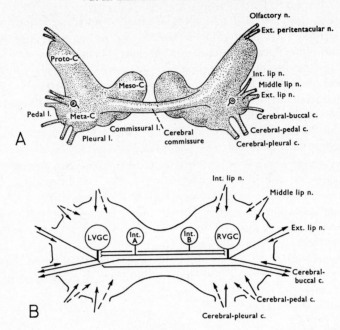

Fig. 16. Giant cells in the cerebral ganglia of *Helix*. There are two large cells in the ganglia (A). They are interconnected functionally as shown in B. (From Kandel and Tauc, 1966a.)

VI. Pharmacological studies on the pulmonate nervous system

There are a number of compounds which have potent actions on pulmonate neurones. Among these compounds are included most of the chemicals which have been postulated as nerve–nerve and/or nerve–muscle transmitter agents in other groups of animals; for example, acetylcholine, dopamine, 5-hydroxytryptamine (5-HT), glutamic acid and GABA. Pharmacological studies have been carried out on a number of pulmonate species and there is considerable similarity amongst the different species: *Helix aspersa*, Kerkut and Walker (1961, 1962); Gerschenfeld and Stefani (1966, 1968); *Helix pomatia*, van Wilgenburg (1970); Sakharov and Salanki (1969a,b); *Lymnaea stagnalis* (L.), Zeimal and Vulfius (1968); Kiss and Salanki (1971).

A. Acetylcholine

Acetylcholine has an action on almost all neurones in the pulmonate central nervous system. In general cells are either depolarized and excited by acetylcholine, termed D cells, or else the cell membrane potential is hyperpolarized and the cell firing rate inhibited, termed H cells. These effects are mediated via a change in membrane permeability to one or more ions or possibly to the stimulation of an electrogenic sodium pump. The H effect can be due to an increase in permeability either to chloride ions (Kerkut and Thomas, 1963; 1964) or potassium ions (Lambert, 1972). In the former case the cells have a very low internal chloride concentration (Kerkut and Meech, 1966). The D effect seen with acetylcholine can be due either to an increase in sodium permeability, or to chloride permeability in cells with a high internal

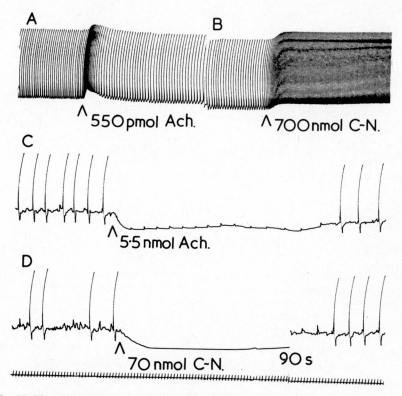

Fig. 17. The effects of acetylcholine and carnitine on the activity of two cells from *Helix aspersa*.

A, 8 mV depolarization to 550 pmol acetylcholine; B, 4 mV depolarization to 700 nmol carnitine nitrile; C, 12 mV hyperpolarization to 5.5 nmol acetylcholine; D, 16 mV hyperpolarization to 70 nmol carnitine nitrile. A, B are from the same cell, action potentials 90 mV amplitude. C, D are from the same cell, action potentials 76 mV amplitude. The time interval is in seconds. In record D there is a gap of 90 s.

concentration of chloride or to an increase in permeability to both ions. Examples of excitation and inhibition due to acetylcholine are shown in Fig. 17.

Evidence for inhibitory postsynaptic potentials mediated via the release of acetylcholine has been obtained by Kerkut and Thomas (1964) and for an excitatory postsynaptic potential by Walker *et al.* (1972).

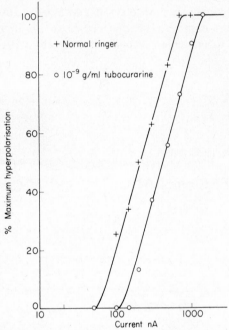

Fig. 18. The effect of tubocurarine on the acetylcholine dose–effect curve of a neurone from *Helix aspersa* which is hyperpolarized by acetylcholine. The acetylcholine was applied iontophoretically from an acetylcholine electrode. (From Newton, 1972.)

This cholinergic excitatory postsynaptic potential can be significantly reduced in size when *Helix* is pretreated with hemicholinium, a compound known to inhibit the synthesis of acetylcholine. Both the responses to acetylcholine and the cholinergic synaptic potentials can be reduced or completely blocked by pretreatment with cholinergic antagonists, for example, tubocurarine, atropine, strychnine and gallamine (Gerschenfeld and Stefani, 1966; Ralph, 1970; Newton, 1972). Figure 18 shows that tubocurarine competitively antagonizes the inhibitory action of acetylcholine, the acetylcholine being applied iontophoretically. Extensive investigations using cholinergic blockers have been undertaken by Zeimal and Vulfius (1968), Vulfius and Zeimal (1968) and Ger *et al.* (1971) on *Lymnaea stagnalis* and *Planorbarius corneus.*

One difference appears to exist between the cholinergic receptors of *Lymnaea* and *Planorbarius* and those of *Helix*. In the former case atropine is a very weak antagonist while in the latter case atropine is normally equipotent with tubocurarine (Walker and Hedges 1967; Newton, 1972). Zeimal and Vulfius (1968) consider that the cholinergic receptor of *Lymnaea* and *Planorbarius* resembles that of the vertebrate somatic myo-neural juction. In general, blocking agents would appear to block both H and D responses equally. Hexamethonium is about 1000 times less potent than tubocurarine in shifting the dose-response curve to the right for acetylcholine (Newton, 1972). Hexamethonium only partially blocks the action of acetylcholine on D cells and recent experiments suggest that hexamethonium may block only the chloride component of the acetylcholine D response.

Attempts have been made to classify the acetylcholine receptors in terms of nicotinic and muscarinic receptors. A wide range of compounds which mimic the action of acetylcholine have similar effects to acetylcholine on pulmonate neurones (Walker and Hedges, 1968; Walker, 1967; Newton, 1972; Zeimal and Vulfius, 1968; Ger *et al.* 1971). Neurones respond to nicotine and this is specifically blocked by tubocurarine, atropine having no effect (Newton, 1972). A few cells respond to low doses of muscarine and would appear to have distinct muscarinic receptors (Walker and Hedges, 1968).

Two groups of compounds have been found to act only on H cells or are far more potent on H cells than on D cells. McN A-343 was found to act only on H cells and was approximately twenty times less potent than acetylcholine (Woodruff and Walker, 1971). Similarly carnitine nitrile is far more potent on H cells than on D cells (Walker *et al.*, 1972); Fig. 17 shows the effect of acetylcholine and carnitine nitrile on an H and on a D cell.

B. Dopamine

This is the major catecholamine present in pulmonates (Cardot, 1963; Sweeney, 1963; Kerkut *et al.* 1966) though recent work has demonstrated the presence of small amounts of noradrenaline (Osborne and Cottrell, 1970). Small amounts of a related compound, octopamine, have also recently been found in the pulmonate nervous system (Walker *et al.*, 1972). In almost all cases where dopamine has an action on a neurone, then this action is inhibitory (Fig. 19). Although there are a

H

few cells in the pulmonate nervous system that are excited by dopamine (Kiss and Salanki, 1971) it is necessary to check that the excitation is direct and not mediated via an interneurone. Often cells which are inhibited by dopamine are excited by both 5-HT and acetylcholine; the action of dopamine is due to an increase in permeability to potassium ions (Kerkut *et al.*, 1969).

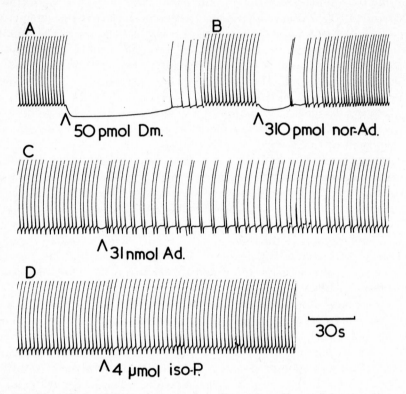

Fig. 19. The effects of dopamine, noradrenaline, adrenaline and isoprenaline on the spontaneous activity of a *Helix aspersa* neurone.
A, 20 mV hyperpolarization to 50 pmol dopamine; B, 10 mV hyperpolarization to 310 pmol noradrenaline; C, slowing of rate to 31 nmol adrenaline; D, no effect with 4 μmol isoprenaline. The action potentials were 80 mV amplitude.

There are at least two types of dopamine mediated inhibitory synaptic events; one is a normal short duration inhibitory postsynaptic potential (Walker *et al.*, 1971b), the other a long-lasting inhibition, termed ILD, that is, inhibition of long duration. Ergometrine has been found to be a potent blocking agent both of the synaptic events and of applied dopamine (Walker *et al.*, 1968) (Fig. 20). Beta blocking agents have no effect on the dopamine response while alpha blocking agents are considerably less potent than ergometrine. Dopamine is generally about 25 times more potent than noradrenaline and over 100 times more potent than adrenaline on pulmonate neurones (Woodruff and Walker, 1969) (Fig. 19). Structure–activity studies indicate that pulmonate neurones possess a distinct dopamine receptor rather than an alpha catecholamine receptor. The structural requirements for potent dopamine-like activity are as follows: (a) two hydroxyl groups on the 3 and 4 position of the benzene ring, (b) the presence of a terminal nitrogen on the two carbon side chain either unsubstituted or with one methyl group. Further substitution greatly reduces potency.

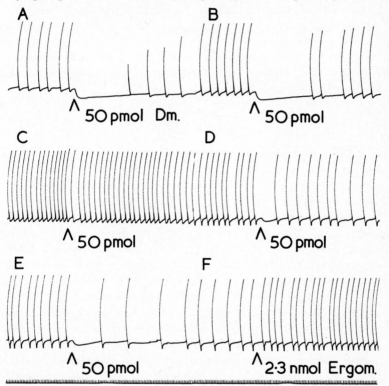

Fig. 20. The effect of ergometrine on the response to dopamine of a neurone from *Helix aspersa*.

A and B, 12 mV hyperpolarization to 50 pmol dopamine; C, 5 min following the addition of 23 pmol ergometrine, the response to 50 pmol dopamine is blocked; D, 45 min after the ergometrine, 8 mV hyperpolarization to 50 pmol dopamine; E, 60 min after the ergometrine, 8 mV hyperpolarization to 50 pmol dopamine; F, direct excitation of the cell in response to 2·3 nmol ergometrine. The time interval is in seconds. The action potentials were 85 mV in amplitude.

There are a few pulmonate neurones that are inhibited by noradrenaline but not by dopamine (Glaizner, 1968a). It is possible that these cells may possess noradrenaline synaptic potentials but these have not yet been identified. Glaizner (1968b) has found that the noradrenaline response is blocked by both alpha and beta blocking agents.

The effect of pretreatment of *Helix aspersa* with various compounds which alter the levels of brain dopamine has been investigated by Walker *et al.* (1971a,b). They found that when *Helix* is pretreated with DOPA 3,4-dihydroxyphenylalanine or the immediate precursor for dopamine, the amplitude of the dopamine inhibitory potentials was significantly increased; while pretreatment with reserpine, alpha-methyl-DOPA or alpha-methyl-*p*-tyrosine all significantly reduced the size of the inhibitory postsynaptic potential suggesting that the pre-synaptic stores of dopamine which are normally available for release had been depleted. It has already been shown that pretreatment with either reserpine or alpha-methyl-DOPA greatly reduced the dopamine level in the *Helix* brain (Kerkut *et al.*, 1966).

C. 5-Hydroxytryptamine (5-HT)

5-HT has been shown to be present in the nervous system of *Helix aspersa* (Kerkut and Cottrell, 1963) and of *Helix pomatia* (Cardot and Ripplinger, 1963; Dahl *et al.*, 1966; van Wilgenburg, 1970). The levels of 5-HT range between 3–14 µg/g wet weight of nervous tissue. Using the technique of fluorescence microscopy the 5-HT has been localized in specific nerve cells (Kerkut *et al.*, 1967; Sedden *et al.*, 1968; Dahl *et al.*, 1966; Marsden and Kerkut, 1970; Sakharov and Salanki, 1969a,b; Osborne and Cottrell, 1971). The levels of 5-HT can be depleted by pretreatment with reserpine and *p*-chlorophenylalanine and increased by pretreatment with 5-hydroxytryptophan (Marsden, 1970).

5-HT has both excitatory and inhibitory actions on snail neurones (Fig. 21). The inhibitory effects are mediated via either an increase in chloride influx or in potassium efflux (Gerschenfeld, 1971). The excitatory effect is due to an increase in sodium influx. The structure–activity relationships of the excitatory 5-HT receptor have been analysed by Walker and Woodruff (1971, 1972) and are as follows: a. the presence of an indole nucleus; b. a hydroxyl group on the five position of the indole ring; and c. a substitution on either the terminal nitrogen or the alpha carbon of the side chain no greater than one methyl group.

There is good evidence for 5-HT as an excitatory transmitter in the pulmonate central nervous system (Gerschenfeld and Stefani, 1966; Cottrell, 1970, 1971; Walker *et al.*, 1971, a, b, 1972). Cottrell has shown

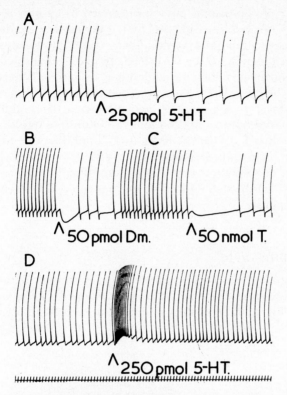

Fig. 21. The effect of 5-HT, dopamine and tryptamine on the activity of two neurones from *Helix aspersa*.
A, 5 mV hyperpolarization to 25 pmol 5-HT; B, 14 mV hyperpolarization to 50 pmol dopamine; C, 6 mV hyperpolarization to 50 nmol tryptamine; D, 8 mV depolarization to 250 pmol 5-HT.
A, B and C are from the same cell and D is from a second cell. Cell D is also inhibited by dopamine and tryptamine. In A, B and C the time trace is in intervals of 2 s. In D the time trace is in intervals of 1 s. The action potentials are 90 mV amplitude in A, B and C while in trace D they are 85 mV.

that when the 5-HT containing cells in the cerebral ganglion of *Helix* are stimulated excitatory post-synaptic potentials can be recorded in cells in the buccal ganglion. Pretreatment with reserpine reduced these synaptic potentials. Walker *et al.*, (1971a,b, 1972) also observed that an excitatory post-synaptic potential due to presynaptic release of 5-HT was reduced in amplitude and fatigued more easily following pretreatment with either reserpine or *p*-chlorophenylalanine. Pretreatment with 5-HTP appeared to increase the presynaptic stores of 5-HT.

A number of compounds have been shown to antagonize the action of 5-HT, these include LSD-25, BOL-148, methylsergide and ergometrine (Gerschenfeld and Stefani 1966; Ralph 1971). However great care must be taken in testing the specificity of blocking agents as several of them also block dopamine and other receptors (Woodruff *et al.*, 1971). Gerschenfeld (1971) has attempted to distinguish between 5-HT receptors which cause excitation and inhibition. He found that tubocurarine blocked both excitation and the inhibition due to an increase in chloride influx but failed to block 5-HT inhibition due to an increase in potassium efflux. Prostigmine had no effect on either 5-HT induced excitation or on inhibition due to potassium efflux but blocked inhibition due to chloride influx. LSD-25, tryptamine and 5-HT blocked all three types of response.

D. Glutamic acid

This compound has both excitatory and inhibitory actions on snail neurones (Kerkut and Walker, 1961; Gerschenfeld and Lasansky, 1964) (Fig. 22). The excitatory effect is due to an increase in permeability to

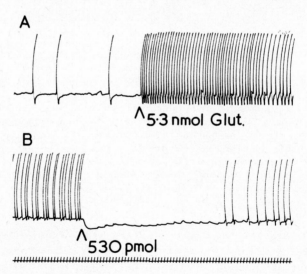

Fig. 22. The effect of glutamic acid on the activity of two neurones from *Helix aspersa*.
A, 4 mV depolarization to 5·3 nmol glutamic acid; B, 8 mV hyperpolarization to 530 pmol glutamic acid.
The action potential amplitude in A is 85 mV and in B it is 80 mV. The time scale is in intervals of one second.

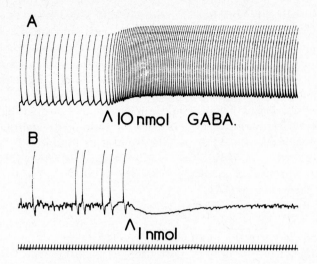

Fig. 23. The effect of gamma-aminobutyric acid (GABA) on the activity of two
neurones from *Helix aspersa*.
A, 6 mV depolarization to 10 nmol GABA; B, 10 mV hyperpolarization to 1
nmol GABA.
The action potential amplitude in A is 80 mV and in B it is 75 mV. The time
scale is in intervals of one second.

sodium ions while the inhibitory effect is due to an increase in permeability to both chloride and potassium ions (Kerkut *et al.*, 1969). The action of glutamic acid is mimicked by aspartic acid and cysteic acid (Gerschenfeld and Lasansky, 1964) and by ibotenic acid (Walker *et al.*, 1971). The inhibitory action of ibotenic acid is also due to an equal increase in permeability to both chloride and potassium ions. There are inhibitory synaptic events in pulmonates which are associated with an increase in permeability to both chloride and potassium ions and it is possible that glutamate is the transmitter agent at these sites.

E. Gamma-aminobutyric acid (GABA)

As in the case of glutamic acid, GABA has both excitatory and inhibitory actions on pulmonate neurones (Kerkut and Walker, 1961; Gerschenfeld and Lasansky, 1964) (Fig. 23). The action of GABA is mimicked by muscimol (Walker, *et al.*, 1971). In general beta-alanine is an inhibitory agent on cells which are inhibited by GABA but on certain cells which are excited by GABA, beta-alanine causes the membrane potential to hyperpolarize (Walker *et al.*, 1972). 3-Aminobutyric acid and 4-amino-3-hydroxybutyric acid mimic GABA on cells which respond to GABA but are considerably less active. Taurine and glycine either have no effect or are very weak antagonists of GABA. Glycine, in high doses, has a non-specific excitatory action on some neurones. No potent agent has yet been found which will antagonize the action of GABA on pulmonate neurones though bicuculline will inconsistently block both excitation and inhibition due to GABA (Walker *et al.*, 1971). There would appear to be considerable hetero-geneity between amino acid receptors of pulmonate neurones. Cells which are excited by GABA may be excited or inhibited by glutamic acid and likewise cells inhibited by GABA may be either inhibited or excited by glutamic acid. Some cells which are sensitive to glutamic acid are insensitive to GABA.

GABA has been shown to be present in the nervous system of *Helix* (Osborne, 1971; Briel *et al.*, 1971) and so is a potential candidate as a transmitter agent in the pulmonate nervous system.

In summary it can be said that there is good evidence for acetyl-choline as both an excitatory and inhibitory transmitter agent between neurones in the pulmonate central nervous system. 5-HT has clearly been shown to be an excitatory transmitter while dopamine has been shown to be an inhibitory central transmitter agent. So far the evidence for amino acids as central transmitter agents in the pulmonates is inconclusive and further investigations are required. Structure–activity studies using agonists and studies using blocking agents have indicated similarities between pulmonate neurone receptors and receptors present on neurones from other groups of animals. However great care must be taken in checking the specificity of blocking agents before they can be used as pharmacological tools in the identification of synaptic trans-mitter agents in the pulmonates.

VII. *Otala lactea*

The Moroccan snail *Otala lactea* (Müll.) has a central nervous system very similar to that of *Helix*. Similar identifiable neurones can be found in the brain of both species (Gainer, 1972a).

In a study of protein synthesis in cell 11 in the right parietal ganglion (equivalent to the giant right parietal cell F1 in *Helix*) it was found that this cell synthesized low molecular weight proteins (5000 daltons) that were not present in other cells. These proteins were associated with the neurosecretory activity of cell 11. Under dry conditions at 21°C *Otala* went into diapause (Gainer 1972b). The low molecular weight proteins in cell 11 were no longer present. There was a 3-fold decrease in the

Fig. 24. Location of specific cell bodies in the central nervous system of *Lymnaea*.
A, Dorsal view of the central nervous system with known nerve cells numbered.
B, Dorsal view of the central nervous system showing large nerve cells. C, Ventral
view of the central nervous system.
(A, After Salanki and Kiss; B and C from Kostyuk, 1967.)

membrane resistance and a slight increase in membrane potential of
cell 11. The cell also lost its normal endogenous pace-maker activity in
diapause.

In the active snail, cell 11 shows a bursting spike pattern with
between 2–20 impulses/burst and a variable interburst interval from

3 to 15 s. There were no EPSP in the cell but IPSP could be driven by stimulation of the peripheral nerve trunks. The pace-maker activity of the cell was dependent upon the sodium ions in the solution. The axon from cell 11 runs along the visceral nerve and terminates in the auricle of the heart. It is suggested that the neurohaemal location of the nerve terminal is specifically related to the cell's secretory activity (Gainer, 1972c).

VIII. *Lymnaea stagnalis*

This aquatic pulmonate has been extensively studied by the Hungarian neurobiologists and they have made a very valuable contribution to our knowledge of the organization of its nervous system.

Gubicza (1970) counted the number of nerve cells in the brain of *Lymnaea stagnalis*. Animals weighing between 14 and 16 g contained between 13 000 and 16 000 nerve cells in their circumoesophageal ganglia. The majority of the large nerve cell bodies were located in the caudal zone of the buccal and cerebral ganglia, and in the oral part of the left parietal ganglion. In other ganglia the large neurones were distributed equally throughout the ganglia. In general most of the paired ganglia contained an equal number of cells in each ganglion of the pair. The main exception was the parietal ganglia where the right parietal ganglion contained three times as many cells as the left parietal ganglion.

It is suggested that the right parietal ganglion may have fused with the right abdominal ganglion to form what we now call the right parietal ganglion. This would explain the larger number of cells in the right parietal ganglion and also some of the cell homologies between the right parietal and the abdominal ganglion It would be more correct to call the right parietal ganglion the right parieto-abdominal ganglion. The abdominal ganglion has the largest number of giant cells, between 4% and 5% of the total number of neurones were larger than 100μm.

Gubicza and S-Rosza (1969) identified the cell bodies in the *Lymnaea* ganglia that have axons emerging in the intestinal nerve trunk. They cut the intestinal nerve and studied the change in staining properties of the neurones in the CNS to malachite green and pyronine. 46 neurones in the brain sent axons to the intestinal nerve. The cerebral ganglia had the largest share (15 cells) then the abdominal ganglion (13 cells) and finally the parietal and pedal ganglia (9 cells each). Gubicza and S-Rosza (1970) studied the connections between the other peripheral nerve trunks and the nerve cell bodies in the CNS by means of the histo-chemical change following denervation.

Salanki and Kiss (1969) studied the electrophysiology of specific neurones in the brain of *Lymnaea* and identified 20 large cells in terms of their topographical position, their electrical activity and their membrane potentials. They found neurones in the abdominal ganglion which were homologous in their electrical activity to neurones in the right parietal ganglion. Kiss and Salanki (1971) mapped the pharma-cological specificity of some of the large neurones in the CNS of *Lymnaea*.

Sakharov and Zs-Nagy (1968) localized two neurones in the cerebral ganglia of *Lymnaea* that fluoresced under u.v. light after treatment with paraformaldehyde vapour. These two neurones probably contain 5-HT. Sakharov (1970) pointed out that these two cells are probably homologous with the pair of giant nerve cells in the brains of *Helix*, *Agriolimax*, and the nudibranch *Dendronotus*. Similarly nerve cells containing dopamine can be identified histochemically from species to species, from pulmonates to opisthobranchs and prosobranchs.

IX. *Archachatina marginata*

The central nervous system and the musculature of this large land snail *Archachatina (Calachatina) marginata* have been studied by Nisbet and his colleagues (Nisbet, 1956, 1960, 1961a,b; Baxter and Nisbet, 1963; Amoroso et al., 1964; Nisbet and Plummer, 1968, 1969). Figure 25 shows a view of the cerebral pedal and visceral ganglia of *Archacha-tina* seen from the right side. The pedal ganglia appear larger than the nerve mass actually is, due to the investment of the ganglia by a mass of tough fibrous tissue. The central nervous system of *Archachatina* is less concentrated than that of *Helix*. Figure 25B is a diagrammatic horizontal section through the cerebral, pedal and pleurovisceral ganglia indicating the relative sizes of the nerve cells and the arrangement of the neuropile and tracts. There is a progressive increase in cell size from the small 8–10μm cells of the cerebral ganglia to the larger cells 300–400μm in the parietal and abdominal ganglia.

Details of the fine structure of the nerve cells (see Figs. 5, 6) are given in Baxter and Nisbet (1963) and Amoroso et al. (1964).

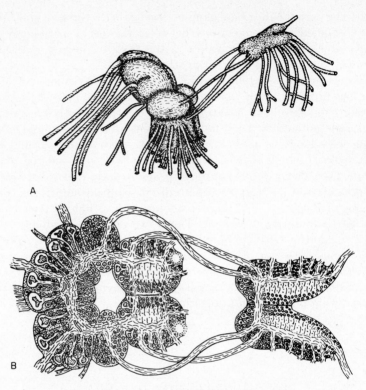

Fig. 25. Brain of *Archachatina*.

A, View from the right side; the cerebral ganglia are to the right and the sub-oesophageal complex to the left. B, Diagrammatic horizontal section through the central nervous system. Note the smaller cells in the cerebral ganglia and the larger cells in the parietal and abdimal ganglia. Note also the double cerebral–suboesophageal connective. (From Nisbet, 1961a.)

X. *Onchidium verruculatum*

The nervous system of this marine pulmonate *Onchidium verrucula-tum* Cuvier has been studied by Japanese neurophysiologists (Hagiwara and Saito, 1959; Oomura *et al.*, 1962).

Katayama (1970) studied the neuronal connections in the ganglia. The ganglionic complex consists of seven ganglia, five of which are easily visible from the dorsal side. Up to 30 large nerve cell bodies are easily distinguished in these ganglia, the largest cell being 300µm in diameter. The pleural and parietal ganglia tend to merge together and for this reason they are described as the pleuroparietal ganglia. By means of stimulating and recording the activity in the cell body and nerve axons (using an averaging technique and a LINC 8) Katayama showed that neurone 1 in the left pleuroparietal ganglion sends branches through the left parietal nerve, the intestinal nerve and the genital nerve. It probably receives excitatory inputs from the left parietal nerve and the genital nerve.

Cell V1 in the visceral ganglion has no clear output to the peripheral nerve trunks but is excited through the left parietal nerve, the intestinal nerve, the left and right genital nerves.

Cell V5 in the visceral ganglion sends a branch out through the left parietal nerve, and the left genital nerve. It receives excitatory inputs from the intestinal nerve and the right and left genital nerves.

It is probable that some of the nerves in *Onchidium* have calcium as the main ion carrying the current (Oomura *et al.*, 1962).

XI. Behavioural studies

Habituation (the relatively permanent waning of a response as a result of repeated stimulation) has been observed in many pulmonates such as *Helix albolabris* Say (Humphrey, 1933), *Physa* (Dawson, 1911; Wells and Wells, 1971), *Lymnaea stagnalis* (Buytendijk, 1921).

Cook (1971) studied the withdrawal response of *Lymnaea stagnalis* to various stimuli (mechanical vibration, mechanical shock, 60W lamp being switched on, moving shadow). With repeated stimulation the amplitude and frequency of the response declined but the latency of the response remained at about 0·6 s. *Lymnaea* appeared to recover from habituation to 20 stimuli after a rest period of about five minutes. This relatively long lasting effect distinguishes habituation from fatigue.

Wells and Wells (1971) showed that the withdrawal response of the freshwater pulmonate *Physa acuta* Drap. varied with mechanical shock and light. However it was not possible to show operant conditioning in the snails (i.e. the animals could not be trained to avoid a light/dark or smooth/rough boundary). Wells and Wells suggested that operant conditioning fails because the animals are unable to recognise some sequences of events in time as more important than other events. Many of the responses of *Physa* could be explained in terms of sensitization (Wells, 1965).

The balance between sensitization and habituation would allow the animals to achieve the level of response that is appropriate to the state of the world around the snail.

There has also been considerable interest in the electrophysiological changes at synaptic level that take place during habituation (Bruner and Tauc, 1966; Peretz, 1970; Castelluci *et al.*, 1970; Veprintsev and Rosanov, 1968).

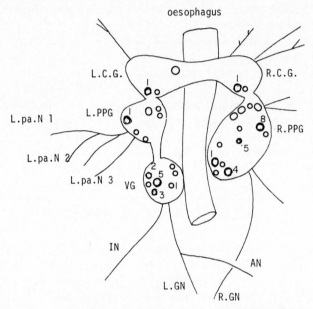

Fig. 26. Diagram of location of large cells in the brain of *Onchidium*. The parieto-pleural ganglia are fused together (PPG). AN, anal nerve; CG, cerebral ganglion; GN, genital nerve; IN, intestinal nerve; pa.N, pallial nerve; VG, visceral ganglion. L, left; R, right. (After Katayama, 1970.)

A. Chemical changes in the CNS

Emson *et al.* (1971) set up two *Helix aspersa* so that when one snail, the experimental animal, put out its tentacle, it received an electric shock. The second snail (control animal) received a shock every time that the first animal received a shock, regardless of the position of the control animal's tentacle. The first snail learned to keep its tentacle in whilst the control animal did not.

Figure 27 shows the result of a training experiment. The experimental animal (shocked when it put out its tentacle) learned in about 20 min to keep its tentacle in. When put on retest, the experimental animal required relatively few shocks to keep its tentacle in. The control animal when put on retest, required a large number of shocks before it could learn to keep its tentacle in, even though it had already received as many shocks as the experimental animal.

The study suggests that *Helix aspersa* can associate tentacular position with shock. Injection of drugs such as cycloheximide, congo red, or acridine orange (Fig. 28) all made the animals learn more slowly and require more shocks. The effect was dose-dependent; the more drug the animal had injected into its body cavity one hour before testing, the greater the effect. Injection of pemoline or amphetamine made the animals learn more quickly. There were also differences in the extent of protein synthesis and rate of incorporation of labelled uridine into the

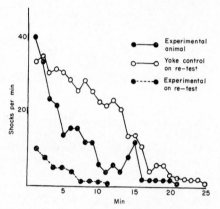

Fig. 27. Learning curve for *Helix aspersa*. The experimental animal (●) learns to associate its tentacle position with shock. On retesting (●- -●- -●- -●) it requires a few shocks compared to the yoked control on retest (○——○——○). (From Emson, Walker and Kerkut, 1971.)

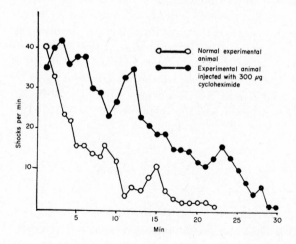

Fig. 28 (a)

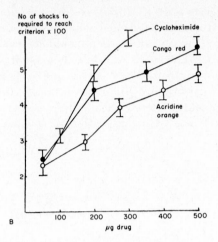

Fig. 28. The effect of drugs on learning in *Helix aspersa*
A, Comparison between learning rate of saline injected animal (O-O-O) and
cycloheximide injected animal (●-●-●). The normal animal learns more quickly
and requires less shocks. (From Emson and Kerkut.) B, Effect of drug injected on
the number of shocks required by snail to learn tentacle retraction. The more
drug is injected, the more shocks the animals require before they can learn. (From
Emson, *et al.*, 1971.)

brains of the animals, it being greater in the experimental animals than
in the control.

There was also a change in the activity of the brain acetylcholine
esterase (AChE). Experimental snails had a higher activity of AChE in
the brains than did the controls. If the animals were left for 24 h. they
forgot their training and the activity of the AChE returned to the initial
level. The enzyme had a different K_m in the experimental animals'
brain ($1{\cdot}33 \times 10^{-4}$M) to that in the control ($5{\cdot}88 \times 10^{-5}$M). It is suggested that ACh is mainly an inhibitory transmitter in the snail brain
and that the increased AChE activity would facilitate the breakdown
of ACh and hence increase the response (Kerkut, 1973).

XII. Vision

A. *The structure of the pulmonate eye*

There have been two recent papers dealing with the structure of pulmonate eyes, that of Newell and Newell (1968) dealing with the eye of

the slug *Agriolimax reticulatus*, and that of Eakin and Brandenburger (1967) on the differentiation of the eye of *Helix aspersa*. In addition there are two further papers on the structure of the eye of *Helix pomatia*, (Rohlich and Török, 1963; Schwalbach *et al.*, 1963). For this reason the description of the structure of the pulmonate eye will be mainly confined to *Helix* and *Agriolimax*.

B. Structure of the eye of Helix

Eakin and Brandenburger (1967) have studied the development of the eye of *Helix aspersa* at various stages of morphogenesis from the initial invagination of the embryonic ectoderm to form an eyecup to the fully differentiated organ in the adult.

The formation of an optic vesicle occurs about day 8 of development by invagination of the ectoderm covering the rudiment of a posterior tentacle. Examination of the tentacular rudiment reveals a V-shaped invagination (Fig. 29A). A day later the optic vesicle is beginning to form from the base of the infolding of the ectoderm on one side of the

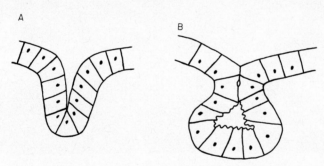

Fig. 29. A. Diagram of invagination of optic ectoderm on side of tentacle anlage of 8-day old embryo of *Helix*. B. Diagram of early optic vesicle of 9-day embryo still connected to tentacular ectoderm. (From Eakin and Brandenburger, 1967.)

tentacular rudiment (Fig. 29B). The base of the former furrow becomes expanded into a small sac, the primordial optic vesicle. The cells are wedgelike and have their nuclei situated at the broad inner ends of the cells. The narrower outer ends, bordering the cavity, bear villi that are longer and more numerous than those in the preceding stage. At this stage presumptive corneal cells contain clusters of dense granules. By the 10th to 12th day the wall of the vesicle is a single layer of columnar cells joined together at their luminal ends by terminal bars (zonulae adhaerentes). Several irregular microvilli, about 2·5 μm in length, project from the luminal border of each cell, which has at least one cilium,

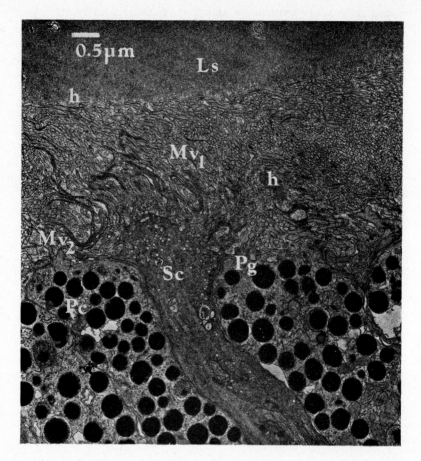

Fig. 30. Distal end of sensory cell (Sc) flanked by two pigment cells (Pc) of 14 day embryo of *Helix*. Microvilli (Mv$_1$) of sensory cell extend to under surface of lens (Ls). Humour (h), short microvilli (Mv$_1$) of pigment cells, mature pigment granule (Pg). $\times$ 15 000. (From Eakin and Brandenburger, 1967.)

though bearing no relationship to the microvilli. Eakin and Brandenburger consider these cilia to be incidental and have no significance so far as photoreception is concerned. They occur because the eye is formed from ciliated embryonic ectoderm. These authors regard the photoreceptors as a rhabdomeric type, that is, an array of microvilli formed directly from the cell membrane independently of cilia, although in another pulmonate, *Onchidium verruculatum*, there is good evidence that in the dorsal eye, the light sensitive organelles are formed from cilia (Yanase and Sakamoto, 1965). The lens at this stage is a prominent dense body which is formed by the aggregation of luminal material. This coalescence may be initiated at several foci and as these centres enlarge they touch and fuse. It is suggested that the luminal material from which the lens is derived originates from highly electron-dense granules present in both presumptive corneal and retinal cells of the optic vesicle. Both the lens and the microvilli of the surrounding cells are bathed by a humour which can be distinguished from the lens itself.

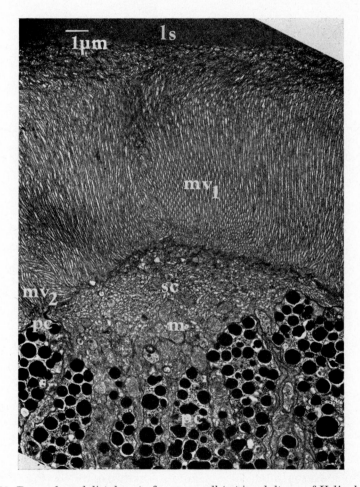

Fig. 31. Dome-shaped distal part of sensory cell (sc) in adult eye of *Helix* showing photoreceptoral microvilli (mv$_1$) extending to the lens (ls). Pigment cells (pc) flank sensory cell and send extensions into supranuclear region of receptor cell subdividing it into vertical columns in which lie many mitochondria (m); short microvilli (mv$_2$) of pigment cell. × 5400. (From Eakin and Brandenburger, 1967.)

By the end of 14 days two types of cell can be observed in the retina, sensory and pigmented cells (Fig. 30). The microvilli which extend from the outer surfaces of the receptor cells are about 4 μm long and 0·09 μm in diameter. Distally the microvilli abut the under surface of the lens and arise basally from the dome-shaped outer end or head of the sensory

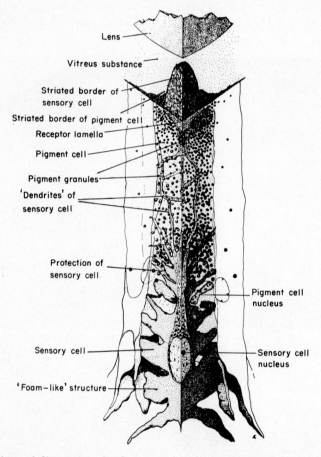

Fig. 32. A semi-diagrammatic electron microscope reconstruction of a section through the retina of *Helix pomatia* to show the arrangement of the sensory and pigment cells. (From Schwalbach *et al.*, 1963.)

cell which overlies adjacent pigment cells like an umbrella (Fig. 32). Pigment granules are abundant in the supportive cells but absent from the sensory cells. The origin of these granules is difficult to determine but it would appear that they can be formed from areas of aggregation which need not be limited to the Golgi region of the cell.

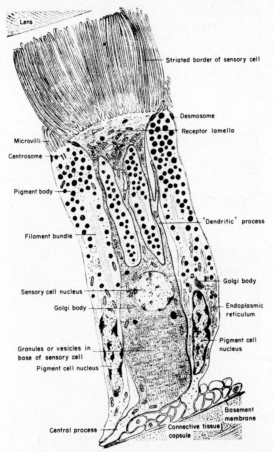

Fig. 33. Diagram of the fine structure of the retina of *Helix pomatia*. A sensory cell is shown surrounded by a pigment cell. The peripheral processes of the sensory cell join to form a receptor lamella from which projects the striated border. The central process arises from the base of the sensory cell. (From Rohlich and Török, 1963.)

The secretory granules which were present in both cell types now disappear from the sensory cells. The base of each receptor cell tapers into a neurite, and the fine structure of the adult *Helix* eye can be seen in Fig. 32. The head or distal part of the sensory cell is dome-shaped (Fig. 31) with its sides overlapping the neigbouring pigmented supporting cells. A corona of long, slender microvilli, about 10 μm long, extends from the convex surface of the receptor cell, their tips touching the lens. The distal end of the receptor cell is filled with mitochondria, microtubules and vesicles. The sensory cell below its apex consists of several columns (Figs 31 and 32) which are rich in mitochondria. The free outer ends of the pigmented cells possess short microvilli. The distal halves of these cells are packed with membrane-bound spheroid pigmented granules, $0.5 - 1.0 \mu$m diameter, which constitute an inner pigmented zone of the retina.

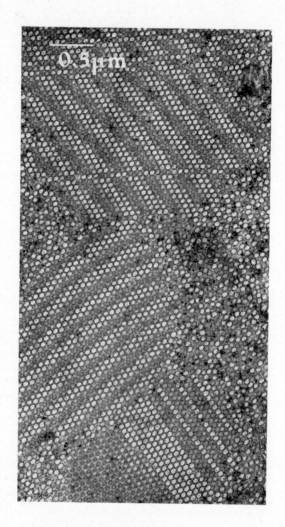

Fig. 34. Paracrystalline arrangement of small clear vesicles in nuclear region of sensory cell of adult *Helix*. Note the dense granules among vesicles. $\times$ 22 400. (From Eakin and Brandenburger, 1967.)

The main feature of the basal part of the sensory cell is a massive body composed of tightly packed small vesicles (85nm) (Figs 33 and 34). Eakin suggests that these vesicles are neurotransmitters produced by the Golgi apparatus of the sensory cells, and that a careful study using light and dark adapted snails might be useful in testing this hypothesis. The neurites in the adult retina and optic nerve show considerable variation in the distribution of organelles and inclusions but Eakin and Brandenburger were unable to determine whether or not there are several types of nerve fibres, though they concluded that there was only one type of vesicle in the optic nerve varying in size from 80 – 130 nm and in appearance from clear to having a dense core. These differences may reflect the physiological state of the animal. These authors looked for the presence of ganglion cells since there is conflicting opinion as to whether they are present or not at the base of the eye, that is, whether the neurites of sensory cells synapse peripherally with ganglion or bipolar neurones or pass directly to the cerebral ganglia. Both Smith (1906) working on *Limax maximus*, *Helix pomatia* and *Planorbis trivolvis* Say and Hesse (1908) working with *Helix pomatia* could find no evidence for second order neurones, but Eisenmann (1920) claimed to have found ganglion cells in the eyes of *Arion* and *Viviparus*. Recently Newell (1965) working not with a pulmonate but with a prosobranch, *Littorina littorea* (L.), has reported the presence of bipolar neurones, one being associated with each visual cell.

Eakin and Brandenburger postulate that the eye of *Helix* is capable of as good resolution as possible in an eye of its size. Their evidence is as follows: the presence of a dioptric apparatus which presumably can focus an image upon the retina, about 4000 slender sensory cells surrounded by pigment-bearing supportive cells, and a direct pathway from each receptor cell to the cerebral ganglia. They suggest that the relatively poor vision of *Helix* compared to cephalopods is largely owing to the small size of its pupil which limits the amount of light entering the eye and the maximal obtainable resolving power. Some details concerning the structure of the cornea and lens of *Helix* have been described by Eakin and Westfall (1964).

The cornea of the adult *Helix aspersa* has a thickness of approximately 57 µm and consists of three distinct layers. The outer stratum is a one-cell thick (10 µm) layer of highly interdigitated epithelial cells, the distal surfaces of which are studded with long, slender microvilli (1·5 µm long and 0·2 µm diameter). The tips of the villi are embedded in a cuticle or layer of jelly. The second stratum is a narrow (7µm) light layer of fibrils the orientation of which is mainly horizontal. Eakin suggests these fibrils may be collagenous in nature. Within this light matrix are denser irregular masses of dark granules, fibres, vesicles, endoplasmic reticulum and mitochondria. The innermost layer is a broad (40 μm) layer of columnar cells. These cells are highly interdigitated along their lateral borders. The cytoplasm is packed with granules. The lower border of the cells possess many long narrow microvilli, similar to those of the outer layer, embedded within the light humour that surrounds the lens.

The lens of the adult eye is a large sphere, about 150 μm in diameter, composed of fine granular, moderately electron-dense material. The position of the lens within the eye is such that it protrudes slightly through the pupil. The exposed part of the lens presents a smooth even arc. The lens is separated from the microvilli of the lower layer of the cornea and from the microvilli of the retinal photoreceptors by a narrow space filled with a fine granular electron-lucent substance.

Recently, Eakin and Brandenburger have carried out a study to try to determine the function of the small 80 nm vesicles which are tightly packed into paracrystalline masses in the basal, perinuclear parts of the photosensory cells. It is possible that these vesicles may transport the photopigment or its precursors from the synthetic centres at the base of the sensory cells to the apical microvilli, and if these vesicles contain a photopigment, then it is likely to be a protein–vitamin A complex. Eakin and his group therefore investigated the presence of vitamin A in the eye and studied the uptake of injected labelled vitamin A into the sensory cells. Using thin layer chromatography they were able to demonstrate the presence of vitamin A in the eye of *Helix aspersa* (unpublished observation quoted in Brandenburger and Eakin, 1970). Vitamin A was found to be selectively taken up by the sensory cells. The next problem was to try to identify the intracellular region of the sensory cells which first selectively incorporated vitamin A and then to determine the pathway of transport from the initial site of uptake to other intracellular regions.

Brandenburger and Eakin found that the first selective incorporation of vitamin A occurs in the synthetic centres of the sensory cells. These are the cytoplasmic areas surrounding the nucleus and adjacent to the large paracrystalline masses of vesicles (region I in Fig. 35). This region contains large numbers of organelles, e.g. rough and smooth endoplasmic reticulum, bound and free ribosomes, and Golgi cisternae and vesicles. It is suggested that in region I, the vitamin A is conjugated with protein and enclosed (or sequestered) in vesicles. The vesicles are then transported into the sensory cell columns (region II) to the apical parts of the cell where they are broken down and the photopigment is incorporated in the microvilli (region IV). Newly formed vesicles instead of moving into the columns may become aggregated into paracrystalline

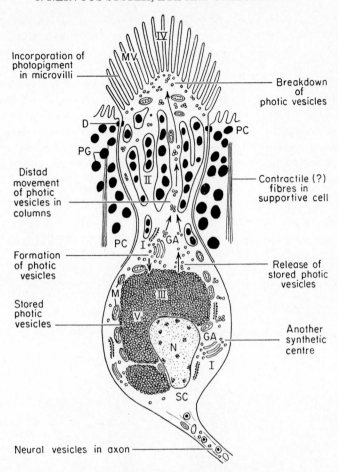

Incorporation of photopigment in microvilli

Breakdown of photic vesicles

Distad movement of photic vesicles in columns

Contractile (?) fibres in supportive cell

Formation of photic vesicles

Release of stored photic vesicles

Stored photic vesicles

Another synthetic centre

Neural vesicles in axon

Fig. 35. Diagram of regions I—IV in a receptor cell and the pathway, indicated by arrows, of transport of photic vesicles possibly containing photopigment from the Golgi body (GA) to microvilli (MV) or alternatively to storage in large masses (V); D, desmosome; M, mitochondria; N, nucleus; PC, supportive pigment cell; PG, pigment granules; SC sensory cell. (From Brandenburger and Eakin, 1970.)

masses which can be regarded as storage centres, region III. Vesicles may leave the storage centres and migrate to the microvilli as required.

Brandenburger and Eakin speculate as to the method of transport of the vesicles to the microvilli. They suggest that there may be a massaging action of the sensory cell columns by the interdigitating folds of the supportive cells. Fine filaments arranged in lengthwise bundles in the supportive cells may be contractile.

Another possibility might be that the 800 Å vesicles migrate into and along the axons of sensory cells to serve as neurotransmitters. Vesicles are found in the neurites and some do resemble those found in the paracrystalline masses, however these authors do not think that these vesicles contain neurotransmitter substances. The vesicles observed in the axons are very variable in size and appearance. These axon vesicles do not occur in clusters, and many of them have dark centres. In contrast the photic vesicles are very uniform in size, never possess dark centres and show a tendency to clump. In addition, the neurites do not selectively take up labelled vitamin A. Thus these authors favour a photic function rather than a neural one for these 800 Å vesicles.

C. Structure of the eye of Agriolimax reticulatus

The eye of *Agriolimax* is a closed vesicle, ovoid in shape, with mean outer dimensions of 140 μm and 180 μm (along the optic axis). The wall

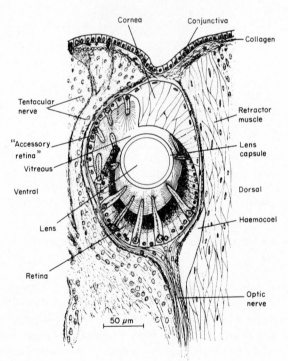

Fig. 36. A diagrammatic longitudinal section through the eye of *Agriolimax* to show its main features. (From Newell and Newell, 1968.)

of this vesicle is made up of a single layer of cells (Fig. 36). At the front surface the cells are large, transparent and have a high refractive index of 1·4. These cells provide the first refractive surface and make up the cornea and refract the light received through the conjunctiva and overlying collagen. In the posterior part of the eye the cells are of two types, sensory cells and pigment cells, their bodies forming the retina which is about 45 μm thick. The sensory cells are large columnar cells which are relatively few in number. These cells are easily recognizable since each at its apex has a projection, the rod (Figs 37 and 38). These outer seg-

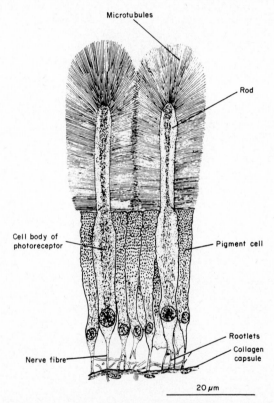

Fig. 37. A vertical section through the retina of *Agriolimax* as seen with the light microscope. (From Newell and Newell, 1968.)

ments or rods have a large number of microtubules or microvilli which project 15–25 μm beyond the ends of the cells. This radial arrangement of the microvilli is in contrast to those already described for *Helix* which arise from the ends of the sensory cells and lie parallel to the direction of the incident light. In *Agriolimax* most of the tubules are at right angles to the light and in this respect resemble the eyes of cephalopods and arthropods. The microvilli are filled with a substance which has an irregular granular appearance in electron micrographs. The central part of each rod is rich in mitochondria and has a well developed endoplasmic reticulum. The regular shape of these sensory cells contrasts with the dendritic arrangement seen in *Helix* (Fig. 32).

As in *Helix* there is no evidence for ciliary structures associated with the sensory cells. From the base of the sensory cell arises a nerve fibre (Figs 37 and 38) which runs in a radial fashion under the connective

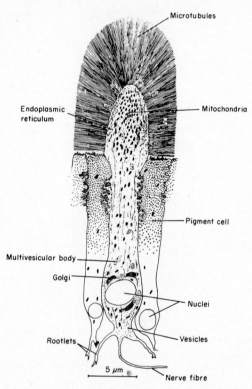

Fig. 38. A diagram to show the general structure of the retinal cells of *Agriolimax* as seen in electron micrographs. The rod of the photoreceptor is proportionately shorter than those shown in the previous light microscope figure. (From Newell and Newell, 1968.)

tissue capsule of the eye before piercing it to contribute to the optic nerve which runs to the cerebral ganglion, there being no evidence for an optic ganglion. The pigment cells are more numerous than the sensory cells. Each pigment cell is packed with spherical granules of dark brown melanin throughout its inner half. Inside the eye lies an approximately spherical lens with a diameter of about 65 μm and a focal length of 100 μm, thus having a radius of curvature of 35 μm and a ratio of focal length/radius of curvature of 2·8. Between the lens and the retina is a layer of vitreous humour into which the sensory rods project.

The dioptric system consists of the transparent cornea formed by the large cells at the front of the eye vesicle (Fig. 36). Taking the refractive index as 1·44 and the focal length of the lens in saline as 100 μm it is possible to construct the approximate path of light rays entering the eye (Fig. 39). From the diagram it can be seen that light from distant

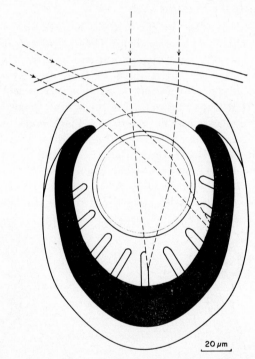

20 μm

Fig. 39. A scale drawing of a longitudinal section through the eye of *Agriolimax* with rays drawn to show their convergence on the retina. (From Newell and Newell, 1968.)

objects can be brought to a focus on the retina. However Newell and Newell consider the eye to be unsuitable for form vision, as the sensory cells are few in number and at the back of the eye are 20 μm apart so that the rods subtend an angle of about 15° at the centre of the lens which indicates the low visual acuity. The fringe of microvilli presents a large surface over which visual pigment is spread and through which light must pass. Since the microvilli of neighbouring sensory cells over-lap, the image will be blurred. Newell and Newell conclude that the eye of *Agriolimax* is structurally adapted for detecting changes in light intensity only and for operating at night. The cornea and lens would function to concentrate light on the retina while the few sensory cells, each with a large area provided by their microvilli largely at right angles to the incident light, would also favour responses at low light intensity. This type of eye would be adequate to perform the simple orientating behaviour associated with *Agriolimax*.

D. *The accessory retina*

A structure absent from the eye of *Helix* is the accessory retina (Fig. 36). Since the accessory retina lies below the optical axis of the eye when the tentacle is extended, light reaching it will not pass through the lens and so will not be concentrated on the sensory cells. It seems unlikely that it can play any part in vision, so what is its function? Newell and Newell suggest that it may be an infrared receptor. It has been observed that slugs will turn and crawl away from a black body heat emitter. If the ends of the tentacles are removed, then slugs lose the ability to avoid a radiant heat source though they continue to crawl normally. Newell and Newell have observed that when a slug is crawling, the tip of first one and then the other tentacle is temporarily withdrawn and a study of sections of a tentacle shows that this has the effect of rotating

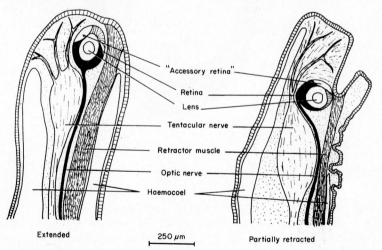

"Accessory retina"

Retina

Lens

Tentacular nerve

Retractor muscle

Optic nerve

Haemocoel

Extended   250 μm   Partially retracted

Fig. 40. Diagrams to show the change in orientation of the accessory retina of *Agriolimax* as the eye is withdrawn into the tentacle. (From Newell and Newell, 1968.)

the eye so that the accessory retina becomes displaced to lie under the outer surface of the tentacle at the base of a pit, thus placing it nearer to an outside source of radiation than the main retina (Fig. 40).

E. Phototropic experiments on whole animals

Wheeler (1921) undertook a series of experiments to determine whether the eyes of *Helix* were functional or not, as earlier work was contradictory on this point. Willem (1892) found that *Helix aspersa* was negatively phototropic while Yung (1911) concluded from his experiments that *Helix pomatia* was totally blind and completely indifferent to light. It is possible that a species variation could exist but Willem (1892) also tested *Helix pomatia* and found these animals to be also influenced by light and tended to gather in brightly illuminated areas. Wheeler found that *Helix aspersa* is negatively phototropic, at least at certain light intensities. The photoreceptors are in the eyes at the ends of the dorsal tentacles since on removal of the tentacles the snails were no longer negatively phototropic. He could find no evidence for skin photoreceptors. These eyes were sensitive only to light falling in a particular direction on the ends of the tentacles and he concluded that they were probably direction eyes only.

Hermann (1968) has recently reinvestigated the problem of visual guidance of snail locomotion using the land snail *Otala lactea*. The visual stimulus used was a slot of relative darkness in a band of uniform brightness encircling the snail. From his experiments Hermann demonstrated that a snail could pick out a 45° × 30° sector of darkness; blinding destroyed this ability. Hermann concluded that *Otala* can see with its eyes but the question of how much it could see remained unanswered.

F. Physiology of the eye

Recently three groups of workers have recorded electrical activity from pulmonate optic nerves and sensory receptors and their observations will now be discussed; Berg and Schneider (1967) worked on *Helix pomatia*, Gillary and Wolbarsht (1967) and Goldman and Hermann (1967) on *Otala lactea* and Gillary (1970) on *Helix aspersa*. Goldman and Hermann (1967) recorded extracellular action potentials from the optic nerve of *Otala lactea* in response to shining light of varying intensity through the lens. In about 30% of their preparations, a step change in white light on the lens resulted in a burst of action potentials within 1·5 s. Increasing the light intensity decreased the latency of the burst and a plot of latency against logarithm of intensity fits approximately to a straight line. When the light was turned off or its intensity reduced, there was no sign of a change in the steady-state firing rate, i.e. there was no "off" response. If the light was left on after the initial burst, the steady-state level of activity following the burst was greater than the steady-state level prior to illumination, this effect being greater at high intensities. In general the latency of response to light of a given intensity was constant in a preparation over several hours. These authors conclude that *Otala* responds to diffuse light with a latency and rise time which is sufficiently rapid to be of use in determining its quicker movements. While the latencies observed are consistent with the ability to perceive motion, the present work does not prove this ability.

Gillary and Wolbarsht (1967) also investigated the electrical responses from the eye of *Otala lactea*. They used an isolated eye-optic nerve preparation at 10°C which would remain viable for at least 40 h.

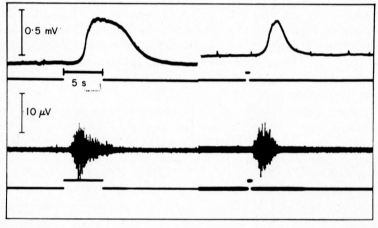

Fig. 41. Records of the electroretinogram (ERG) and spike activity from *Otala* evoked by light. The upper records were obtained by d.c. recording across the eye and the lower by a.c. recording from the optic nerve. The traces under the records indicate the light stimulation. The upper and lower records were not obtained simultaneously. (From Gillary and Wolbarsht, 1967.)

When the eye was stimulated with light, a slow cornea-negative electroretinogram (ERG) was evoked from the intact eye, together with spike activity in the optic nerve (Fig. 41). The amount of spike activity in the nerve was proportional to the amplitude of the ERG. During prolonged stimulation, after a fairly long latency, both the amplitude of the ERG and the intensity of spike activity rose rapidly to a maximum and then declined to a more steady level. When illumination ceased the ERG returned to the baseline level. There was no evidence for an "off" component in either the ERG or the spike activity which agrees with the observation of Goldman and Hermann (1967). The long latency before the onset of the ERG is typical of other photoreceptors, e.g. the caudal photoreceptor of the crayfish (Kennedy, 1958). The time course of the *Otala* ERG is similar to those recorded from cephalopod retinas (Tasaki *et al.*, 1963). The absence of an "off" response suggests the absence of inhibitory interaction between the receptors. "Off" responses are also absent from the optic nerve fibres of *Onchidium* (Dennis, quoted by Gillary and Wolbarsht, 1967).

Gillary and Wolbarsht quantitatively investigated the maximum amplitude of the ERG and the latency from the onset of stimulation to 50% maximum rise. Above threshold the response amplitude increased

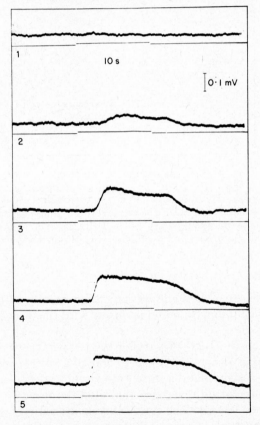

Fig. 42. Recordings from *Otala* taken during stimulation with different intensities of light. Five intensities, each a log unit greater than the preceding one, were tested consecutively on a single preparation. The lower trace of each record indicates the stimulus. (From Gillary and Wolbarsht, 1967.)

approximately in proportion to the log of the intensity (Fig. 42). The linear range varied between one to four log units. The latency was found to decrease in proportion to the log intensity (Fig. 43).

Illumination of the eye prior to a test stimulus reduced the response and decreased the latency as compared with the dark-adapted preparation, while repeated test stimulation progressively light adapted

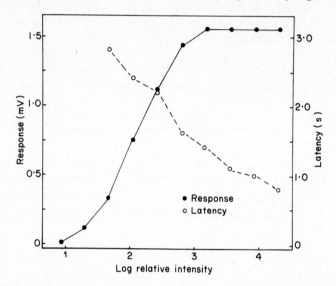

Fig. 43. Response amplitude and latency vs. stimulus intensity. Increasing intensities of 490 nm light were tested on a dark-adapted *Otala* preparation. Response is the maximum amplitude of the ERG. Latency is measured from the onset of stimulation to 50% of maximum rise. (From Gillary and Wolbarsht, 1967.)

the eye. The degree of light adaptation varied directly with the intensity of the adapting illumination and inversely with the recovery time in the dark.

The effect on the ERG of varying the temperature between 10 and 30°C is shown in Fig. 44. Between 10 and 20°C, increasing temperature increased the response and reduced the latency. Between 20 and 30°C, increasing temperature reversibly inhibited the response but had little further effect on latency. The constant response action spectrum for the *Otala* eye had a peak at around 490 nm (Fig. 45). The general shape is somewhat sharper than the absorption spectrum of a rhodopsin-like pigment as plotted from Dartnall's nomogram (Dartnall, 1953). Gillary and Wolbarsht tried to fractionate the action spectrum by adapting the

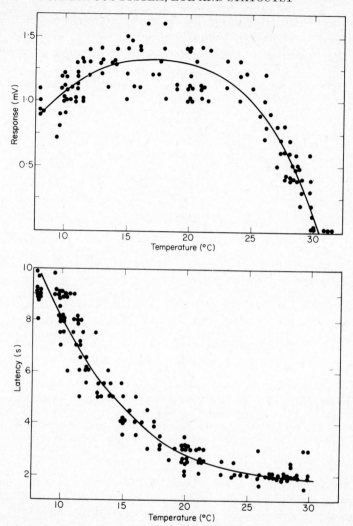

Fig. 44. The effect of temperature on the ERG from *Otala*. The temperature of a single preparation was raised and lowered through four cycles while stimulating every minute for 5 seconds with 490 nm light whose intensity was below that which saturated the response. Response refers to the maximum amplitude of the ERG. Latency is measured from the onset of stimulation to 50% of maximum rise. (From Gillary and Wolbarsht, 1967.)

preparation with blue or red light but were unsuccessful. The shape of the action spectrum was not altered when the temperature was varied between 10 and 20°C. The invariance in the shape of the action spectrum during chromatic adaptation indicates only a single photopigment which absorbs maximally at 490 nm. This is similar to the situation in other molluscan photopigments (Dennis, 1965).

Gillary and Wolbarsht consider that the *Otala* ERG reflects the initial electrical events in the outer segments of a homogeneous population of primary photoreceptors in the retina. This interpretation would agree

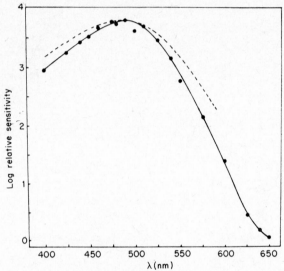

Fig. 45. Constant response action spectrum from *Otala*. The response amplitude vs. intensity curves at each of the 18 different wavelengths were determined in random order. The constant response was above threshhold but well below saturation. The dotted curve is the absorption spectrum of a pigment (max = 490 nm) plotted from Dartnall's nomogram (Dartnall, 1953). (From Gillary and Wolbarsht, 1967.)

with the conclusions of Tasaki *et al.* (1963) and Hagins (1965) concerning similar structures in the cephalopod retina. Gillary and Wolbarsht suggest that the inhibitory effect of high temperature on the ERG amplitude indicates a thermolabile step in the excitatory process. At higher temperatures the concentration of the visual pigment may be reduced by inactivation or alternatively regeneration is less efficient. The visual pigment may be thermolabile. When the preparation was electrically stimulated near the eye a compound action potential was produced which was conducted along the optic nerve with a velocity of about 0·1 m/s. The initial phase was negative when a single recording electrode was used. Above threshold, the amplitude of the action potential increased with the intensity of the electric current. When the stimulating current was near threshold, the amplitude of the action potential was greatly increased by illumination of the eye (Fig. 46). Above threshold illumination and below a saturating intensity of light, this light facilitating effect increased approximately with the log of the relative light intensity. This light facilitating effect was not seen when the optic nerve was illuminated. Gillary (1970) has extended his study

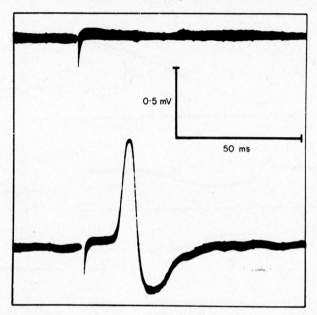

Fig. 46. Records from the optic nerve of *Otala* during electrical stimulation in darkness and light. For the upper record this was in darkness; for the lower one, it was illuminated. The artifact indicates the stimulating pulse which was just below threshold for the unilluminated preparation The action potential is initially negative at the recording electrode. (From Gillary and Wolbarsht, 1967.)

to *Helix*. He recorded both an ERG from the surface of the eye using a suction electrode and a compound action potential from the optic nerve of *Helix aspersa*, intra-ocular recordings being made with 2 M KCl glass microelectrodes. The ERG evoked by a flash of light was usually 1–2 mV in amplitude though amplitudes of up to 5 mV were recorded. ERGs recorded from the cornea were negative and monophasic while those from the back of the eye were positive and monophasic. ERGs recorded from the side of the eye were often biphasic. Examples are shown in Fig. 47. Gillary only observed "on" responses, there being no evidence for "off" responses. Each ERG waveform can be regarded as a summed potential from an array of receptors whose receptor potentials have amplitudes and latencies which depend on the local light intensity.

Increasing the temperature from 6 to 34°C decreased the latency and duration of the ERG, the ERG amplitude being maximal around 20°C. In addition to evoking an ERG, stimulation of the eye with light caused a concomitant asynchronous discharge of action potentials which were propagated centripetally along the optic nerve (Fig. 48). This discharge

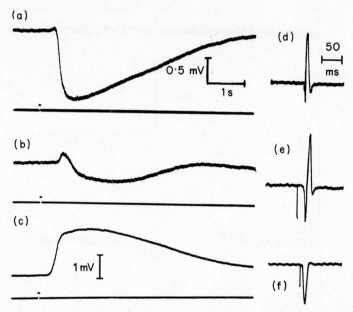

Fig. 47. Recordings of the electroretinogram (a)—(c) and the electrically evoked compound action potential (d)—(f) through suction electrodes placed on the back (a) and (d), side (b) and (e), and the corneal surface (c) and (f) of the eye. The lower trace in (a), (b) and (c) monitors a flash of light. Vertical calibrations for (a), (b), (d) (e) are identical and for (c), (f) are identical. Horizontal calibrations for (a), (b), (c) are identical and for (d), (e), (f) are identical. A negative deflection is upwards. (From Gillary, 1970.)

is due to activity in many small fibres each about $0 \cdot 1$ µm in diameter. Gillary was unable to record activity in single units.

Electrical stimulation of the optic nerve evoked a compound action potential composed of synchronous action potentials in many fibres. It contained a single peak which propagated along the nerve in both directions at a velocity of around 5 cm/s. The Q_{10} of the conduction velocity was about 2. The waveform of the compound action potential recorded from the eye varied with the position of the recording electrode (Fig. 47). Gillary suggests that the large negative phase recorded with electrodes on the back and sides of the eye indicates that the action

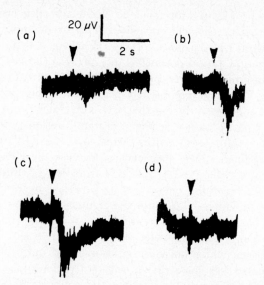

Fig. 48. d.c. Recordings with a suction electrode from the optic nerve of an eye-nerve preparation. (a)—(c) is activity evoked by a flash of light recorded 4, 2·5 and 2 mm respectively from the eye; (d) the electrode position was the same as in (c) but the optic nerve was crushed at its point of exit from the eye. (From Gillary, 1970.)

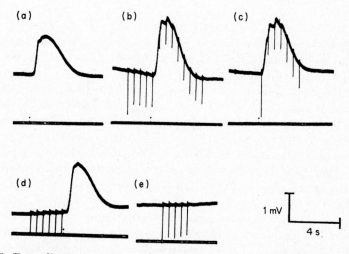

Fig. 49. Recordings via suction electrodes from the corneal surface of a single preparation. The lower trace in each record monitors a flash of light. The slow potential changes are electroretinograms. The rapid transients are compound action potentials electrically evoked at 2/s. (From Gillary, 1970.)

potential was propagated into the eye. When the optic nerve was stimulated and the ERG recorded via a suction electrode on the cornea, there was no effect on the baseline voltage. However a train of electrically evoked compound action potentials did increase the ERG amplitude if the last impulse immediately preceded the flash or it if occurred during the ERG (Fig. 49). The increment was greatest if the train began before the flash and continued during the ERG. The increase in the ERG amplitude could be up to 90%. The compound action potentials had no effect on either the latency or the duration of the ERG. Compound action potentials were never observed to decrease the ERG amplitude.

The results of Gillary were obtained from several hundred units recorded with extracellular electrodes. It would be of interest to investigate the way in which gross potential changes evoked in the eye correspond to events occurring in individual receptors. However the receptors in *Helix* are small and the cells are embedded in a tough capsule of connective tissue which surrounds the eye and optic nerve. The connective tissue makes penetration with microelectrodes very difficult. Gillary attempted partly to dissolve the connective tissue with pronase but this did not assist penetration.

XIII. Statocysts

The two statocysts, one on the left side and the other on the right, lie laterally between the pleural and pedal ganglia. The statocysts are spherical in shape and approximately 200 µm in diameter (Fig. 50). They are hollow; the inside is liquid filled and contains many plate like statoliths (Pfeil, 1922; Fretter and Graham, 1962; Laverack, 1968). The fine structure of the statocysts of *Arion* and *Limax* have been studied by Wolff (1969) (Fig. 51). There is an outer sheath of connective tissue. Within is the sensory epithelium. This contains two main types of cells: (1) large hair cells which have cilia and microvilli projecting into the cavity. There are 12–14 large hair cells; the cilia have basal plates that project centrifugally, and (2) small cells. These have microvilli and a few cilia.

The hair cells are in close connection with the nerve axons and there are whorl-like structures in both the hair cells and the axons. The statocyst nerve is in two bundles; one large bundle (14µm in cross section) and a small bundle (3 µm in cross section). The large bundle contains 13 axons. It would appear that there is one axon for each large hair cell. The statocyst nerve runs from the sensory epithelium to the cerebral

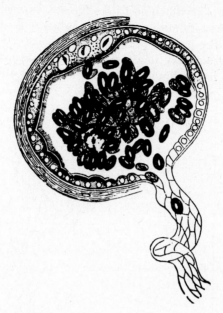

Fig. 50. General diagram of the structure of the snail statocyst.

ganglion. There are also some small, possibly efferent axons, below the statocyst cells.

Wolff (1970a) recorded electrical activity from the statocyst nerve of *Helix, Arion,* and *Limax* (Figs 52 and 53). When the animal is rotated the maximal impulse frequency occurs in the statocyst nerve when the

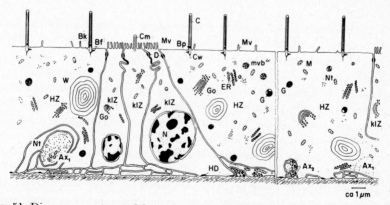

Fig. 51. Diagram constructed from electronmicrographs of the sensory epithelium of the statocyst of *Helix.* Note the large hair cells (HZ) and the smaller epithelial cells. (klZ.) (From Wolff, 1969.)

animal is in the upside down position (Fig. 52). In this position, hair cells
are stimulated that are not normally in contact with the statoliths when
the animal is in the normal upright position.

The impulse frequency increases with the rotation velocity. Differences in output frequency of a single cell rotated about different axes

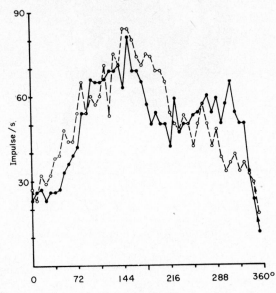

Fig. 52. The effect of rotation through 360° on the electrical activity of the
statocyst nerve. There is a maximal activity when the preparation is at 180°.
(From Wolff, 1970a.)

with the same rotation velocity appear to be due to the size of the receptive area stimulated. The statocyst receptors are of the phasic-tonic
type. The sense organs are mainly gravity receptors.

Wolff (1970b) developed a simple preparation of the circumpharyngeal
ring of *Arion*, *Limax* and *Helix* where he recorded from a single statocyst nerve (Fig. 53). Rotating the preparation caused an increase of activity in the nerve. If the nerve was cut between the ganglia and the
statocyst and the activity recorded on the statocyst side, the pattern of
activity changed with rotation. If the nerve to one statocyst was cut and
the recording made on the gangliar side of the nerve, the activity (efferent activity from the ganglia) increased when the brain was rotated. This
reaction to movement stopped if the other statocyst was removed. It
would appear that the activity from the statocysts enters the brain and
that some of this is relayed as efferent activity to the other statocyst.

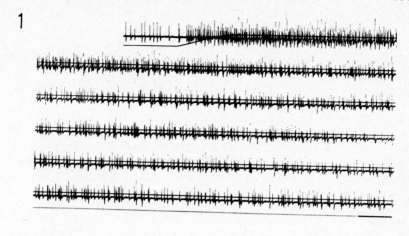

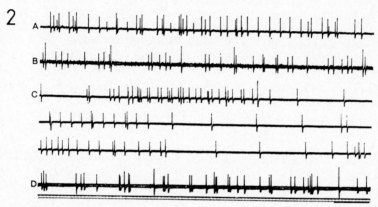

Fig. 53. Electrical activity in the statocyst nerve of *Limax*, *Arion* and *Helix*. 1, change in activity as the preparation is rotated. The movement is indicated by the continuous line. 2, activity in the nerves of A, *Arion*; B, *Limax maximus*; C & D, *Helix pomatia*. (From Wolff, 1970a.)

The situation is similar to the vertebrate situation with efferent fibres going to the sensory cells of Corti's organ (Iturato, 1962).

XIV. Acknowledgements

We should like to thank the authors of papers who have sent us reprints of their work and also in many cases copies of the illustrations and electron micrographs.

XV. References

Amoroso, E. C., Baxter, M. I., Chiquoine, A. D. and Nisbet, R. H. (1964). *Proc. R. Soc.* B **160**, 167–180.

Baker, P. F., Hodgkin, A. L. and Ridgway, E. B. (1971). *J. Physiol. Lond.* **218**, 709–755.

Bang, T. (1917). *Zool. Anz.* **48**, 281–292.

Barondes, S. (1967). *Neurosci. Res. Program Bull.* **5**, 307–419.

Barrantes, F. J. (1970). *Z. Zellforsch. mikrosk. Anat.* **104**, 208–212.

Baxter, M. I. and Nisbet, R. H. (1963). *Proc. malac. Soc. Lond.*, **35**, 167–177.

Benjamin, R. L. and Peat, A. (1968). *Nature, Lond.* **219**, 1371–1372.

Berg, E. and Schneider, G. (1967). *Naturwissenschaften* **54**, 591–592.

Boer, H. H. (1965). *Arch. néerl. Zool.* **3**, 313–386.

Brandenburger, J. L. and Eakin, R. M. (1970). *Vision Res.* **10**, 639–653.

Briel, G., Neuhoff, V. and Osborne, N. N. (1971). *Int. J. Neurosci.* **2**, 129–136.

Bruner, J. and Tauc, L. (1966). *Nature, Lond.* **217**, 880.

Buytendijk, F. J. J. (1921). *Arch. néerl. Physiol.* **5**, 458–466.

Cardot, J. (1963). *C.r. hebd. Séanc. Acad. Sci. Paris* **257**, 1364–1366.

Cardot, J. and Ripplinger, J. (1963). *J. Physiol. Paris* **55**, 217–218.

Castellucci, V., Pinsker, H., Kaupfermann, I. and Kandel, E. R. (1970). *Science, Wash.* **167**, 1745.

Chamberlain, S. G. and Kerkut, G. A. (1967). *Nature, Lond.* **216**, 89–90.

Chamberlain, S. G. and Kerkut, G. A. (1969). *Comp. Biochem. Physiol.* **28**, 787–803.

Chiarandini, D. J. and Gerschenfeld, H. M. (1967). *Science, Wash.* **156**, 1595–1596.

Chiarandini, D. J., Stefani, E. and Gerschenfeld, H. M. (1967). *Science, Wash.* **156**, 1597–1599.

Cook, A. (1971). *Anim. Behav.* **19**, 463–574.

Cottrell, G. A. (1970). *J. Physiol. Lond.* **208**, 28–29 P.

Cottrell, G. A. (1971). *Experientia* **27**, 813–815.

Cottrell, G. A. and Osborne, N. N. (1970). *Nature, Lond.* **225**, 470–472.

Curtis, D. J. and Kerkut, G. A. (1969). *Comp. Biochem. Physiol.* **30**, 835–840.

Dahl, E., Falck, B., Mecklenburg, C.von, Myhrberg, H. and Rosengren, E. (1966). *Z. Zellforsch. mikrosk. Anat.* **71** 489–498.

Dartnall, H. J. A. (1953). *Brit. med. Bull.* **9**, 24–30.

Dawson, J. (1911). *Behav. Monogr.* **1**, No. 4, 1–120.

Dennis, M. J. (1965). *Am. Zool.* **5**, 651.

Duncan, C. J. (1961). *Comp. Biochem. Physiol.* **3**, 42–51.

Eakin, R. M. and Brandenburger, J. L. (1967). *J. ultrastruct. Res.* **18**, 391–421.

Eakin, R. M. and Westfall, J. A. (1964). *Z. Zellforsch. mikrosk. Anat.* **62**, 310–332.

Eisenmann, H. (1920). *Zool. Anz.* **51**, 143–158.

Emson, P., Walker, R. J. and Kerkut, G. A. (1971). *Comp. Biochem. Physiol.* **40B**, 223–239.

Falck, B. (1962). *Acta physiol. scand.* **56**, (Suppl. **197**) 1–25.

Fernandez, J. (1966). *J. comp. Neurol.* **127**, 157–182.

Fernandez, J. (1971). *Z. Zellforsch. mikrosk. Anat.* **119**, 512–524.

Fretter, V. and Graham, A. (1962). "British Prosobranch Molluscs." The Ray Society, London.

Gainer, H. (1972a). *Brain Res.* **39**, 369–385.

Gainer, H. (1972b). *Brain Res.* **39**, 387–402.

Gainer, H. (1972c). *Brain Res.* **39**, 403–418.

Geduldig, D. and Junge, D. (1968). *J. Physiol. Paris* **142**, 516–453.

Ger, B. A., Zeimal, E. V. and Kvitko, I. J. (1971). *Comp. gen. Pharmac.* **2**, 225–246.

Gerasimov, V. D. (1964). *Fiziol. Zh. U.S.S.R.* **50**, 457–463. (In Russian). English translation: (1965) Federation of American Societies for Experimental Biology; *Fed. Proc. Transl. Suppl.* **24** No 2 (II) 371–374.

Gerasimov, V. D., Kostyuk, P. G. and Maisky, V. A. (1965). *Biofizika*, **10**, 447–453. (In Russian). English translation: *Biophysics* **10**, 494–502.

Gerasimov, V. D. and Maisky, V. A. (1963). *Fiziol. Zh. U.S.S.R.* **49**, 1099–1104 (In Russian).

Gerschenfeld, H. M. (1963). *Z. Zellforsch. mikrosk. Anat.* **60**, 258–275.

Gerschenfeld, H. M. (1971). *Science, Wash.* **171**, 1252–1254.

Gerschenfeld, H. M. and Lasansky, A. (1964). *Int. J. Neuropharmacol.* **3**, 301–314.

Gerschenfeld, H. M. and Stefani, E. (1966). *J. Physiol. Lond.* **185**, 684–700.

Gerschenfeld, H. M. and Stefani, E. (1968). *Pharmacol.* **6A**, 369–392.

Gillary, H. L. (1970). *Vision Res.* **10**, 977–991.

Gillary, H. L. and Wolbarsht, M. L. (1967). *Rev. Can. Biol.* **26**, 125–134.

Glaizner, B. (1968a). *In* "Neurobiology of Invertebrates" (J. Salanki Ed.) pp. 267–284. Plenum Press, New York.

Glaizner, B. (1968b). *J. Physiol. Lond.* **197**, 70P.

Goldman, T. and Hermann, H. (1967). *Vision Res.* **7**, 533–537.

Gubicza, A. (1970). *Ann. Biol. Tihany* **37**, 3–15.

Gubicza, A. and S-Rozsa, K. (1969). *Ann. Biol. Tihany* **36**, 3–10.

Gubicza, A. and S-Rozsa, K. (1970). *Ann. Biol. Tihany* **37**, 17–31.

Hagins, W. A. (1965). *Cold Spring Harb. Symp. quant. Biol.* **30**, 403–418.

Hagiwara, S. and Saito, Y. (1959). *J. Physiol. Lond.* **148**, 161–179.

Hanström, B. (1925). *Acta zool., Stockh.* **6**, 183–215.

Hesse, R. (1908). "Das Sehen der Niederen Tiere" Fischer, Jena.

Hermann, H. T. (1968). *Vision Res.* **8**, 601–612.

Hiripir, L. (1970). *Ann. Biol. Tihany* **37**, 33–41.

Hodgkin, A. L. and Huxley, A. F. (1945). *J. Physiol. Lond.* **104**, 176–195.

Humphrey, G. (1933). "The Nature of Learning." Kegan Paul, New York.

Iturato, S. (1962). *Exp. Cell Res.* **27**, 162–164.

Joose, J. (1964). *Arch. néerl. Zool.* **16**, 1–103.

Kandel, E. R., Frazier, W. T., Waziri, R. and Coggeshall, R. E. (1967). *J. Neurophysiol.* **30,** 1352–1376.

Kandel, E. R. and Kupfermann, I. (1970). *A. Rev. Physiol.* **32,** 193–258.

Kandel, E. R. and Tauc, L. (1966a). *J. Physiol. Lond.* **183,** 269–286.

Kandel, E. R. and Tauc, L. (1966b). *J. Physiol. Lond.* **183,** 287–304.

Katayama, Y. (1970). *Jap. J. Physiol.* **20,** 711–724.

Kehoe, J. S. (1969). *Nature, Lond.* **221,** 866–868.

Kehoe, J. S. (1973). *In* "Drug Receptors" (H. P. Rang Ed.) pp. 63–85 Macmillan, London.

Kennedy, D. (1958). *Am. J. Ophthalmol.* **46,** 21.

Kerkut, G. A. (1973). *Br. med. Bull.* **29,** 100–103.

Kerkut, G. A. and Cottrell, G. A. (1963). *Comp. Biochem. Physiol.* **8,** 53–63.

Kerkut, G. A., French, M. C. and Walker, R. J. (1970). *Comp. Biochem. Physiol.* **32,** 681–690.

Kerkut, G. A. and Gardner, D. R. (1967). *Comp. Biochem. Physiol.* **20,** 147–162.

Kerkut, G. A., Horn, N. and Walker, R. J. (1969). *Comp. Biochem. Physiol.* **30,** 1061–1074.

Kerkut, G. A., Lambert, J. D. C. and Walker, R. J. (1973). *In* "Drug Receptors" (H. P. Rang Ed.) pp. 37–44 Macmillan, London.

Kerkut, G. A., Lambert, J. D. L., Loker, J., Gayton, R. and Walker, R. J. (1974). *Comp. Biochem. Physiol.* **50A,** 1–25.

Kerkut, G. A. and Meech, R. W. (1966). *Comp. Biochem. Physiol.* **19,** 819–832.

Kerkut, G. A., Ralph, K., Walker, R. J., Woodruff, G. N. and Woods, R. (1970). *In* "Excitatory Synaptic Mechanisms" (Per Andersen and Jan Jansen Eds.) pp. 105–117. Universitetsforlaget, Oslo.

Kerkut, G. A., Sedden, C. B. and Walker, R. J. (1966). *Comp. Biochem. Physiol.* **18,** 921–930.

Kerkut, G. A., Sedden, C. B. and Walker, R. J. (1967). *Comp. Biochem. Physiol.* **23,** 159–162.

Kerkut, G. A., Shapira, A. and Walker, R. J. (1967). *Comp. Biochem. Physiol.* **23,** 729–748.

Kerkut, G. A. and Thomas, R. C. (1963). *Comp. Biochem. Physiol.* **8,** 39–45.

Kerkut, G. A. and Thomas, R. C. (1964). *Comp. Biochem. Physiol.* **11,** 199–213.

Kerkut, G. A. and Walker, R. J. (1961). *Comp. Biochem. Physiol.* **3,** 143–160.

Kerkut, G. A. and Walker, R. J. (1962). *Comp. Biochem. Physiol.* **7,** 277–288.

Kerkut, G. A., Woodhouse, M. and Newman, G. R. (1966). *Comp. Biochem. Physiol.* **19,** 309–311.

Kerkut, G. A. and York, B. (1969). *Comp. Biochem. Physiol.* **28,** 1125–1134.

Kerkut, G. A. and York, B. (1971). "The Electrogenic Sodium Pump." Scientechnica, Bristol.

Kiss, I. and Salanki, J. (1971). *Annal. Biol. Tihany* **38,** 39–52.

Kostyuk, P. (1968). *In* "Neurobiology of Invertebrates" (J. Salanki Ed.) pp. 145–167. Plenum Press, New York.

Kostyuk, P. G., Sorokina, Z. A. and Kholodova, Yu.D. (1969). *In* "Glass Microelectrodes" (Lavalee, M., Schanne, A. and Herbert, N. C. Eds.) pp. 322–348. Wiley, New York.

Krishtal, O. A. and Magura, I. S. (1970). *Comp. Biochem. Physiol.* **35**, 857–866.

Kristensson, K. and Olsson, Y. (1971). *Brain Res.* **29**, 363–365.

Kunze, H. (1917). *Zool. Anz.* **49**, 123–137.

Kunze, H. (1921). *Z. wiss. Zool.* **118**, 25–203.

Lambert, J. D. C. (1972). "Drug response and membrane potential in identified molluscan neurones." Ph.D. Thesis, Southampton University.

Lane, N. J. (1962). *Q. Jl microsc. Sci.* **103**, 211–226.

Laverack, M. S. (1968). *Symp. zool. Soc. Lond.* **23**, 299–326.

Lever, J. (1965). *Arch. néerl. Zool.* **13**, 194–201.

Loker, J. (1973). "The identification of catecholamine and 5-HT containing neurones in the snail brain." Ph.D. Thesis, Southampton University.

Maisky, V. A. (1963). *Fiziol. Zh. U.S.S.R.* **49**, 1468–1474. (In Russian.) English translation: *Fed. Proc. Transl. Suppl.* 23 No. 6 II, 1173–1176 (1964). Federation of American Societies for Experimental Biology.

Magura, I. S. (1969). *Neurophysiology*, **1**, 109–117. (In Russian).

Marsden, C. A. (1969). "The cellular localisation of monoamines in *Helix aspersa, Planorbis corneus* and *Hirudo medicinalis*." Ph.D. Thesis, Southampton University.

Marsden, C. A. and Kerkut, G. A. (1969). *In* "Experiments in Physiology and Biochemistry" (G. A. Kerkut Ed.). **2**, pp. 327–360. Academic Press, London and New York.

Marsden, C. A. and Kerkut, G. A. (1970). *Comp. gen. Pharmac.* **1**, 101–116.

Meves, H. (1966). *Pflügers Arch. ges. Physiol.* **289**, R10.

Meves, H. (1968). *Pflügers Arch. ges. Physiol.* **304**, 215–241.

Moreton, R. B. (1968). *Nature, Lond.* **219**, 70–71.

Nabias, B. de (1898). *Trav. Lab. Soc. Sci. Stat. Zool. Arcachon* **3**, 43–72.

Newell, G. E. (1965). *Proc. zool. Soc. Lond.* **144**, 75–86.

Newell, P. F. and Newell, G. E. (1968). *Symp. zool. Soc. Lond.* **23**, 97–111.

Newman, G., Kerkut, G. A. and Walker, R. J. (1968). *Symp. zool. Soc. Lond.* **22**, 1–18.

Newton, L. C. (1972). (Studies of invertebrate acetylcholine receptors.) Ph.D. Thesis, Southampton University.

Nisbet, R. H. (1956). *J. Physiol. Lond.* **135**, 34–35.

Nisbet, R. H. (1960). *J. Physiol. Lond.* **151**, 15–17P.

Nisbet, R. H. (1961a). *Proc. R. Soc.* B **154**, 267–287.

Nisbet, R. H. (1961b). *Proc. R. Soc.* B **154**, 309–331.

Nisbet, R. H. and Plummer, J. M. (1968). *Symp. zool. Soc. Lond.* **22**, 193–211.

Nisbet, R. H. and Plummer, J. M. (1969). *Experientia* Suppl. **15**, 47–68.

Nolte, A. (1968). *In* "Neurobiology of Invertebrates" (J. Salanki Ed.) pp. 123–133. Plenum Press, New York.

Oomura, Y. (1963). Quoted in "The Synapse" by J. C. Eccles. Springer, Berlin

Oomura, Y. and Maeno, T. (1963). *Nature, Lond.* **197**, 358–359.

Oomura, Y., Maeno, T., Ozaki, S. and Nakashima, Y. (1962).*Seitai No Kagaku* 13, 83–90.

Oomura, Y., Ooyama, H. and Sawada, M. (1965). *Int. Cong. Physiol. Sci.* Tokyo, 5, 389 Abstract 913.

Oomura, Y., Ooyama, H. and Sawada, M. (1966). *Symp. Cell Chem.* 17, 233–241. (In Japanese).

Oomura, Y., Ooyama, H. and Sawada, M. (1968). *Int. Cong. Physiol. Sci.* Washington 7, 330 Abstract 901.

Osborne, N. N. (1971). *Comp. gen. Pharmac.* 2, 433–438.

Osborne, N. N. and Cottrell, G. A. (1970). *Comp. gen. Pharmac.* 1, 1–10.

Osborne, N. N. and Cottrell, G. A. (1971). *Z. Zellforsch. mikrosk. Anat.* 112, 15–30.

Pentreath, V. W. and Cottrell, G. A. (1970). *Z. Zellforsch. mikrosk. Anat.* 111, 160–178.

Peretz, B. (1970). *Science, Wash.* 169, 379.

Pfeil, E. (1922). *Z. wiss. Zool.* 119, 79–113.

Ralph, K. L. (1970). "The mechanism of excitatory and inhibitory potentials in the snail brain." Ph.D. Thesis, Southampton University.

Rogers, D. C. (1968). *Z. Zellforsch. mikrosk. Anat.* 89, 80–94.

Rogers, D. C. (1969a). *Z. Zellforsch. mikrosk. Anat.* 102, 113–128.

Rogers, D. C. (1969b). *Z. Zellforsch. mikrosk. Anat.* 102, 99–112.

Rohlich, P. and Török, L. J. (1963). *Z. Zellforsch. mikrosk. Anat.* 60, 348–368.

Sakharov, D. A. (1970). *Ann. Rev. Pharmacol.* 10, 335–352.

Sakharov, D. A. and Salanki, J. (1969a). *Acta Physiol. Acad. Sci. Hung.* 35, 19–30.

Sakharov, D. A. and Salanki, J. (1969b). *Acta Biol. Hung.* 19, 391–406.

Sakharov, D. A. and Zs-Nagy, I. (1968). *Acta Biol. Hung.* 19, 145–157.

Salanki, J. and Kiss, I. (1969). *Annal. Biol. Tihany* 36, 63–75.

Sanchis, C. A. and Zambrano, D. (1969a). *Experientia* 25, 385–386.

Sanchis, C. A. and Zambrano, D. (1969b). *Z. Zellforsch. mikrosk. Anat.* 94, 62–71.

Sawada, M. (1969). *J. physiol. Soc. Japan,* 31, 491–504.

Schlote, F. W. (1957). *Z. Zellforsch. mikrosk. Anat.* 45, 543–568.

Schultze, H. (1879). *Arch. mikrosk. Anat.* 16, 57–110.

Schwalbach, G., Lickfeld, K. G. and Hahn, M. (1963). *Protoplasma,* 56, 242–273.

Sedden, C. B., Walker, R. J. and Kerkut, G. A. (1968). *Symp. zool. Soc. Lond.* 22, 19–32.

Smith, G. (1908). *Bull. Mus. comp. Zool. Harv.* 48, 231–283.

Sorokina, Z. A. and Zelenskaya, V. S. (1967).*J. Evol. Biochem. Physiol.* 3, 25–30. (In Russian).

Sorokina, Z. A. and Kholodova, Yu.D. (1968). *J. Evol. Biochem. Physiol.* Suppl. 3, 76–84.

Stretton, A. O. W. and Kravitz, E. A. (1968). *Science, Wash.* 162, 132–134.

Sweeney, D. (1963). *Science, Wash.* 139, 1051.

Tauc, L. (1958). *Arch. Ital. Biol.* 96, 78–110.

Tauc, L. (1966). *In* "Physiology of the Mollusca" (K. M. Wilbur and C. M. Yonge Eds) **21**, pp. 187–454. Academic Press, New York.

Tauc, L. (1967). *Physiol. Rev.* **47**, 522–593.

Tauc, L. and Gerschenfeld, H. M. (1962). *J. Neurophysiol.* **25**, 236–262.

Tasaki, K., Oikawa, T. and Norton, A. C. (1963). *Vision Res.* **3**, 61–73.

Van Wilgenburg, H. (1970). "An electrophysiological analysis of neurones in the visceral and parietal ganglia of *Helix pomatia* L." Ph.D. Thesis, Amsterdam.

Veprintzev, B. N. and Rosanov, S. I. (1968). *In* "Neurobiology of Invertebrates" (J. Salanki Ed.) pp. 413–421. Plenum Press, New York.

Vulfius, E. A. and Zeimal, E. V. (1968). *J. Evol. Biochem. Physiol.* Supplement **3**, 92–100.

Wald, F. (1972). *J. Physiol. Lond.* **220**, 267–281.

Walker, R. J. (1967). *In* "Symposium of Invertebrate Neurobiology" (J. Salanki, Ed.) pp. 227–253. Plenum Press, New York.

Walker, R. J., Azanza, M. J. and Woodruff, G. N. (1972) (in preparation).

Walker, J. L. and Brown, A. M. (1971). *Science, Wash.* **167**, 1502–1504.

Walker, R. J., Crossman, A. R., Woodruff, G. N. and Kerkut, G. A. (1971). *Brain Res.* **33**, 75–82.

Walker, R. J. and Hedges, A. (1967). *Comp. Biochem. Physiol.* **23**, 977–989.

Walker, R. J. and Hedges, A. (1968). *Comp. Biochem. Physiol.* **24**, 355–376.

Walker, R. J., Kato, G., Morris, D. and Woodruff, G. N. (1973). *Comp. gen. Pharmacol.* **4**, 65–74.

Walker, R. J., Ralph, K. L., Woodruff, G. N. and Kerkut, G. A. (1971a). *Experientia* **27**, 281–282.

Walker, R. J., Ralph. K. L. Woodruff, G. N. and Kerkut, G. A. (1971b). *Comp. gen. Pharmacol.* **2**, 15–26.

Walker, R. J., Ralph, K. L., Woodruff, G. N. and Kerkut, G. A. (1972). *Comp. gen. Pharmacol.* **3**, 52–60.

Walker, R. J., Ramage, A. G. and Woodruff, G. N. (1972). *Experentia* **28**, 1173–1174.

Walker, R. J. and Woodruff, G. N. (1971). *Brit. J. Pharmacol.* **43**, 415–416.

Walker, R. J. and Woodruff, G. N. (1972). *Comp. gen. Pharmacol.* **3**, 27–40.

Walker, R. J., Woodruff, G. N., Glaizner, B., Sedden, C. B. and Kerkut, G. A. (1968). *Comp. Biochem. Physiol.* **24**, 455–469.

Walker, R. J., Woodruff, G. N. and Kerkut, G. A. (1971). *Comp. gen. Pharmacol.* **2**, 168–174.

Wells, M. J. (1965). *Adv. mar. Biol.* **3**, 1–59.

Wells, M. J. and Wells, J. (1971). *Anim. Behav.* **19**, 305–312.

Wheeler, G. C. (1921). *J. comp. Psychol.* **1**, 149–154.

Willem, V. (1892). *Arch. Biol.* **12**, 57–98.

Wolff, H. G. (1969). *Z. Zellforsch. mikrosk. Anat.* **100**, 251–270.

Wolff, H. G. (1970a). *Z. vergl. Physiol.* **69**, 326–366.

Wolff, H. G. (1970b). *Z. vergl. Physiol.* **70**, 401–409.

Woodruff, G. N. and Walker, R. J. (1969). *Int. J. Neuropharmacol.* **8**, 279–289.

Woodruff, G. N. and Walker, R. J. (1971). *Eur. J. Pharmacol.* **14**, 81–85.

Woodruff, G. N., Walker, R. J. and Kerkut, G. A. (1971). *Eur. J. Pharmacol.* **14,** 77–80.

Yanase, T. and Sakamoto, S. (1965). *Zool. Mag. Tokyo* **74,** 238–242.

Yung, E. (1911). *Rev. suisse Zool.* **19,** 339–382.

Zeimal, E. V. and Vulfius, E. V. (1968). *In* "Neurobiology of Invertebrates" (J. Salanki, Ed.) pp. 255–265. Plenum Press, New York.

Zelenskaya, V. S., Oleynikova, T. N. and Sorokina, Z. A. (1968). *J. Evol. Biochem. Physiol.* Suppl. **3,** 84–92.

Chapter 6

Endocrinology

H. H. Boer and J. Joosse

Department of Zoology, Free University, Amsterdam, The Netherlands

I. Introduction

In the Pulmonata a number of endocrine and supposed endocrine structures occur. Firstly, in the central nervous system various types of neurosecretory cells have been described. Secondly, two structures which are associated with the nervous system, viz. the optic tentacle and the cerebral gland have been supposed to be endocrine organs. Furthermore some non-nervous organs are considered to be endocrine structures, viz. the dorsal bodies and the gonad. Also the organ of Semper of the Stylommatophora has originally been considered as an endocrine organ (e.g. Lane 1964). Later research has shown, however, that this organ is an exocrine gland (Van Mol, 1967; Renzoni, 1969).

Recent surveys on the endocrinology of pulmonates are those of Gabe (1966), Simpson *et al.* (1966b), Durchon (1967), Martoja (1968, 1972), Highnam and Hill (1969), Joosse (1972) and Wendelaar Bonga (1972).

In the present chapter on the endocrinology of the Pulmonata first the morphological and cytological aspects of the endocrine and supposed endocrine structures will be dealt with (Section II). In Section III attention will be focussed on the functional significance of the structures concerned.

II. Morphology of the endocrine structures

A. Neurosecretion

The first indication of the occurrence of neurosecretory cells in Pulmonata was presented by Gabe in 1954. The first extensive study on neurosecretory phenomena in a pulmonate is that of Lever (1957), who investigated the basommatophore *Ferrissia shimekii* Pilsbry. In the central nervous system (CNS) of this species various types of neurosecretory cells were distinguished. Since these first reports, the literature on neurosecretion in the Pulmonata has rapidly accumulated (Lever, 1958a, b; Lever *et al.*, 1959; Krause, 1960; Van Mol 1960a, b, 1967; Lever and Joosse, 1961; Lever *et al.*, 1961b; Jungstand, 1962; Quattrini, 1962, 1963; Boer, 1963, 1965; Joosse, 1963, 1964; Kuhlmann, 1963, 1970; Nolte, 1963, 1964, 1965; Röhnisch, 1964; Lever *et al.*, 1965; Cook, 1966; Krkac and Mestrov, 1970; for further references see Gabe, 1966; Simpson *et al.*, 1966b; Martoja 1968, 1972; Wendelaar Bonga, 1970b).

1. Neurosecretory cells and systems

One of the problems in the study of neurosecretion in general, which has also affected and still affects that in pulmonates, is how to define a neurosecretory cell. In his book, Gabe (1966) goes into this question in some detail. He elucidates that from the first discovery of neurosecretory cells (Scharrer, 1928), the concept of neurosecretion has been a histological concept. In Gabe's opinion it still seems "justifiable to apply the term neurosecretory cell, just as it was done in the past, to elements having the characteristics of neurones and also exhibiting the morphological signs of glandular activity – the intracellular elaboration and discharge from the cell of a formed, histologically detectable secretory product" (p. 6). This outline by Gabe seems quite justified. However, in the case of gastropods it seems desirable to define the criteria in more detail, especially as in these animals neurones have been mistaken for neurosecretory cells when criteria have been too loosely applied (see p. 248). A useful addition to light microscopy in this connexion would be an ultrastructural verification of the evidence. The electron microscope not only makes possible a correct identification of neurosecretion, but it usually also makes the distinction between neurotransmitters* (chemical intermediaries of nerve transmission, like acetylcholine and noradrenalin) and neurosecretions possible.

Taking the above mentioned facts into account, we could then define the conditions for evidence of neurosecretion as follows:

1. Histochemical evidence from pulmonates among others (Boer, 1963, 1965; Kuhlmann, 1963; Röhnisch, 1964; Wendelaar Bonga, 1970b), shows that neurosecretions consist mainly of proteins. In fact, the widely used Gomori methods for the demonstration of neurosecretory cells, the chrome-haematoxylin (CH) and the paraldehyde fuchsin (PF) staining techniques, most probably demonstrate the presence of protein-bound cysteine (e.g. Landing et al., 1956; Boer, 1965). This apparently also holds for the pseudoisocyanin method of Sterba (1963). Furthermore, not only the "Gomori-positive" neurosecretions, but also neurosecretions not stainable with CH and PF ("Gomori-negative" neurosecretions) are mainly proteinaceous in character (e.g. Boer, 1965; Wendelaar Bonga, 1970b).

2. At the ultrastructural level neurosecretions appear to be composed of membrane-limited, electron-dense elementary granules with a mean diameter of 100–300nm. This criterion should, however, be used with caution, since probably neurotransmitter granules may also extend into this size range (see Boer et al., 1968a), although they usually have a diameter of less than 100nm.

3. In contrast to neurotransmitters, which are released at synapses, neurosecretions are discharged from axone terminals which end non-synaptically in a neurohaemal organ. In these organs, which may have either a compact structure or be rather extensive, as in pulmonates (neurohaemal area, e.g. Joosse, 1964; Nolte, 1965), the neurosecretory axone endings are in close contact with blood or lymph, into which they discharge their secretion. Although this criterion is not always applicable, as there are reports of neurosecretory axones ending with "synaptoid" junctions on target cells ("neurosecretomotor innervation",

---

* The literature on neurotransmitters is not considered here, but is covered by Sakharov (1970) and Osborne and Cottrell (1971).

e.g. Knowles, 1965), it seems of special importance, as release of a neuronal product into the blood strongly indicates the hormonal nature of this product. Because the knowledge of many neurosecretory cells in the Pulmonata is still fragmentary, it appears justifiable to attach special value to the establishment of the location and the ultrastructure of neurohaemal areas. In this respect it seems useful to introduce the term *neurosecretory system* for the description of neurosecretory phenomena in the Pulmonata, and to agree that this term be used only if the location of the neurohaemal area of the "neurosecretory cells" under discussion is known. Although neurosecretory cells have been distinguished in most of the central ganglia of the Pulmonata, the term neurosecretory system can be applied only in a few cases. In most of the species investigated, it is only in the cerebral ganglia that such a system has been described. In *Lymnaea stagnalis* (L.) alone have other systems been recognized; in this species at least two systems have been found in the cerebral ganglia and others have been described in the ganglia of the visceral ring (Joosse, 1964; Wendelaar Bonga, 1970b).

Summarizing the above considerations, it may be stated that a neurone can be recognized as a neurosecretory cell on the basis of morphological criteria. It should, however, be emphasized that as many criteria as possible have to be taken into account to avoid misinterpretations. This statement can easily be illustrated from the literature on the Pulmonata. In the early investigations (Lever, 1957; Krause, 1960; Kuhlmann, 1963) neurones have been called neurosecretory solely on the basis of their stainability with CH and PF. Later ultrastructural and histochemical evidence has shown that in a number of cases this stainability is due to the presence in these cells of what we now call secondary lysosomes (Boer, 1965; Nolte *et al.*, 1965; Boer *et al.*, 1968a). Another example is that of neurones of *Helix aspersa* Müll and *Planorbis trivolvis* Say which were considered to be neurosecretory because in the cell bodies elementary granules (probably neurotransmitters) were observed (Lane, 1966). However, the presence of such granules in non-neurosecretory neurones of *Lymnaea stagnalis* (Boer *et al.*, 1968a) has emphasized the statement that "their presence is, by itself, not diagnostic for neurosecretion" (Knowles and Bern, 1966).

Of course, only endocrinological experiments, such as extirpations and implantations of cells, can prove that a neurosecretory cell indeed produces a neurohormone. In this respect it should be admitted that such "definite neurosecretory cells" (Bern, 1962) are still scarce in molluscs, although the function of some neurosecretory cells is known. Firstly, the bag cells of the abdominal ganglion of the opisthobranch *Aplysia californica* Cooper, are involved in the regulation of ovulation

(Kupfermann, 1972) the same holds for the caudo-dorsal cells of the cerebral ganglia in the Casommatophore *Lymnaea stagnalis* (Geraerts and Bohlken, in prep.) In this species the light green cells of the cerebral ganglia are involved in growth (Geraerts, 1973) and that certain cells in the pleural and parietal ganglia play a role in osmoregulation (Hekstra and Lever, 1960; Lever *et al.*, 1961a; Wendelaar Bonga, 1971b).

Concerning the definition of a neurosecretory cell two more points should be discussed briefly.

Firstly, the term "Gomori-positive" cell is widely used for cells staining with PF and CH, which suggests that these cells are similar to a great extent. Obviously this is not so. After staining with PF differences in colour have been observed between "Gomori-positive" cells, e.g. in *Succinea putris* (L.) (Cook, 1966; see also Simpson, 1969; Schooneveld, 1970). In *Lymnaea stagnalis* Wendelaar Bonga (1970b) showed with the alcian blue-alcian yellow staining method and with electron microscopy that the class of "Gomori positive" cells consists of seven different types.

The second point concerns the question as to whether it is possible to make statements on the activity of a neurosecretory cell on the basis of light microscope observations. In this respect it should be mentioned that filled cells have often been considered to be active and empty cells inactive. The validity of this supposition has been questioned by several authors (Highnam, 1965; Mordue, 1967; Highnam and Hill, 1969), who argued that no definite conclusions about the secretory activity can be drawn on the basis of the amount of secretory substance present in the cells, unless the rates of formation and release of the material are taken into account. This can be done by various methods, e.g. by morphometrical methods which may be applied at the light as well as at the electron microscope level (Wendelaar Bonga, 1971a, b).

2. Location of neurosecretory systems and cells

a. Stylommatophora

In most of the Stylommatophora investigated CH and PF positive neurosecretory cells have been found in the mesocerebrum (*Arion ater rufus* (L.), *Achatinella fulgens* Newcomb, *Succinea putris*, *Succinea elegans* Risso, Van Mol, 1960, 1967; in 8 helicid species: *Cepaea nemoralis* L, (L.), *C. hortensis* (Müll.) *Helix pomatia* L., *Helix aspersa*, *Helicigona lapicida* (L.), *Theba pisana* (Müll.) *Eobania vermiculata* (Müll.)

Sphincterochila candidissima (Draparnaud) and in the species *Stropho-cheilusoblongus* Müll., Kuhlmann, 1963; in *Succinea putris,*Cook,1966;in *Helix aspersa*, Guyard, 1971). These neurosecretory cells are, like most neurones in pulmonates, relatively large, obviously polyploid cells (Boer, 1965). Their exact number is not known. The cells occur in groups. In all species investigated there is a large group situated in the dorsal part of the mesocerebrum apposed to the perineurium close to the origin of the cerebral commissure (medio-dorsal cells). Kuhlmann (1963) reports that in addition two other groups of cells are present in the central and dorsal parts of the mesocerebrum of the Helicidae (Fig. 1). The axones

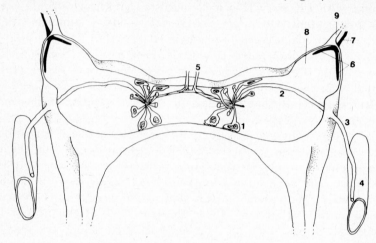

Fig. 1. Schematic representation of the mesocerebral neurosecretory system of the cerebral ganglia of Helicidae. 1, Mesocerebral neurosecretory cells; 2, main tract of neurosecretory axones; 3, nerve of the cerebral artery; 4, cerebral artery; 5, intercerebral nerve; 6, intracerebral part of the cerebral gland; 7, extracerebral part of the cerebral gland; 8, procerebrum; 9, olfactory nerve. (Redrawn after Nolte and Kuhlmann, 1964.)

of all these mesocerebral neurosecretory cells form a bundle, which runs into the central neuropile of the ganglion, where it branches into a larger and a smaller one. The larger bundle of each ganglion runs to the lateral part of the ganglion and forms the thin nerve of the cerebral artery, which leaves the ganglion close to the origin of the median lip nerve. It then runs to the cerebral artery and branches in the wall of this vessel. The areas of branching are considered to be neurohaemal areas of the mesocerebral cells. The smaller bundles of axones run into the cerebral commissure and form, at the dorsal side of the commissure, the very thin, paired intercerebral nerves (Fig. 1). These nerves branch

in the connective tissue surrounding the cerebral ganglia and the commissure. This area is apparently a second neurohaemal area of the
mesocerebral neurosecretory cells. (According to Van Mol axones of the
mesocerebral cells are additionally present in the subcerebral commissure of *Achatinella fulgens*.)

It seems doubtful whether this scheme is applicable to all Stylommatophora, as possibly *Succinea putris* (Cook, 1966) and *Strophocheilus oblongus* (Kuhlmann, 1963; Nolte, 1965) lack nerves of the
cerebral arteries. The larger axone bundles originating from the mesocerebral neurosecretory cells would, as in the Basommatophora (see
below), run primarily into the median lip nerve and form a neurohaemal
area in the periphery of this nerve. Van Mol (1967), on the other hand,
holds the opinion that nerves of the cerebral arteries also exist in
Succinea putris. As in *Achatinella fulgens* they do not originate from the
cerebral ganglia as separate nerves, but from the median lip nerves at
some distance from the ganglia. A second deviation from the general
scheme is the fact that intercerebral nerves were not found in every
species studied. In *Succinea putris, Achatinella fulgens, Sphincterochila
candidissima* and *Strophocheilus oblongus* these nerves are absent
(Nolte, 1965; Cook, 1966; Van Mol, 1967). (On the other hand, there are
indications that such nerves exist in the basommatophore *Helisoma
tenue* Philippi (Simpson, 1969; see also Wendelaar Bonga, 1970b).
Probably the periphery of the cerebral commissure of those Stylommatophora lacking intercerebral nerves serves as a neurohaemal area
(cf. Nolte, 1965). It is interesting to mention in this respect that the
periphery of the cerebral commissure of the basommatophore *Lymnaea
stagnalis* is primarily the neurohaemal area of certain Gomori-negative
neurosecretory cells (caudo-dorsal cells) of the cerebral ganglion (Joosse,
1964; Boer *et al.*, 1968a; Wendelaar Bonga, 1970b, 1971a; see below).

Histochemical and ultrastructural information on the mesocerebral
neurosecretory system of the Stylommatophora (Guyard, 1971) is
extremely limited. Kuhlmann (1963) concludes on the basis of the results
of some histochemical reactions that the secretion product in Helicidae
is lipoproteinaceous in nature. The same holds for *Succinea putris*
(Cook, 1966). The function of the mesocerebral neurosecretory system is
still unknown.

The intercerebral nerves branch in the connective tissue dorsal to the
commissure. In this tissue cell groups of the medio-dorsal body are
located. Nolte (1965), Kuhlmann (1966) and Van Mol (1967) suggest that
this morphological situation reflects the neurosecretomotor innervation
of the endocrine medio-dorsal body. This question will be discussed
further in the section on the dorsal bodies.

In addition to the mesocerebral neurosecretory system neurosecretory cells have been described in most other ganglia of the CNS of the Stylommatophora. In the buccal ganglia of *Arion ater rufus* Gomori-positive cells were found sending their axones into the pharyngeal nerve (Herlant-Meewis and Van Mol, 1959). In Helicidae, Gomori-positive cells were described in the postcerebrum of the cerebral ganglia, the buccal, pedal and visceral ganglia (Krause, 1960; Kuhlmann, 1963); no neurohaemal areas were found. This also holds for Gomori-positive cells

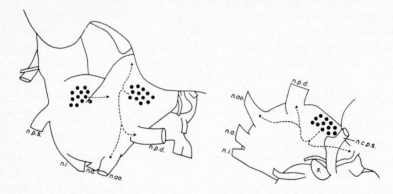

Fig. 2. Diagrammatic representation of the visceral complex (dorsal and right lateral view) of *Succinea putris* showing the phloxinophilic neurones and the route of transport of the phloxinophilic material. n.a., anal nerve; n.ao., aortal nerve; n.c.p.s., second cutaneous pedal nerve; n.i., intestinal nerve; n.p.d., right pallial nerve; n.p.s., left pallial nerve; s., statocyst. (From Cook, 1966.)

in the buccal ganglia of *Succinea putris* (Cook, 1966). In this species furthermore a group of Gomori-positive cells was observed in the ventral part of the procerebrum. These 'possible' neurosecretory cells have staining properties which differ from those of the mesocerebral cells. They are histochemically similar to a group of 13–15 Gomori-positive cells present at the right side of the visceral complex (in *Succinea* the pleural, parietal and visceral ganglia have fused into a complex). No associated neurohaemal areas were found. On the other hand, indications of the presence of neurohaemal areas were obtained for two other groups of Gomori-positive cells and especially for two groups of Gomori-negative cells (phloxinophilic cells) occurring in the visceral complex of *Succinea* (Cook, 1966). The groups of Gomori-positive cells (17–24 cells per group) are located on either side of the complex dorsal to the origins of the pleuropedal connectives. Histochemically these cells differ from the two other types of Gomori-positive cells already mentioned. It seems possible that the cells transport their secretion to various nerves of the complex, viz. to the pallial nerve, to the anal nerve and to the intestinal nerve, since in these nerves axone bundles were found with the same staining properties as the cell bodies in the visceral complex. The groups of phloxinophilic cells (each of 10–15 cells) are located in the dorso-lateral parts of the visceral complex (Fig. 2). The axones of the cells of the right group could be traced into the right nerve of the pharynx retractor muscle, the right pallial nerve and the aortic nerve. The axones of the cells of the left group run in the direction of the right cell group, and could not be traced further. Although in these cases it is difficult to speak of neurosecretory systems, as no definite neurohaemal areas were found, the organization of the cells and their axones indicates that the neurosecretory material is transported into various nerves. This is interesting, since in the basommatophore *Lymnaea stagnalis* a similar diffuse organization was found (Wendelaar Bonga 1970b; see below).

b. Basommatophora

The first extensive study on neurosecretion in a basommatophore is that of Lever (1957) on *Ferrissia shimekii*. After that the following

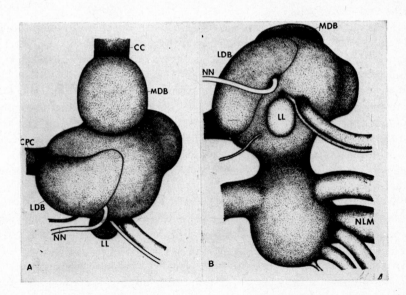

Fig. 3. Dorsal (A) and lateral (B) view of the right cerebral ganglion of *L. stagnalis* showing the position of the medio- (MDB) and latero-dorsal body (LDB). CC, cerebral commissure; CPC, cerebro-pleural connective; LL, lateral lobe; NLM, median lip nerve; NN, nuchal nerve. (After Joosse, 1964.)

species received most attention: *Lymnaea stagnalis* (e.g. Altmann and Kuhnen-Clausen, 1959; Lever *et al.*, 1961b; Joosse, 1964; Boer, 1965; Boer *et al.*, 1968a, b; Brink and Boer, 1967; Wendelaar Bonga, 1970a, b, 1971a, b), *Planorbarius corneus* (L.) (Röhnisch, 1964; Nolte, 1965; Kuhlmann, 1970), *Helisoma tenue* (Simpson *et al.*, 1966a, b; Simpson, 1969) and *Biomphalaria glabrata* Say (Lever *et al.*, 1965). The most comprehensive investigation is that of Wendelaar Bonga on *L. stagnalis*. The author extended the earlier work by using a new staining method, the alcian blue–alcian yellow technique (AB/AY).

In *L. stagnalis* the technique appeared to be highly selective for various types of Gomori positive cells, which had until then only been identified by their position (Lever *et al.*, 1961b; Joosse, 1964). With AB/AY seven types of CH-positive cells can be distinguished, as they stain in different shades of green. The colour differences obtained are supposed to reflect different ratios of strong and weak acid groups in the secretion materials (cf. Peute and Van de Kamer, 1967). Furthermore in *L. stagnalis* 4 types of phloxinophilic Gomori-negative cells occur, of which three stain with alcian yellow. Not only the cell bodies, but also the axones take the colour characteristic for the cells. This finding facilitates the search for the neurohaemal areas of the different cells and cell groups.

Most of the cell types were also studied with the electron microscope. The observations revealed that the secretions distinguished with AB/AY and with phloxin consist of elementary granules which differ in size and/or appearance, indicating that the histochemical differences observed are borne out at the ultrastructural level (see Fig. 8). The various granule types could also be discerned in the neurohaemal areas. The results of the investigations are summarized in a map of the CNS of *L. stagnalis*, showing the location and staining characteristics of the neurosecretory cells and systems (Fig. 4). In all ganglia, except for the buccal and the pedal ganglia, neurosecretory cells were found.

In the cerebral ganglia of *L. stagnalis* two types of Gomori-positive cells (light green cells, bright green cells; Figs. 5a, b, 6a) and two types of Gomori-negative cells (caudo-dorsal cells, see Joosse, 1964; and SBB cells, see Boer, 1965) are present.

Of the light green cells in each ganglion two large groups occur (mean number of cells per group fifty; maximum cell diameter 80 μm) in the medio- and latero-dorsal parts of the ganglia, under the endocrine medio- and latero-dorsal bodies (see also section on dorsal bodies). On the basis of this location they have been called medio- and latero-dorsal cells, respectively (Joosse, 1964). In other species similar cells

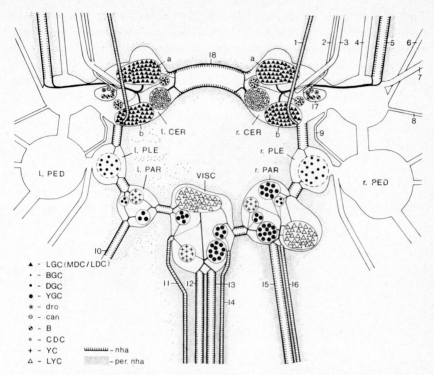

Fig. 4. Diagram of the location of the neurosecretory cell groups and their neurohaemal areas (nha) in the central nervous system of *Lymnaea stagnalis* (dorsal view). The mottled areas (per. nha) represent parts of the perineurium and of the connective tissue which are traversed by small nerves containing neurosecretory axones. The pedal ganglia and the ventral parts of the cerebral ganglia are turned to the sides. The medio- and latero-dorsal bodies are not indicated. CER, cerebral ganglia; PLE, pleural ganglia; PAR, parietal ganglia; VISC, visceral ganglion; PED, pedal ganglia; LGC, light green cells; MDC, medio-dorsal cells (a); LDC, latero-dorsal cells (b); BGC, bright green cells; DGC, dark green cells; YGC, yellow green cells; dro, droplet cells; can, canopy cell; B, B-cells; CDC, caudo-dorsal cells; YC, yellow cells; LYC, light yellow cells; 1, nuchal nerve; 2, optic nerve; 3, tentacular nerve; 4, superior frontal lip nerve; 5, median lip nerve; 6, cerebrobuccal connective; 7, penial nerve; 8, subcerebral commissure; 9, statocyst nerve; 10, left pallial nerve; 11, cutaneous pallial nerve; 12, anal nerve; 13, intestinal nerve; 14, genital nerve; 15, right internal pallial nerve; 16, right external pallial nerve; 17, lateral lobe (the follicle gland is not indicated); 18, cerebral commissure. (From Wendelaar Bonga, 1970b.)

have been found (e.g. in *Planorbarius*, Röhnisch, 1964; Kuhlmann, 1970). In species lacking latero-dorsal bodies, only one group of cells, the medio-dorsal cells, is present as in *Helisoma tenue* (Simpson, 1969). The axones of the dorsal cells form a main bundle which runs into the median lip nerve. Here they branch repeatedly. In *L. stagnalis* the periphery of the entire nerve serves as a neurohaemal area (Joosse, 1964; Fig. 7b). In addition to this main bundle small numbers of light green axones were observed in part of the cerebral commissure and in tiny nerves in the perineurium surrounding this commissure and the median

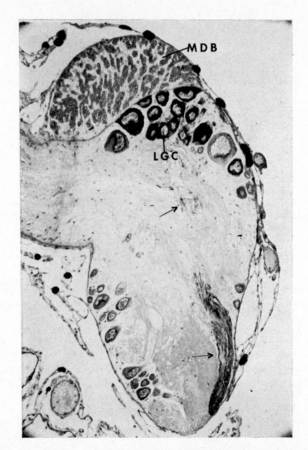

Fig. 5. *Lymnaea stagnalis*. Transverse section through a cerebral ganglion, showing the medio-dorsal light green neurosecretory cells (LGC) and the axone tract (arrows) running into the median lip nerve. MDB, medio-dorsal body. × 100.

lip nerve (Wendelaar Bonga, 1970b; see also Röhnisch, 1964; Nolte, 1965; Simpson, 1969). Furthermore light green-stained nerves were found in low numbers in the perineurium around the cerebral ganglia and the medio- and latero-dorsal bodies (see also Sanchiz and Zambrano, 1969).

With the electron microscope axones containing elementary granules characteristic of the light green cells were found in all these places (Fig. 8). Moreover, signs of release of the NSM were observed. Thus, the neurohaemal area of the light green cells is rather diffuse. Apparently the occurrence of diffuse neurohaemal areas is, as already mentioned, a general phenomenon in the Pulmonata (Nolte, 1965; Wendelaar Bonga, 1970b; see Figs 9, 10). It is conceivable that the neurosecretory materials easily reach the general circulation, since the perineurium of the CNS in both Stylommatophora and Basommatophora is well provided with blood vessels and blood spaces (Boer and Lever, 1959; Pentreath and Cottrell, 1970; Bekius, 1972).

It has been suggested that the dorsal cell–median lip nerve neurosecretory system of the Basommatophora is homologous with the mesocerebral system of the Stylommatophora (Nolte, 1965). The systems have similar locations and staining properties. Differences were, however, observed at the ultrastructural level between different basommatophores. For example, the mean diameter of the elementary neurosecretory granules in *L. stagnalis* and *Planorbarius corneus* is 200 nm (Boer *et al.*, 1968a; Kuhlmann, 1970), whereas the granules measure 150 nm in *Helisoma tenue* and 170 nm in *Biomphalaria glabrata* (Simpson, 1969; Boer and De Morree, unpublished observations). This finding throws doubt on the supposed homology of the systems in different species and groups.

The function of the neurosecretory system of the dorsal cell–median lip nerve has been investigated thoroughly in *L. stagnalis* (Joosse and Geraerts, 1969). Very probably the cells produce a growth hormone. As argued before, histologically discernible changes in the amount of secretion are difficult to interpret in functional terms. Yet it should be mentioned that Joosse (1964) has presented histological evidence suggesting that the dorsal cells are most active during spring, when the animals grow rapidly, so that the histological findings may be taken to sustain the supposition that the cells produce a growth hormone. Interestingly Wendelaar Bonga (1971a) found with quantitative electron microscopy that the light green cells show a diurnal rhythmicity. During the night, the release of neurosecretory material, which is low during the day, increases considerably. In this respect the observations of Röhnisch (1964) on *Planorbarius corneus* should also be mentioned. The dorsal cells of this species appeared to accumulate secretion if the animals were kept in continuous light. It is not yet clear how these observations relate to the function of the cells.

The second type of Gomori-positive cell in the cerebral ganglia of *L. stagnalis* stains bright green with AB/AY (Fig. 6a). The cells are relatively small (diameter 15–25 μm) and occur in two groups of 10–25 cells in each ganglion. One group is located near the medio-dorsal cells between the anterior lobe and the cerebral commissure; the other lies adjacent to the neuropile between the medio- and latero-dorsal cells (Fig. 4). No ultrastructural data are available, because these cells could not be recognized beyond doubt in the electron microscope. Their neurohaemal area is not known since their axons could only be traced over a small distance, because they intermingle with the main tract of the medio- and latero-dorsal cells.

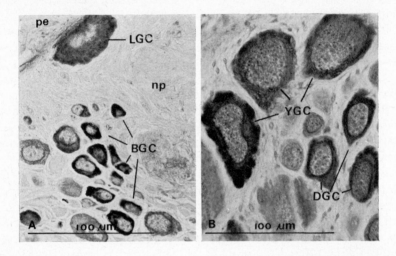

Fig. 6. *L. stagnalis*. A group of bright green cells (BGC) in the right cerebral ganglion. LGC, light green cell; np, neuropile; pe, perineurium; B, parietal ganglion with dark green cells (DGC) and yellow green cells (YGC). AB/AY staining. (From Wendelaar Bonga, 1970b.)

The first type of Gomori-negative cells in the cerebral ganglia are the phloxinophilic caudo-dorsal cells. They lie caudo-dorsally in groups of 20–40 cells in the left and of 50–100 cells in the right ganglion and transport their neurosecretory material to the periphery of the cerebral commissure, the neurohaemal area of the cells (Joosse, 1964). The cells are not stained by AB/AY. The neurosecretory material consists of fairly electron-dense elementary granules with a mean diameter of 150 nm (Fig. 7b; Boer *et al.*, 1968a). Like the light green cells, the caudo-dorsal cells show a diurnal rhythm, release being high during the night and low during the day (Fig. 13). The function of these cells is not yet clear.

Of the second type of Gomori-negative cells in the cerebral ganglion of *L. stagnalis* only little is known. They occur in two small groups of one to three cells located near the pleural connective and near the origin of the statocyst nerve, respectively (Boer, 1965). They stain intensely with phloxin, Sudan black B and alcian yellow. Their axones could be traced only over a short distance. Therefore their neuro-haemal area is not known, but may well be located in the periphery of the median lip nerve, as along the peripheral part of this nerve Sudan black B positive material was observed*. The cells have not

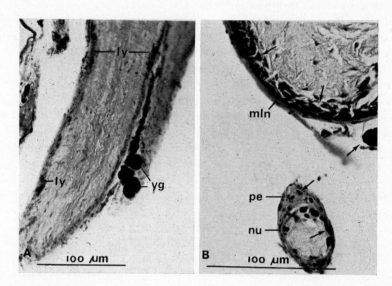

Fig. 7. *L. stagnalis*, neurohaemal areas. A. Oblique section through the right internal pallial nerve. Neurosecretory axone endings of the light yellow type (ly) occur at the periphery of the nerve. Distended yellow green axones (yg) are present in the perineurium. AB/AY staining. B. Median lip nerve (mln) and nuchal nerve (nu). Gomori-positive axone terminals occur in the nerves and in the perineurium (pe). PF staining. (From Wendelaar Bonga, 1970b.)

been recognized in electron microscope preparations. Their position is not indicated in Fig. 4.

In the cerebral ganglia of the Basommatophora investigated so far

* The periphery of the median lip nerve, the main neurohaemal area of the medio- and latero-dorsal cells is very probably the neurohaemal area for at least four types of neurosecretory cells. This can be concluded from the fact that at the ultrastructural level three types of neurosecretory elementary granules were distinguished in addition to those of the medio- and latero-dorsal cells (Wendelaar Bonga, 1970b).

K

Gomori-negative cells have only been reported for *L. stagnalis*. Nevertheless, in the periphery of the cerebral commissure of *Planorbarius corneus* and *Helisoma tenue* axones carrying elementary granules, which are different from those of the dorsal Gomori-positive cells, are present (Nolte, 1965; Simpson, 1969). These findings indicate that in these species as in *L. stagnalis*, the periphery of the commissure has to be considered as a neurohaemal area. Very likely the cells from which the neurosecretory axones originate are not stained by the routine staining methods (Simpson, 1969).

In the Basommatophora a small lobe is attached to the lateral part of the cerebral ganglion. This lateral lobe has to be considered homologous with the procerebrum of the Stylommatophora (Van Mol, 1967). In most species investigated neurosecretory cells have been described in this lobe (e.g. Lever, 1958b; Lever and Joosse, 1961). No neurosecretory cells were found in the lateral lobes of *Helisoma tenue* (Simpson, 1969), *Lymnaea auricularia* (L.) (Nishioka *et al.*, 1964) and *Planorbarius corneus* (Röhnisch, 1964; Nolte, 1966). In those of *L. stagnalis* and of *B. glabrata* (Lever *et al.*, 1965) three types of Gomori-positive cells were observed: one very large "canopy cell", two droplet cells and one large and several small B cells (Fig. 15; Brink and Boer, 1967). In *L. stagnalis* the cells stain with AB/AY brownish green, faint light green and brilliant green, respectively (Wendelaar Bonga, 1970b). The mean diameter of the elementary granules is 205 nm in the canopy cells, 135 nm in the droplet cells and 110 nm in the B cells. The location of the neurohaemal areas of these cells is not known. The neurosecretory axones leave the lateral lobe via the posterior connective with the cerebral ganglion. They could only be traced up to the central neuropile of this ganglion. In the lateral lobe a vesicular structure, the cerebral gland, also occurs. The question whether or not this gland is the neurohaemal organ of the B-cells – these cells are bipolar and send one projection to the gland – will be discussed in the section on the cerebral gland.

In the ganglia of the visceral ring of *L. stagnalis* two types of Gomori-positive cells (dark green cells, yellow green cells) and two types of Gomori-negative cells (light yellow cells, yellow cells) occur (Wendelaar Bonga, 1970b; Fig. 4).

In the pleural ganglia dark green cells are the only type of neurosecretory cells. The cells are medium-sized and occur (10–15 in the left ganglion, 15–20 in the right) scattered between the ordinary neurones. Furthermore dark green cells are found in small groups in the parietal ganglia (4–10 cells in the right and 2–5 cells in the left) near the parietopleural connectives (Fig. 6b). The neurohaemal area of the dark green

cells is rather diffuse. Part of the axones end in the periphery of the
cerebro- and parieto-pleural connectives. Dark green axones were also
found in the periphery of the nuchal nerve and in tiny nerves originating from the pleural ganglia and their connective. The tiny nerves form
a network between blood spaces in the perineurium (Fig. 9). The mean
diameter of the elementary granules of the dark green cells is 200 nm
(Fig. 8c). There is strong experimental evidence indicating that the
dark green cells are involved in osmoregulation (Hekstra and Lever

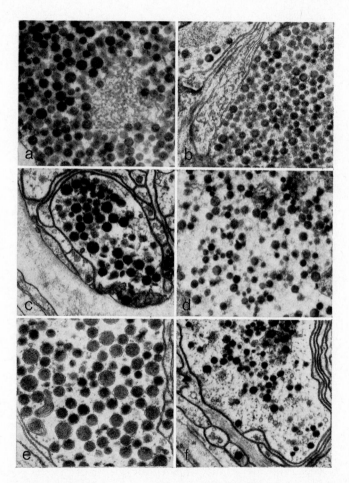

Fig. 8a–f. *L. stagnalis*. Neurosecretory elementary granules in axone endings of
six cell types. a, light green type; b, type of the caudo-dorsal cells; c, dark green
type; d, yellow green type; e, light yellow type; f, yellow type. × 18 000. (From
Wendelaar Bonga, 1970b.)

1960; Lever *et al.*, 1961a; Wendelaar Bonga, 1971b; see sections on synthesis and release of neurosecretory material and on osmoregulation).

The second type of Gomori-positive neurosecretory cells in the visceral ring are the yellow green cells (Fig. 6b). They occur in several groups in the left and right parietal and in the visceral ganglion, in the regions near the origins of the connectives and nerves (Fig. 4; see also Lever *et al.*, 1961b). In the right parietal ganglion 10–20 of these

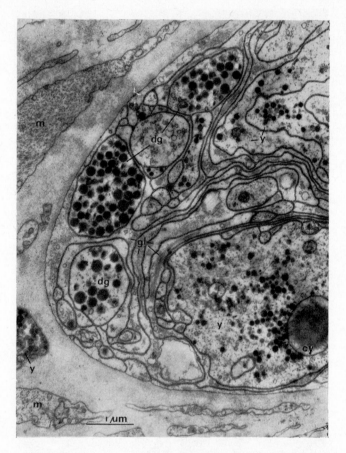

Fig. 9. *L. stagnalis.* Cross section of part of a small nerve in the perineurium near the pleural ganglion, containing many neurosecretory axones. y, axone containing the yellow type of neurosecretory elementary granules; dg, axone containing the dark green type of elementary granules; gl, filamentous glial cell processes; m, muscle fibre; cy, cytosome. (From Wendelaar Bonga, 1970b.)

large cells (length of the perikarya: 25–70 μm) are intermingled with dark green cells near the parietopleural connective, while 10–30 are concentrated in the region where the pallial nerves leave the ganglion. About two to five cells are situated near the parietovisceral connective. The left parietal ganglion contains only one to six cells of this type. In the visceral ganglion yellow green cells occur in two groups of five to ten cells near the connectives with the parietal ganglia. In addition to these approximately ten to twenty cells were found dispersed in the region from which the visceral nerves originate. Like those of the dark green cells the axones of the yellow green cells do not form prominent axonal tracts. In general, the axones proceed to the nearest connective or nerve, many endings being found in the peripheries of the right parietal and visceral nerves. There are also numerous axones traversing the neuropile. They were often traced to their endings in more distant parts of the nervous system, e.g. in adjoining ganglia and also in the nuchal nerve. In the perineurium surrounding the visceral ring and the pallial, visceral and nuchal nerves a network of yellow green axones is present. So, the neurohaemal area for the yellow green cells is also very diffuse (Figs 7a, 9, 10). The mean diameter of the elementary granules is 165 nm. Yellow green cells were also found outside the CNS proper, viz. in the visceral and pallial nerves. Up till now there is no indication of the function of the yellow green neurosecretory system.

In other Basommatophora no neurosecretory cells have been established in the pleural ganglia. When compared with *L. stagnalis* the number of neurosecretory cells in the parietal ganglia of *Ferrissia shimekii* (Lever, 1957), *Planorbarius corneus* (Röhnisch, 1964), *Biomphalaria glabrata* (Lever *et al.*, 1965) and *Helisoma tenue* (Simpson *et al.*, 1966a; Simpson, 1969) is rather limited. Also in the visceral ganglion of most species the number of Gomori-positive cells is low when compared with that of *L. stagnalis*. There are also indications that in the other species the Gomori-positive cells of the visceral ring differ from those of the cerebral ganglia. For example, Kuhlmann (1970) reports that the mean diameter of the elementary granules of cells in the visceral ganglion is 150 nm in *Planorbarius corneus*, whereas that of the granules of the dorsal cells in the cerebral ganglia is 200 nm (see also Simpson, 1969). Furthermore, the cells had different staining affinities.

Only little is known about the neurohaemal areas of the Gomoripositive cells in other Basommatophora. Simpson (1969) observed neurosecretory axone terminals in the periphery of the proximal part of the intestinal nerve of *Helisoma*. No terminations were found in

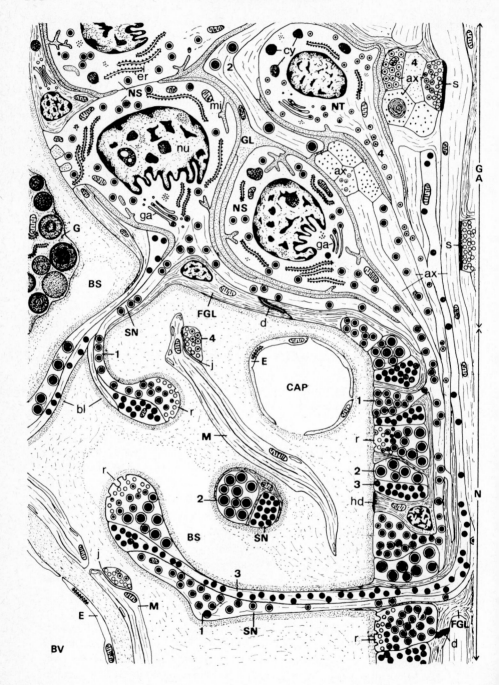

the other nerves of the visceral ganglion or in the pallial nerves. In *Planorbarius corneus* Röhnisch (1964) detected fuchsinophilic secretory tracts ending in the pallial nerves.

The first type of Gomori-negative cells in the visceral ring of *L. stagnalis* are the light yellow cells. Cells of this type occur in two large groups. The first group, consisting of 30–50 cells which measure 20–60 μm in length, is located in a lobe-like protrusion in the ventral part of the right parietal ganglion. The same number of cells forms a group in the dorsal part of the visceral ganglion (Fig. 4). Light yellow cells are absent from the left parietal ganglion. The cells stain moderately with phloxin and Sudan black B. The axones and axone endings of these cells were encountered in the same areas as described for the yellow green cells. The elementary granules of the light yellow cells have a mean diameter of 230 nm (Fig. 8e). The function of the light yellow neurosecretory system is still unknown.

The second type of Gomori-negative cells in the visceral ring are the yellow cells (Fig. 4). These cells, which stain strongly with Sudan black B, occur in groups of five to ten cells in the dorsal parts of both parietal ganglia and in the ventral part of the visceral ganglion (Fig. 4). The length of the cell bodies varies between 20 and 40 μm. The distribution of the axone endings containing the yellow secretory material has the same pattern as that of the yellow green and the light yellow cells (Fig. 9). Furthermore many yellow axone terminals were found in the connective tissue of the kidney. The mean diameter of the elementary granules is 140 nm (Fig. 8f). Experimental evidence suggests that these cells, like the dark green cells, are involved in osmoregulation (Wendelaar Bonga, 1970a, 1972).

Gomori-negative cells have up till now not been described in the ganglia of the visceral ring of other Basommatophora. However, preliminary observations on *Biomphalaria glabrata* have shown that

Fig. 10. *L. stagnalis.* Diagrammatic representation of neurosecretory cells and their neurohaemal areas, in particular of the group of yellow green neurones near the origin of the right pallial nerves. GA, ganglion; NS, neurosecretory neurones; NT, neurotransmitter neurone; N, nerve with neurosecretory axone terminals at the periphery; SN, small nerve in the perineurium; FGL, filamentous glial cell; GL, glial cell forming the trophospongium; CAP, blood capillary; BS, blood space; BV, large blood vessel; E, endothelium; M, muscle fibre; G, granular cell; nu, nucleus; ga, Golgi apparatus; cy, cytosomes; er, granular endoplasmic reticulum; mi, mitochondrion; s, synapse; j, neuromuscular junction; d, desmosome; hd, hemi-desmosome; r, release site in neurosecretory axone ending; bl, basement membrane; ax, axones; 1, axone of the yellow green type; 2, axone of the light yellow type; 3, neurosecretory axone of the yellow type; 4, axone with neurotransmitter granules. (From Wendelaar Bonga, 1970b.)

in this species with AB/AY two types of such cells (light yellow cells, dark yellow cells) are present in the parietal and visceral ganglia (Boer and De Morree, unpublished data).

3. Synthesis and release of neurosecretory material

As already mentioned statements about the activity of secretory cells can only be made if the rates of synthesis and release of the products are known. To study these processes quantitatively several techniques

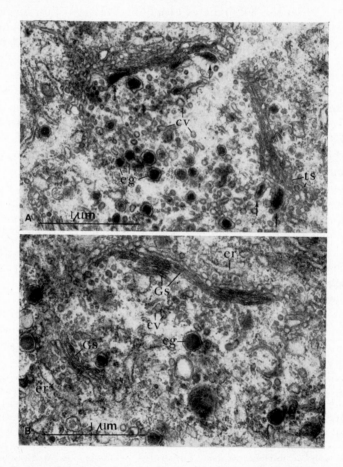

Fig, 11. *L. stagnalis*, caudo-dorsal cells. A, "Active" Golgi field. In distal parts the Golgi saccules contain accumulations of electron-dense secretory material (arrows). cv, clear vesicles; eg, elementary granules. Fixation: 22.00 h. B, "Inactive" Golgi field. Secretory material is absent from the Golgi saccules (Gs). Fixation: 14.00 h. (From Wendelaar Bonga, 1971a.)

may be applied such as autoradiography (Bierbauer *et al.*, 1967; Schooneveld, 1970) cytophotometry (Drawert, 1968) or morphometrical methods (Olivereau, 1970). Morphometry can be applied at the ultrastructural level (Wendelaar Bonga, 1971a, b).

Since neurosecretory material is primarily proteinaceous it can be assumed that in particular the granular endoplasmic reticulum (GER) and the Golgi apparatus are involved in the formation of the secretory granules (Novikoff and Holtzman, 1970). As a measure for the activity of the GER the length of the membranes of this organelle

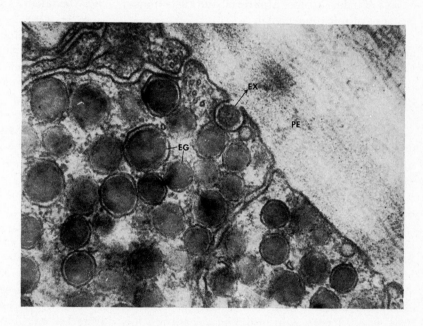

Fig. 12. *L. stagnalis*. Axone ending of a caudo-dorsal cell in the periphery of the intercerebral commissure. Exocytosis (EX) of the contents of an elementary granule (EG). PE, perineurium. × 100 000.

per unit surface area cytoplasm can be taken. This length can be determined by lineal integrative analysis (Loud *et al.*, 1965). As a measure for the activity of the Golgi apparatus the relative number of active Golgi fields can be taken. A Golgi field is considered to be active if electron-dense material is observed in the Golgi cisternae, since the presence of this material indicates that the Golgi field was forming secretion granules at the moment of fixation (Fig. 11) (Wendelaar Bonga, 1971a).

Parameters can also be found to measure release activity. Although several release mechanisms have been proposed (Nolte, 1965; Boer et al., 1968; Scharrer, 1968; Normann, 1969), it seems evident that in L. stagnalis neurosecretions are released by exocytosis, i.e. the membrane of an elementary granule fuses with the plasma membrane of the axone terminal so that the product can be released (Fig. 12). The membranes of the elementary granules then apparently transform into a number of small clear vesicles. The presence of indentations in the plasma-membrane of the axone terminal and of small clear vesicles indicates release of neurosecretory material and their frequency may be taken as a measure of release activity.

Other morphological parameters, e.g. the number of lysosomes and mitochondria, can provide additional information on cell activity. Furthermore, quantitative data on transport of elementary granules through the axones and storage of neurosecretory material in the axone terminals can be obtained.

So far, quantitative electron microscopy on synthesis and release of neurosecretory material has, in pulmonates, only been carried out in L. stagnalis (Wendelaar Bonga, 1971a, b, 1972). Particular attention was paid to the caudo-dorsal cells of the cerebral ganglia and to the dark green cells of the pleural ganglia.

Preliminary investigations had indicated that the caudo-dorsal cells show a diurnal rhythmicity in their activity. To test this hypothesis the CNS of a group of animals was fixed every 4 h during a period of 24 h and the synthetic and release activities of the cells were determined with quantitative electron microscopy. The relative number of active Golgi fields was taken as a measure of the synthetic activity only, since the variation in length of the GER per unit surface area cytoplasm was too small to be taken as a parameter. As can be seen from Fig. 13 both synthesis and release are low during the day, but during the night a considerable increase in the rate of these processes occurs. A similar curve was obtained for the release of elementary granules of the light green cells in the median lip nerve. Four other types of neurosecretory cells (dark green cells, yellow green cells, light yellow cells and yellow cells) showed no rhythmicity in their release activity.

The pleural ganglia of L. stagnalis produce an osmoregulatory hormone (Hekstra and Lever, 1960, Lever et al., 1961a; see also Chaisemartin, 1968). Since dark green cells are the only neurosecretory cells in these ganglia it may be assumed that the hormone is being produced by these cells. To test this hypothesis animals were kept in osmotically different media (deionized water, 0·1 M NaCl) and the synthetic and release activities of the dark green cells were analysed. The results clearly confirmed the hypothesis. In animals kept in deionized water the length of the GER and the number of active Golgi fields increased significantly. Also the number of release sites in the neurohaemal zones increased when compared with control snails. In animals kept in 0·1 M NaCl inactivation of the cells occurred. Change of the osmolarity of the medium had no effect on the release activity

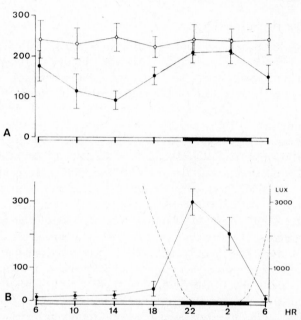

Fig. 13. *L. stagnalis.* A, Determination of the number of active Golgi fields in the cell bodies of the caudo-dorsal cells of animals fixed at different times during a period of 24 h. The horizontal black bar indicates the period between sunset and sunrise. Ordinate: Number of Golgi zones per unit of surface area of cytoplasm. Open circles: total number of Golgi zones; black dots: active Golgi zones. B, Number of axones showing release sites in the neurohaemal area of the caudo-dorsal cells, the cerebral commissure. Ordinate: number of releasing axones per 3 cross sections of the cerebral commissure. Broken line (right scale): light intensity. (From Wendelaar Bonga, 1971a.)

of other neurosecretory cell types, except for the yellow cells (Wendelaar Bonga, 1970a, 1971b, 1972; see also the section on osmoregulation, III C, p. 298).

The above results show that it is possible to study various aspects of the process of neurosecretion with quantitative electron microscopy. One of these aspects, which is now under investigation in this laboratory in *L. stagnalis*, is how neurosecretory cells are regulated in their secretory activity. The first results obtained by combining implantation of pleural and parietal ganglia and quantitative electron microscopy, indicate that the reaction of dark green cells of implanted ganglia and of intact ganglia to media of different osmolarity is not different (Roubos, 1973). Apparently the cells can react without being stimulated by nervous imput from other ganglia of the CNS.

4. Peripheral neurosecretion

In most cases neurosecretory material is released in neurohaemal organs or regions which are located in the vicinity of the CNS. There are, however, also reports showing that neurosecretory material is released peripherally, close to the target organ (Cottrell and Osborne, 1969; Madrell, 1967; Jariol and Scudder, 1970; Scharrer, 1969; Scharrer and Weitzmann, 1970). In the Pulmonata peripheral neurosecretion has been investigated to some extent in *Helix pomatia* (Foh and Schlote, 1965) and in *L. stagnalis* (Wendelaar Bonga, 1970a, 1972). In *H. pomatia* PF-positive axone tracts were observed in the penis retractor muscle and in the heart. In *L. stagnalis* the presence of neurosecretory fibres was established in the connective tissue around the kidney sac and ureter. They originate from the yellow cells of the parietal and visceral ganglia. The release of the yellow neurosecretory material occurs not only peripherally but also in the proximal parts of some nerves and in connectives (see Fig. 4). This release can be influenced by placing snails in osmotically different media, as has been established by quantitative electron microscopy (Wendelaar Bonga, 1970a, 1972). Apparently the axones release a hormone involved in osmoregulation (see section on Osmoregulation, III C, p. 298).

Peripheral neurosecretion has in a preliminary study furthermore been found in some other organs in *L. stagnalis* (Wendelaar Bonga and Plesch, unpublished results). The number of these axone endings was always small when compared to that of "yellow" axones in the kidney. First, axone terminals of the yellow green cells were observed between the lobes of the digestive gland. These axones were not found in the connective tissue around the acini of the ovotestis. This location might indicate that the yellow green cells produce a factor involved in the digestion process. This supposition would not, however, be in accordance with the observation that axone endings of the yellow green cells occur also in the connective tissue around the vagina, and in this latter region axone endings of the light yellow cells were also found. No neurosecretory axone endings were observed in the connective tissue of the other parts of the reproductive tract.

B. Cerebral gland

In an extensive comparative study of the morphology and histology of the cerebral ganglia of the pulmonates Van Mol (1967) pays special attention to the procerebrum. This part of the brain, which is absent in the other subclasses of the Gastropoda, develops during embryogenesis from the distal part of a special ectodermal invagination, the cerebral

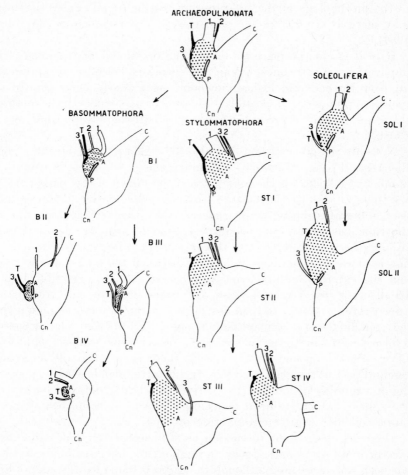

Fig. 14. Stages in the transformation of the procerebrum in the course of the evolution of the Pulmonata. Dots and circles indicate small and large cells in the procerebrum. 1, tentacular nerve; 2, peritentacular nerve; 3, optic nerve; C, cerebral commissure; Cn, cerebropleural connective; A, anterior procerebral connective; P, posterior procerebral connective; T, cerebral gland. (From Van Mol, 1967.)

tube. According to Van Mol the procerebrum (in Basommatophora often called "lateral lobe", see above) is "un lobe d'association en rapport avec les organes sensoriels du tentacule" (Van Mol, 1967, p. 127). Its significance and development in the course of evolution would be closely related to the chemoreceptive function of the tentacle. In terrestrial species this function is much more pronounced than in aquatic forms. Along with differences in tentacular function differences in the morphology of the procerebrum can be found (Fig. 14). These may serve as criteria for the phylogeny of the pulmonates (Van Mol, 1967).

Here it is not the place to discuss the taxonomic and phylogenetic implications of this concept. Van Mol's work is, however, also interesting from the endocrinological viewpoint, as it clearly shows that in all groups of pulmonates a possibly endocrine or neuroendocrine structure, the cerebral gland, is present in association with the procerebrum (Lever, 1958b).

Not only the procerebrum, but also the cerebral gland develops from the cerebral tube. Its lumen – or lumina: the gland may split up into a number of follicles – is the remainder of the lumen of the distal part of the tube, i.e. the part associated with the procerebrum. The proximal part obliterates during embryogenesis after the external opening of the tube has closed. In this respect *Ovatella myosotis* (Draparnaud) (Ellobiidae) is an exception. In this species the tube persists throughout life, as does the open communication between the lumen of the tube and the external environment (Lever et al., 1959; Van Mol, 1967). In all other investigated species, the tubes involute, but not to the same degree. In many of them, especially in the Basommatophora, the only remainder of a cerebral tube is one well developed oval or round follicle, located in the periphery of the procerebrum (follicle gland, see Lever, 1958b). In many other species there is an intra- and an extra-cerebral part of the gland (see Fig. 1), the latter glands usually consisting of a number of follicles. In "active" cerebral glands the lumina of the follicles, which are lined by a simple brush-border epithelium, contain a colloidal or granular secretion product.

There has been some discussion as to whether cerebral glands are present in all pulmonate species throughout the lifespan. This question has been raised because, although the gland is found in all stages of life in Basommatophora, observations suggested that it disappears in Stylommatophora in the course of life (Pelseneer, 1901; see also Lever et al., 1959). However, several authors found the glands to be present not only in young, but also in adult Stylommatophora (Van Mol, 1960, 1967; Nolte and Kuhlmann, 1964; see for further references Van Mol,

1967 p. 105). The discrepancy may be explained by Van Mol's (1967) work on the slug *Arion ater rufus*. In this species the cerebral gland consists of an extra- and an intra-cerebral part. The glandular activity, as judged by the presence of secretion material in the follicles, varies in different periods of the lifespan of the animals. During the first 4 months after hatching the intra-cerebral part is well developed and shows active secretion. In adult specimens, on the other hand, this part of the gland has almost disappeared. The extra-cerebral gland becomes active in the fourth month and remains so throughout the reproductive period. In older specimens this part of the gland reduces too. In *Arion rufus*, at least, the activity of the cerebral gland seems to be age-dependent. These results are, however, not supported by Cook (1966) for *Succinea putris*, as he did not find large size differences during the lifespan.

In considering the function of the cerebral gland on the basis of histological evidence, it is important to establish the histochemical nature and the source of the secretion product of the gland. Of particular interest in this respect is the question whether or not neurosecretory cells contribute to the secretion product (Lever, 1958b; Lever *et al.*, 1959). If so, then the gland should be considered as a neurohaemal organ.

The hypothesis that neurosecretory cells contribute to the secretion product of the cerebral gland was based on the observation that in quite a number of Basommatophora one of the processes of the bipolar neurosecretory B cells of the procerebrum contacts the lumen of the gland (Fig. 15). Moreover, these neurosecretory cells and the colloidal secretion product of the cerebral gland show similar staining reactions with the neurosecretory stains CH and PF (Lever, 1958b; Lever *et al.*, 1959, 1965; Lever and Joosse, 1961; Boer, 1965; Brink and Boer, 1967). The neurosecretory B cells would not be the only source of the colloid, the epithelial cells contributing as well: in the apices of these cells small secretion droplets were often observed (Lever, 1958b).

Although the involvement of neurosecretory cells in the elaboration of the secretion product of the cerebral gland has not yet been ruled out unequivocally, doubt has been thrown upon this concept in four respects: 1. In some Basommatophora and in all Stylommatophora studied so far, no neurosecretory cells were observed in the procerebrum (Nishioka *et al.*, 1964; Nolte and Kuhlmann, 1964; Nolte, 1965; Simpson *et al.*, 1966a, b; Cook, 1966; Van Mol, 1967; Simpson, 1969; Kuhlmann, 1970). 2. The histochemistry of the follicular material differs from that of the neurosecretory cells (Nolte and Kuhlmann, 1964; Boer, 1965; Cook, 1966). The observations on three Basommatophora (*Lymnaea stagnalis, Physa fontinalis* (L.) and *Biomphalaria glabrata*) showed that in these species the follicular secretion consists of acid mucopolysaccharides, whereas that of the neurosecretory B cells is a lipoprotein (Boer, 1965). 3. In electron microscope studies no signs of release of neurosecretory elementary granules into the cerebral gland were observed (Nishioka *et al.*, 1964; Nolte, 1966; Brink and Boer, 1967). 4. Active cerebral glands were observed in young snails, which do not possess active neurosecretory cells (Sanchez and Bord, 1958; Van Mol, 1960; Joosse, 1964; Boer, 1965).

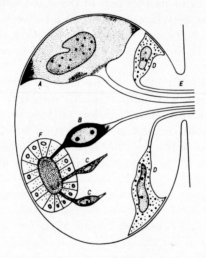

Fig. 15. Diagram of a transverse section through the lateral lobe of the cerebral ganglion of *Lymnaea stagnalis* (modified from Lever and Joosse, 1961). A, canopy cell; B, large B-cell; C, small B-cells; D, droplet cells; E, posterior lobe connection; F, follicle, with cytoplasmic granules in the brush-border epithelial cells. (From Boer, 1965.)

In conclusion it can be stated that it is improbable that neurosecretory cells contribute to the secretion product of the cerebral glands. The involvement of the epithelial cells in the production of this material (Lever, 1958b; Lever *et al.*, 1959; Van Mol, 1960, 1967; Boer, 1965; Cook, 1966) is better documented, especially since electron microscope observations have confirmed this hypothesis (Brink and Boer, 1967). The question may be raised as to whether in all species every epithelial cell has to be considered as a gland cell, since sometimes more than one epithelial cell type can be distinguished. For example, Van Mol (1967) found in the cerebral glands of *Siphonaria pectinata* (L.) and *Succinea putris* large gland cells in addition to ordinary epithelial cells. These large cells have their necks interposed between epithelial cells and their cell bodies lying beneath the epithelium.

Interpretation of these histological observations leads to the conclusion that the cerebral gland is apparently not a neurohaemal organ. It may, on the other hand, well be considered as an endocrine organ (Lever, 1958a; Nolte and Kuhlmann, 1964; Boer, 1965; Nolte, 1965, 1966; Brink and Boer, 1967), which has histologically some resemblance to the thyroid gland as the secretion product is first secreted into closed follicles and secondarily into the haemolymph (e.g. Lever et al., 1959). There are also indications that epithelial cells of cerebral glands secrete their products directly into the haemolymph (Van Mol, 1967).

The function of the gland is still unknown. As the gland shows different periods of activity during life in *Arion rufus*, Van Mol (1960) has suggested that it might be involved in the regulation of growth. However, in his publication of 1967 Van Mol expressed the opinion that the cerebral gland "semble être un organe endocrine dont la fonction nous échappe encore totalement" (p. 120). Nolte and Kuhlmann (1964) rejected the possibility that the gland is involved in osmoregulation, as they found active glands in hygrophilic as well as in aridophilic species. This opinion is sustained by experimental work on *Lymnaea stagnalis* (Boer, McLeod and Lokhorst, unpublished results): exposure of snails to media of various osmolarities or injecting them with saline solutions had no significant effect upon the volume of the glands.

The third possible function, that of a receptor organ, was in fact the first mentioned in the literature (de Nabias, 1899). On the basis of electron microscope studies of *Lymnaea auricularia* and comparison of the morphology of the cerebral gland of this species with the structure of the epistellar organ of *Octopus* Nishioka et al. (1964) share this opinion and suggest the gland is a vestigial photoreceptor (see also Simpson et al., 1966a). The authors found no secretion material in the gland and ascribed the observed fuchsinophilia to accumulations of microvilli present at the apices of the epithelial cells. Furthermore, they observed that non-neurosecretory neuronal processes, which are interposed between the epithelial cells, project cilia into the lumen of the gland. These latter findings are corroborated by electron microscope work of Simpson et al. (1966a) on *Helisoma tenue* and of Brink and Boer (1967) on *Lymnaea stagnalis, Planorbarius corneus* and *Ancylus fluviatilis* Müll. Since this organization is quite similar to the situation in the epidermis, where neurosensory cells send ciliated processes to the epidermal surface (Zylstra, 1972), the possibility that the gland is a receptor organ cannot be excluded.

Much experimental work is still needed to elucidate the function of the cerebral gland.

C. Optic tentacle

There is experimental evidence suggesting that the optic tentacles of Stylommatophora produce a hormone that is involved in the regulation of gametogenesis (Pelluet and Lane, 1961; Pelluet, 1964; Berry and Chan, 1968; Gottfried and Dorfman, 1970a; Bailey, 1973). This evidence will be discussed in the section on reproduction. Here we will discuss whether or not certain cells of the optic tentacles can be considered, on histological criteria, to secrete hormones.

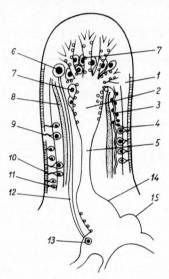

Fig. 16. Schematic representation of the optic tentacle of Stylommatophora. 1, dermato-muscular layer; 2, tentacular ganglion; 3, lateral tentacular fibres; 4, ganglion cells; 5, tentacular nerve; 6, eye; 7, collar cells; 8, retractor muscle; 9, lateral processed cell A; 10, lateral processed cell B; 11, gland cells of the dermato-muscular layer; 12, optic nerve; 13, postcerebrum; 14, procerebrum; 15, mesocerebrum. (From Bierbauer and Török, 1968.)

The optic tentacles of quite a number of Stylommatophora have been studied by histological and cytological techniques (Tuzet *et al.*, 1957; Sanchez and Bord, 1958; Lane, 1962, 1964; Smith, 1966; Bierbauer and Török, 1968; Rogers, 1969; see for further references, Bierbauer and Vigh-Teichmann, 1970). From these studies the following general picture emerges (Fig. 16). In the tentacle the tentacular ganglion is located close to the eye. It is connected with the procerebrum by the tentacular nerve and it gives rise to six branches running to the tip of the tentacle. When searching for hormone-producing cells primarily the classic neurosecretory stains CH and PF have been used. With these and other staining methods no neurosecretory cells were found in the tentacular ganglion (e.g. Lane, 1962; Bierbauer, 1967; Bierbauer et al., 1967). On the other hand, four cell types, occurring in different places in the tentacles gave a strong positive reaction with neurosecretory stains, viz. the collar cells, two types of lateral cells and the gland cells of the dermato-muscular layer (Fig. 16) (Lane, 1962; Bierbauer et al., 1965; Bierbauer and Török, 1968). Primarily on the basis of the staining reaction with the neurosecretory stains it has been assumed that the cells possibly are neurosecretory. However, later histochemical and electron microscope work strongly indicated that all 4 cell types are subepidermal gland cells, which release their secretion at the epidermal surface (e.g. Lane, 1964; Röhlich and Bierbauer, 1966; Rogers, 1969). This is especially clear for the lateral cells and the gland cells of the dermato-muscular layer (Bierbauer and Vigh-Teichmann, 1970). Concerning the collar cells there still consists a controversy. These cells have an intimate morphological relation to the tentacular ganglion (Fig. 16). It has been stated that they send a fine cytoplasmic projection into the ganglion, which would indicate that they are neurosecretory cells (Lane, 1964; Bierbauer and Török, 1968). On the other hand, the observation of Lane (1962) that the cells send a process towards the epidermis is sustained by the electron microscope work of Rogers (1969), who observed this process passing between the cells of the epidermal epithelium. Since moreover the ultrastructure of the secretion granules of the collar cells (Lane, 1964; Röhlich and Bierbauer, 1966; Rogers, 1969) differs greatly from that of characteristic neurosecretory elementary granules, it seems highly unlikely that the collar cells are neurosecretory elements.

So, it can be concluded that, on the basis of the available histological data, at the moment no cell type in the optic tentacle can be considered as a hormone source. Since, on the other hand, the experimental results are in favour of the production of a hormone by the optic tentacle a reinvestigation of this structure with new staining techniques and electron microscopy is feasible. In such a study special attention should be paid to the tentacular ganglion itself, as the ultrastructure of the neurones has hardly been investigated (Guyard, 1971).

D. Dorsal bodies

Dorsal bodies (DB) are structures associated with the cerebral ganglia. Their occurrence is not restricted to the Pulmonata since apparently similar structures have been reported for Opisthobranchia and Diotocardia (Martoja, 1965a, b, c, 1968). In pulmonates they are, up till now, the only non-nervous structures, of which the endocrine nature has been proved beyond doubt (Joosse, 1964, 1972; Joosse and Geraerts, 1969). They are involved in the regulation of reproduction (see section III B, reproduction).

In the earlier embryological literature the DBs are an important source of misunderstandings (Joosse, 1964), primarily because they were originally found only in Basommatophora. (Only since 1965 have DBs also been recognized in Stylommatophora.) Since they are well defined organs in Basommatophora, which are closely attached to the cerebral ganglia, many authors have taken them to be nervous structures, homologization of which with part of the cerebral ganglia of the Stylommatophora caused difficulties (Joosse, 1964). In 1931 it was Böhmig who expressed the opinion that they develop not as part of the cerebral ganglia, but as separate organs and that they are either of ectodermal or of mesodermal origin. Since electron microscope observations have shown that DB cells are not delimited by a basement membrane from tissue derived from the mesoderm (fibroblasts, muscle cells), a mesodermal origin of the DBs is feasable (Boer et al., 1968b).

In Basommatophora two types of DBs can be distinguished according to their location on the cerebral ganglia – medio-dorsal bodies (MDBs) and latero-dorsal bodies (LDBs). MDBs have been found in all Basommatophora studied. Additional LDBs occur only in a number of genera (*Ferrissia, Gundlachia, Planorbarius, Physa, Lymnaea*, see Joosse, 1964; *Biomphalaria*, Lever et al., 1965). In many genera two distinctly separate MDBs are present (*Ovatella, Physa, Lymnaea, Ferrissia, Ancylus*, see Lever, 1958a). They are located upon the medio-dorsal parts of the cerebral ganglia and cover also part of the commissure (Fig. 3). In other genera (*Carychium, Biomphalaria* and especially *Planorbarius*, see Lever, 1958a; Röhnisch, 1964; Lever et al., 1965; Boer et al., 1968b) the two MDBs are jointly encapsulated, extending over the entire commissure, so that it seems as if only one MDB is present. The LDBs, if present, are located upon the latero-dorsal parts of the cerebral ganglia.

In the Stylommatophora the situation is quite different. Compact MDBs, directly attached to the cerebral ganglia, have only been observed in *Succinea putris* (Lever, 1958a; Cook, 1966) and *Succinea elegans* (Van Mol, 1967) and possibly in *Strophocheilus oblongus* (Kuhlmann, 1966). In other species the DBs consist of cell groups occurring primarily dispersed in the dorsal part of the relatively thick layer of connective tissue surrounding the cerebral ganglia. This diffuse location

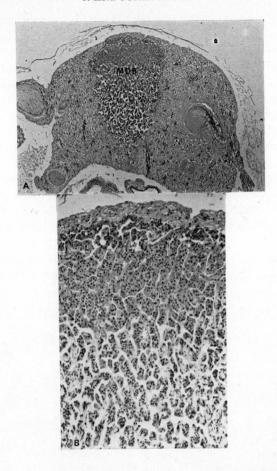

Fig. 17. A. Frontal section through the connective tissue sheath surrounding the cerebral ganglia of the helicid snail *Acavus phoenix* Pfeiffer, showing the location of the dorsal body (DB). B. Detail of the dorsal body. A, ×35; B, ×110

explains why the general occurrence of DBs in Stylommatophora has only recently been established (Fig. 17; Nolte, 1965; Cook, 1966; Kuhlmann, 1966; Van Mol, 1967).

In Basommatophora the histology of the LDBs is not essentially different from that of the MDBs. The DBs consist of small cells arranged in radially oriented groups, which are separated from each other by loose connective tissue. They send a process towards the site of attachment of the DB to the cerebral ganglion. This arrangement means that a cortex, consisting of cell bodies, and a medulla, mainly composed of processes of DB cells, can often be distinguished. The dorsal capsule of DBs is usually well developed and consists of fibro-blasts, collagen fibrils and muscle cells; in addition other connective tissue cells may be present. This dorsal capsule is continuous with the outer layers of the perineurium covering the cerebral ganglia and the intercerebral commissure. Ventrally the DBs are closely apposed to the cerebral ganglia. From the endocrinological point of view the structure of this site of attachment is of special interest, as it has been supposed that at this site morphological contacts exist between the neurosecre-tory cells in the cerebral ganglion and the DBs (Lever, 1958a; Simpson et al., 1966a). This hypothesis was primarily forwarded on the basis of light microscope observations. In the cerebral ganglia the dorsal neuro-secretory cells (MDC, LDC, light green cells, see above) are located just underneath or slightly lateral to the DBs. In addition to the main axones, which run to the median lip nerve, these cells send pro-cesses towards the site of attachment of the DBs. These processes stain positively with CH and PF (Fig. 18). Since, furthermore, positively

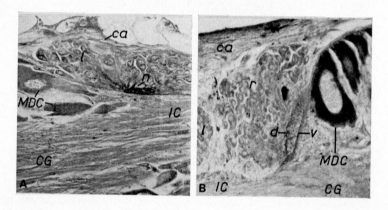

Fig. 18. Photo micrographs of cross sections through the cerebral ganglion – MDB complex in *Ancylus* (Fig. A) and *Biomphalaria* (Fig. B). Paraldehyde-fuchsin stain. l, left MDB; r, right MDB; CG, cerebral ganglion; IC, cerebral commissure; ca, MDB-capsule; MDC, "Gomori-positive" medio-dorsal cell; d, dorsal disc-lamella; v, ventral disc-lamella; n, "network". A, × 330; B, × 260. (From Boer *et al.*, 1968b.)

staining fibres are observed radiating from the attachment area into the DB ("network", Lever, 1958a; "neurosecretory neuropile", Simpson et al., 1966a), it has been supposed that neurosecretory fibres pass through the perineurium into the DBs, where the neurosecretory material is "possibly stored or modified, and finally excreted into blood lacunae'' (Lever, 1958a, p. 200). This possible morphological relationship might in Stylommatophora be reflected in the presence of the commissural neurosecretory nerves (see above), which branch in the connective tissue dorsal to the commissure, the area where groups of DB cells are located (Nolte, 1965; Kuhlmann, 1966; Van Mol, 1967). So, histological observations suggest that the DBs are neurohaemal organs or at least innervated by neurosecretory nerve fibres. This view is not shared by Joosse (1964) for *Lymnaea stagnalis*, as in this species the DBs are always well delimited from the cerebral ganglia by the perineurium.

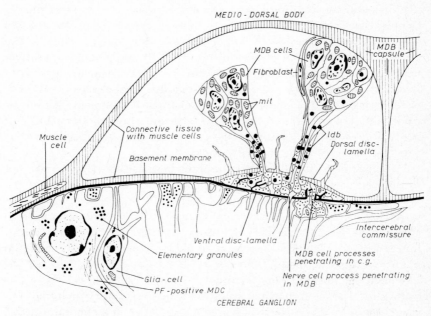

Fig. 19. Schematic drawing of the morphological relation between cerebral ganglion and medio-dorsal body in *Planorbarius* and *Biomphalaria*. (From Boer et al., 1968b.)

To investigate the assumed morphological contacts in more detail, histochemical and ultrastructural work has been carried out on *Lymnaea stagnalis*, *Biomphalaria glabrata*, *Planorbarius corneus* and *Ancylus fluviatilis* (Boer et al., 1968b) and on *Helisoma tenue* (Simpson et al., 1966a; Simpson, 1969). The investigations were restricted to MDBs. The observations showed that the supposed neurosecretory fibres radiating into the MDB differ histochemically from the processes of the neurosecretory cells in the ganglia, which indicates that the network of fibres in the MDBs is not derived from the neurosecretory cells (see also Röhnisch, 1964). This conclusion was sustained by the ultrastructural investigations. Ventrally the MDBs are delimited from the cerebral ganglia by the inner layer of the perineurium. In *Lymnaea stagnalis* no special differentiation of this layer at the site of attachment of the MDBs was observed, the layer being composed of a basement membrane directly apposed to the nervous tissue and of fibroblasts, collagen fibrils and an occasional muscle cell. In *Biomphalaria* and *Planorbarius* the perineurium splits over an area centrally under the MDB, so that a small disk is formed (Fig. 19). The dorsal lamella of this disk consists of fibroblasts and collagen fibrils, the ventral lamella of only the basal lamina. Within the disk processes of MDB cells are present. They penetrate through the dorsal disk lamella. Thus, in the disk area MDB cells are separated from the cerebral ganglion by only a basement membrane. The latter also holds for the attachment sites of the MDBs of *Ancylus* and *Helisoma*, but in these species no "disks" were observed. In *Planorbarius* the separating basement membrane may show holes (Böhmig, 1931). Penetration through these holes of nerve fibres into the MDB as well as of processes of MDB cells into the ganglion was observed, but only rarely.

It seems rather difficult to decide on the basis of this evidence whether or not the neurosecretory cells of the cerebral ganglion influence the MDB cells. Although neuronal and MDB cell processes are in rather close contact with each other, they are normally separated by at least a basement membrane. This indicates that direct neurosecretomotor innervation of the MDB cells is unlikely. Neurosecretory material might, on the other hand, influence MDB cells by diffusing through the basement membrane. This possibility can not be excluded, as it was shown for the neurohaemal areas that neurosecretory material can easily traverse the perineurium (see above). Yet it seems improbable that neurosecretory material is released into the MDB, since no signs of exocytosis of neurosecretory elementary granules were observed in this particular region. Another possibility is that the MDB is innervated by small neurosecretory nerves, which have sometimes been found in the MDB (Nolte, 1965; Boer *et al.*, 1968b; Wendelaar Bonga, 1970b). These nerves are, however, not numerous and they occur mainly in the connective tissue between the MDB cells. So there is no firm basis for considering DBs as neurohaemal organs.

The morphology of the contact between cerebral ganglion and MDB might also reflect an influence of the MDB cells upon the neurosecretory cells (Böhmig, 1931). This possibility seems unlikely. Secretory material of the MDB cells released into the haemolymph bathing the loose

connective tissue of the MDB (see below) probably reaches the haemocoel more easily than the cerebral ganglion (Wendelaar Bonga, 1970b;
Bekius, 1972).

As already mentioned, the DB cells are pear-shaped and send a
process in the direction of the attachment site of the DB to the cerebral
ganglion. At the ultrastructural level the perikarya of the DB cells are
characterized by the presence in their cytoplasm of great numbers of
well developed mitochondria (Fig. 20). The Golgi complex is usually

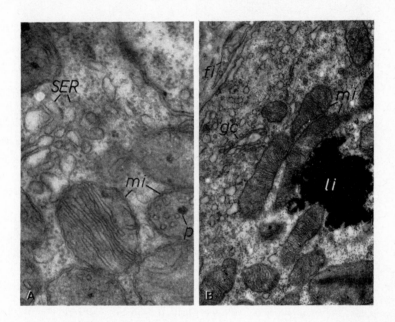

Fig. 20. A. MDB cell of *Biomphalaria*. Note the well developed cristae and the
dark particles (p) in the mitochondria (mi). In the centre of the electron micrograph smooth endoplasmic reticulum (SER) can be observed. × 54 000. B. Part
of a MDB cell of *Planorbarius*. Note the particular arrangement of the mitochondria (mi) around a lipid droplet (li). gc, Golgi complex; fi, extension of a
fibroblast. × 19 000. (From Boer *et al.*, 1968b.)

inconspicuous. The endoplasmic reticulum is mainly of the agranular
type. Free ribosomes and polyribosomes are numerous. Furthermore
often large dense bodies, probably secondary lysosomes, were observed. It is possibly these bodies which are responsible for the positive
staining reactions of the DB cells with paraldehyde-fuchsin, Sudan
black B and other histochemical methods (Röhnisch, 1964; Boer, 1965;
Nolte, 1965; Kuhlmann, 1966; Boer *et al.*, 1968b; Simpson, 1969).

Small, moderately electron-dense granules (diameter 70–90 nm) were infrequently found in the perikarya. Such granules are, on the other hand, abundant in the cell processes, especially in their endings, which occur intertwined in the area just dorsal to the attachment site of the DBs (Fig. 21). The endings are quite numerous, as a result of the

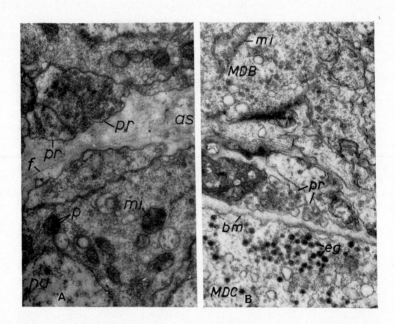

Fig. 21. A. Electron micrograph of part of the MDB of *Lymnaea*, showing MDB cells containing mitochondria (mi) with small electron-dense particles (p). In the spaces (as) between the cells, collagen fibrils (f) can be observed. The MDB-cell processes (pr) contain small granules. nu, nucleus. × 18 000. B. *Ancylus fluviatilis*. Electron micrograph showing a Gomori-positive cell (MDC) in the cerebral ganglion, which is separated from the MDB by a basement membrane (bm). The MDC contains elementary granules (eg). Processes (pr) of MDB cells contain small granules. In the MDB cells similar granules and mitochondria (mi) can be seen. × 18 000. (From Boer *et al.*, 1968b.)

frequent branching of the DB cell processes. Obviously the small granules contain the secretion product of the DBs, which is probably released in the process endings by exocytosis. The ultrastructural observations as yet neither allow a statement on the mode of formation of the secretory granules nor on the chemical nature of the secretion.

Finally, in this review of the DBs the seasonal variation of their histology as described for *Lymnaea stagnalis* (Joosse, 1964) should be mentioned. Evidence was obtained that the total volume of the DBs increases throughout the lifespan (Böhmig, 1931). The cells show, however, a cyclic activity. From October to April they gradually increase in volume. This hyperplasmic stage is followed by a return to the initial hypoplasmic stage during the next two months (Fig. 22).

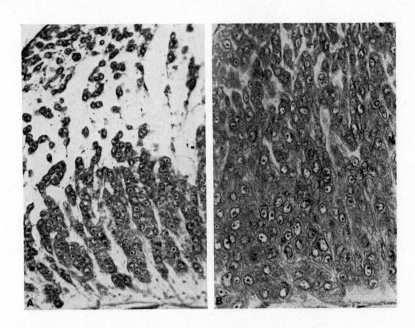

Fig. 22. Periodicity of the medio-dorsal bodies of *L. stagnalis*. A. Inactive stage (January). B. Active stage (April). × 455. (From Joosse, 1964.)

In summary it can be stated that DB cells produce secretory granules with a diameter of 70–90 nm. The secretion is apparently released by exocytosis into the loose connective tissue surrounding the groups of DB cells. These cells show hyperplasia during spring. Direct morphological contacts between processes of neurosecretory cells and MDB cells have not been found. These observations strongly suggest the DBs are endocrine organs. (See section on reproduction, IIIB, p. 289.)

E. The gonad

There are indications from experimental work that the gonad of the pulmonates produces one or more hormones (see section on

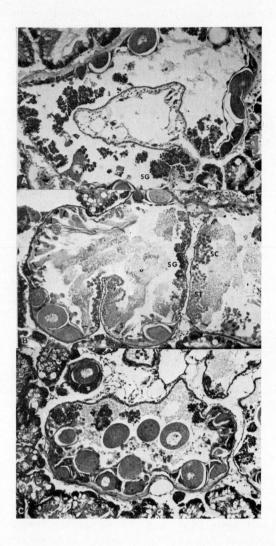

Fig. 23. A and B: Sections of the ovotestis of specimens of *Lymnaea stagnalis* collected in the field and fixed in March (A) and May (B). In March the male cells are represented by spermatogonia (SG) and primary spermatocytes (SC); the later stages of spermatogenesis are absent. In May all stages of spermatogenesis are present including spermatids (ST) and sperms (S). C. Sections of the ovotestis of a specimen of *L. stagnalis* bred in the laboratory at 20 °C until the adult stage and afterwards kept at 8 °C for 70 days. Spermatogenesis is blocked at the meiotic stage (cf. Fig. A), whereas oogenesis is continuing: ripe female cells accumulate in the acini. × 125. CH stain. (A and B after Joosse, 1964.)

reproduction, IIIB, p. 289). Histological studies might give clues as to the possible source(s) of the hormone(s).

The gonad of various pulmonates has been studied by light micro scopy (Barth and Jansen, 1961; Joosse and Reitz, 1969). Electron microscopy has also been carried out (Quattrini and Lanza, 1965; Guyard, 1971; Starke, 1971). The hermaphrodite organ consists of a number of acini in which oocytes as well as sperm ripen. The different stages of sperm formation are connected with Sertoli cells. The later stages of the oocytes are surrounded by nurse cells. In addition to these cell types epithelial cells having a microvillous border and ciliated epithelial cells occur in the acini (Joosse and Reitz, 1969). The gonad is embedded in connective tissue (Fig. 23).

Up till now no definite secretory phenomena have been described in any of the cell types of the ovotestis. In the connective tissue secretory cells occur (Sminia, 1972), but these cells are also found in other parts of the body, so that it seems unlikely that they produce a gonadial hormone.

There are two cell types which might most readily be expected to be involved in the production of hormones: the Sertoli cells (Lofts, 1972) and the nurse cells. These cells have in this laboratory been investigated to some extent in the Basommatophora *L. stagnalis* and *Biomphalaria glabrata* (de Jong-Brink, de Wit, Hommes and Bolhuis, unpublished results). In the younger stages of follicle formation around the growing oocytes, the nurse cells have ultrastructurally an inactive appearance. In the later stages the GER and the Golgi apparatus of the cells show signs of the synthesis of a secretion product. However, the pictures rather suggest that this product contributes to the formation of the cleft which is formed between the oocyte and the nurse cells, than that it represents a hormonal product. In most stages of spermatogenesis the Sertoli cells have an inactive appearance. However, during spermiogenesis a moderately developed smooth endoplasmic reticulum (SER) is present. This might reflect the production of a steroid hormone, since also in other cells producing steroid hormones, including Sertoli cells, a well developed SER occurs. Histochemical demonstration of key enzymes of the biosynthesis of steroid hormones failed, however, although various modifications of the techniques were applied.

So, in conclusion it can be stated that there is to date no histological evidence for the production of hormones in the gonad of the pulmonates.

III. Function of the endocrine structures

A. Introduction

The application of the classical methods of endocrine research (extirpations, implantations and injections of homogenates of supposed endocrine structures) to the pulmonate gastropods has raised a number of purely technical difficulties for the investigators. Some of these difficulties are worth mentioning in this survey.

Touching the body with surgical instruments causes in the terrestrial species violent contractions of the body wall, retraction of the body into the shell and a copious production of mucus. The freshwater pulmonates show an immediate retraction into the shell and the release of the greater part of the blood (Lever and Bekius, 1965). This type of reaction demonstrates that the animals are in a stress situation. To avoid or suppress these reactions, narcotization techniques have been developed. A survey of these methods is given by Runham *et al.* (1965). Recently a new technique for slugs has been introduced by Bailey (1969). Procedures for carrying out operations on centrally located structures (e.g. the central nervous system) have been described for *Lymnaea stagnalis* (Joosse and Lever, 1959; Joosse ,1964).

In studies on the endocrine control of reproduction castration can give much information about the endocrine function of the gonad. Removal of the gonad in slugs can easily be performed by removal of the rear portion of the body (e.g. in certain Arionidae and Limacidae, Laviolette, 1954a, b). However, in shelled pulmonates the gonad is located in the apex of the shell surrounded by the digestive gland. Complete castration is therefore hardly possible, for it causes severe damage and results in a high mortality (Goddard, 1960). In this respect the planorbid snails have the advantage that the gonad is located behind the digestive gland in the apex of the shell, so that in species of this family castration can be carried out experimentally (Harry, 1965).

To evade the experimental difficulties mentioned above, the organ culture technique has been introduced in molluscan endocrinology. This technique opens the possibility for *in vitro* studies of hormonal effects. Two media for the culture of pulmonate organs have now been published: a solid culture medium for *Helix aspersa* tissues (Guyard, 1971), and a fluid medium for organs of *Agriolimax reticulatus* (Müll.) (Bailey, 1973).

Endocrine control of reproduction will be further discussed in section IIIB. The second main subject in gastropod endocrinology is the study of the control of osmoregulation. The results on this subject will be dealt with in section IIIC. Knowledge of the endocrine control of growth is as yet still extremely limited (section III,D).

B. Reproduction

The literature concerning the endocrine control of reproduction in gastropods is rapidly increasing during recent years. However, a general picture is as yet not available. This may be caused by species specificity in the control systems, but also by a lack in uniformity of the techniques employed. For recent reviews of the field, see Joosse (1972) and Martoja (1972).

From an endocrinological point of view (Charniaux-Cotton, 1965) distinction should be made between the differentiation of gametes and the further development of these cells (gametogenesis). In the studies on reproduction of Pulmonata, this has not always clearly been done, so that the interpretation of the results is sometimes difficult.

1. Differentiation of the sex cells

In the hermaphrodite gonad of the Pulmonata the male and female cells differentiate from the germinal epithelial cells. A sexually indifferent intermediate stage, called the protogonium, has been described in *Helix aspersa* by Guyard (1971).

There are several arguments in favour of an endocrine control of sexual differentiation in pulmonate gastropods. During postembryonic development a protandric period of varying length is found in the gonad of all species. Moreover, a seasonally determined phase of protandry (early spring) was demonstrated in several species (e.g. in *Helix aspersa*, Guyard, 1971). This phase is followed by a simultaneous hermaphrodite stage in Basommatophora. In Stylommatophora sex reversal occurs as in *Arion ater* (L.) (Abeloos, 1944; Lūsis, 1961; Smith, 1966) and in *Agriolimax reticulatus* (Runham and Laryea, 1968).

The first report about the endocrine control of gameto-genisis in pulmonate gastropods was published by Pelluet and Lane (1961), and later extended by Pelluet (1964). These studies focussed the attention of many investigators on the optic tentacles and the cerebral ganglia ("brain") as the sites of production of gonadotrophic hormones involved in differentiation of sex cells and/or gametogenesis in gastropods.

The experimental work on differentiation of sex cells has primarily been carried out with the organ culture technique. For pulmonate organs this technique was introduced by Gomot and Guyard (1964) and Guyard and Gomot (1964). Their reports were followed by a number of others (Gomot, 1971; Gomot and Guyard, 1968; Guyard, 1967, 1969a, b,

1970, 1971). The authors studied the differentiation of the sex cells of *Helix aspersa*. Under anhormonal conditions the non-differentiated germinal cells of a gonad anlage transform only into female cells. Guyard (1969b) called this phenomenon the autodifferentiation of the female sex cells (Streiff, 1967). The addition of a cerebral ganglion to the culture medium with the gonad anlage caused differentiation of male sex cells. Guyard considers this differentiation to be the result of the stimulating effect of an androgenic factor produced by the cerebral ganglia. This factor also suppresses the autodifferentiation of the protogonia to female cells and stimulates the mitotic activity of the male cells after their differentiation. The presence of the androgenic factor in the haemolymph of adult snails was demonstrated in the spermatogenic active phase in early spring: adding haemolymph of these animals to the culture medium caused a differentiation of male cells in the gonadal tissue. Later on in spring the gonad enters the female phase. In the haemolymph of animals in this phase the androgenic factor appeared to be absent.

The addition of an optic tentacle to the culture medium with gonadial material suppressed the autodifferentiation of the female cells. However, in contradistinction to the findings of Pelluet and Lane (1961) no positive effect on the differentiation of male cells was found (Guyard, 1971).

The production centres of the androgenic factor of the cerebral ganglia and of the factor from the tentacles that suppresses the differentiation of the female cells have as yet not been identified (see also the section on the morphology of these structures).

Bailey (1973) reported the first results of the *in vitro* culture of gonadial material of *Agriolimax reticulatus*. He did not find any differentiation of sex cells in an anhormonal medium. Close association of a cerebral ganglion and an optic tentacle with gonadial material caused the onset of spermatogenesis. The effect on the differentiation of female cells was not clear.

Summarizing the data it can be stated that the production of an androgenic factor by the cerebral ganglia of terrestrial pulmonates is highly probable. There is good evidence that this factor is present in all gastropods: an androgenic factor also has been demonstrated in various prosobranchs (Choquet, 1971; Streiff, 1967). Apparently the same general conclusion holds for the phenomenon of the autodifferentiation of the female sex cells (Streiff, 1967; Choquet, 1969, 1971; Griffond, 1969; Lubet and Streiff, 1969). Furthermore, there is evidence that the optic tentacles exert an inhibiting influence on the autodifferentiation of the female cells.

2. Gametogenesis (spermatogenesis and oogenesis)

As mentioned the first reports of experimental work on the endocrine control of gametogenesis in gastropods are from Pelluet and Lane (1961) and Pelluet (1964). They extirpated the optic tentacles of *Arion ater* and *Arion subfuscus* (Draparnaud) and/or injected the slugs with homogenates of these tentacles and of the cerebral ganglia. They concluded from the results that the optic tentacles produce a hormone that inhibits oogenesis, whereas the cerebral ganglia produce a hormone that stimulates oogenesis. Moreover, a hormone from the optic tentacles would have a stimulating effect on spermatogenesis.

Since then a great number of investigators have carried out tentacle extirpations in various species of Stylommatophora. Sanchez and Sablier (1962) extirpated the optic tentacles of *Helix aspersa*, but could not confirm the results of Pelluet and Lane. Moreover, it was observed that the experimental animals deteriorated physiologically due to malnutrition caused by the absence of the optic tentacles. Guyard (1971) reported the results of a detailed study on the same species. No changes in gametogenesis were observed after optic tentacle extirpations. The extirpations were repeated frequently, since the regenerative capacity of the tentacles is high (Chétail, 1963). Kuhlmann and Nolte (1967) and Renzoni (1969) also obtained negative results with *Helix pomatia* and *Vaginula borelliana Colosi*, respectively. However, changes in the ovotestis were observed by some authors. Berry and Chan (1968) found an increase in the production of oocytes and egg capsules in *Achatina fulica* Bowdich after optic tentacle extirpation, whereas on the other hand, Gottfried and Dorfman (1970a) observed an increase in spermatogenic activity in *Ariolimax californicus* (Cooper) upon removal of the optic tentacles.

From these reports it is clear that optic tentacle extirpations have presented contradictory results with respect to the control of gametogenesis. These contradictions can be expected by considering that not only different species were used, but that also the experimental conditions were not uniform and, moreover, in some instances rather unphysiological.

There may be important species specific differences in the gonadotrophic function of the optic tentacles. Furthermore, this function may change with the seasons, so that the results of experiments can only be compared if the experiments are carried out in the same season. From the available literature this is hardly likely. Moreover, if the optic tentacles are important for daylight perception and orientation for food, a significant indirect effect of the extirpation of optic tentacles on the reproductive activity obviously cannot be excluded. Next, some

L

authors have not given attention to the regeneration of the tentacles. Finally, it is very difficult to quantify exactly the number of sex cells in an ovotestis and thus to assess the changes in number of these cells. For all these reasons it is not surprising that this type of experiment has met with criticism in the literature (Sanchez and Sablier, 1962; Durchon, 1967; Guyard, 1971; Joosse, 1972). Apparently, the results obtained so far need to be confirmed by using other experimental techniques.

a. Spermatogenesis

As already mentioned the androgenic factor of the cerebral ganglia of *Helix aspersa* is not only involved in the differentiation of male cells, but also in the mitotic activity of the various stages of spermatogenesis. All stages following upon the spermatogonial stage degenerate in an anhormonal culture medium. This means that the androgenic factor is necessary for the maintenance of the entire spermatogenic activity. In contrast to these results, all stages of spermatogenesis survived in an *in vitro* culture of the gonad of *Arion ater rufus* under anhormonal conditions (Badino, 1967).

From observations on specimens of *Lymnaea stagnalis* taken from the field Joosse (1964) demonstrated a clear annual cycle in the spermatogenic activity (see Fig. 23). During winter only the spermatogonia and primary spermatocytes persist in the gonad. It appeared that spermatogenesis is temperature dependent. In animals kept at temperatures below about 10°C spermatogenesis was blocked at meiosis (Joosse and Veld, 1972). According to the latter authors the results strongly suggest that spermatogenesis is controlled by a hormone which exerts its effect at the meiotic stage.

b. Vitellogenesis

The autodifferentiation of the female cells in *in vitro* cultures is not followed by vitellogenesis. In prosobranch snails the addition of a cerebral ganglion of a specimen with a gonad in the active vitellogenic stage induced vitellogenesis in autodifferentiated oocytes in organ cultures (Streiff, 1967; Choquet, 1971). Guyard (1971), however, did not succeed in inducing vitellogenesis in *Helix aspersa* oocytes *in vitro* by addition of cerebral ganglia of snails in the active female phase.

Joosse (1964) observed a stop in the production of egg capsules in adult specimens of *L. stagnalis* after extirpation of the dorsal bodies. But the proportions of the stages of the female cells (previtellogenic, vitellogenic and degenerating oocytes) were not different from those in sham-operated controls. So, the effect of the dorsal bodies on oogenesis was not apparent. However, in a more detailed study by Joosse and

Geraerts (1969) the function of the dorsal bodies with regard to the control of reproduction was elucidated. Extirpation and subsequent implantation of dorsal bodies was performed in adult as well as juvenile laboratory bred specimens of *Lymnaea stagnalis*. The juvenile

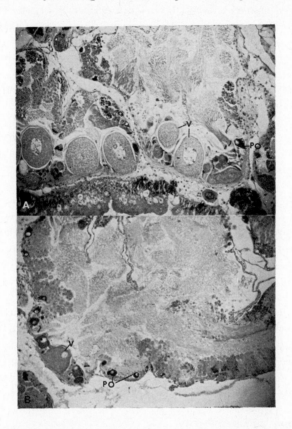

Fig. 24. Endocrine control of vitellogenesis by the dorsal bodies in *L. stagnalis*. A: Ovotestis of a sham-operated juvenile specimen, fixed 90 days after the operation. All stages of oogenesis are present (PO, previtellogenic oocyte; V; vitellogenic oocyte). B. Ovotestis of a specimen fixed 90 days after the cauterization of the dorsal bodies. Previtellogenic oocytes are abundant, whereas vitellogenic oocytes are scarce (cf. Fig. 25). × 125. CH stain.

snails had not yet started oviposition. As in the earlier experiments it was found that oviposition is not possible after extirpation of the dorsal bodies. The proportions of the stages of female cells in the juvenile snails clearly demonstrated that vitellogenesis had stopped (Figs 24, 25). The differentiation of new female cells had not been affected. Subsequent implantation of dorsal bodies restored vitello-genesis, and later on also ovipository activity. This effect on vitello-genesis could not be determined from the data in the adult snails.

The interpretation of these results could be as follows. The experi-ments proved that the dorsal bodies are endocrine organs producing a hormone which stimulates vitellogenesis. Thus, after extirpation of the dorsal bodies, vitellogenesis stops as, consequently, does oviposition. Apparently the hormone does not affect the autodifferentiation of the female cells, since dorsal body extirpations had no effect on the pro-portions of the stages of female cells in adults. The increase of the number of the previtellogenic female cells is apparently more rapid in juvenile snails than in adults.

From the data obtained with morphometrical methods on the annual periodicity of the dorsal bodies of *L. stagnalis* (see section on the morphology of the dorsal bodies), it was concluded that these bodies are most active in spring, i.e. just before and during the reproductive period (Joosse, 1964). In view of the results of Joosse and Geraerts (1969) this active period of the bodies can be related to the production of ripe oocytes. The activity of the dorsal bodies is not dependent on the temperature of the medium in which the snails are kept. At tem-peratures below 10°C vitellogenesis is not inhibited: juvenile snails kept at temperatures below 10°C showed a clear increase in number of "vitellogenic" oocytes (Joosse, unpublished results). Apparently (an) other factor(s) than temperature are involved in the regulation of the activity of the dorsal bodies in *L. stagnalis* and hence in vitellogenic activity (day length?).

In contrast to these findings, in *Helix aspersa* the vitellogenic oocytes appeared not to persist in the gonad at low temperatures during winter (Guyard, 1971).

3. Ovulation and oviposition

In the pond snail *L. stagnalis* ovulation can be stimulated by supply-ing oxygen to the water (Lever: see Woerdeman and Raven, 1946). This finding has been used in experiments dealing with synchronization of oviposition in *L. stagnalis* (Joosse and Dijkstra, unpublished results). Every third day, fresh oxygenated water was supplied to the animals. In this way up to 70% of the animals could be stimulated to oviposi-tion. Observations under a dissecting microscope showed that ovulation occurs rather rapidly after applying the stimulus, since many oocytes had reached the end of the spermoviduct within 10 min after stimula-tion with oxygen. From these experiments it could not be concluded whether the direct stimulus for ovulation had either a nervous or a

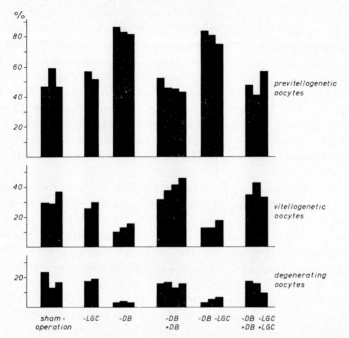

Fig. 25. Percentages of three distinct stages of oocytes in the ovotestis of juvenile specimens of *L. stagnalis*, fixed 90 days after various types of operations and implantations. The implantations were carried out 60 days after the operation. The columns correspond vertically and each represent the data of one specimen. Vitellogenesis is inhibited in the absence of the dorsal bodies. After subsequent implantation of the dorsal bodies vitellogenesis is restored. For further conclusions see text. −LGC: cauterized light green neurosecretory cells; −DB+DB: cauterized dorsal bodies and subsequent inplantation of one pair of medio-dorsal bodies.

hormonal character. However, the results obtained in the opisthobranch species *Aplysia californica* (Kupfermann, 1972) strongly suggest that in gastropods ovulation is under neuroendocrine control. In this opisthobranch species a special cell type, the bag cell, in the abdominal ganglion, produces an ovulation hormone.

4. Endrocine control of the growth and differentiation of the reproductive tract

In several gastropods an influence of the gonad upon the differentiation and functioning of the reproductive tract has been demonstrated. The first and excellent studies on this subject are those of Abeloos (1943) and Laviolette (1954). They castrated various limacid and arionid slugs by simply removing the rear portion of the body of the slugs in which only the ovotestis is situated. As a result of this operation the albumen gland, and the hermaphrodite duct retain the same size as at the time of operation, or even regress in size. Moreover, Laviolette has transplanted juvenile gonads or reproductive tracts into the haemocoel of normal or castrated adults in which they lie free in the blood of the host. It appeared that the differentiation and growth of the albumen gland and the hermaphrodite duct are controlled by the gonad via a blood-borne factor. Gonadial implants of adults into juveniles induced a precocious development of the reproductive tract of the juvenile hosts.

Recently, Runham et al. (1973) have extended these data in experiments with the slug Agriolimax reticulatus. They transplanted undifferentiated tracts into the haemocoel of adult hosts which were in various stages of the sexual cycle. Transplantation of undifferentiated tracts into animals in the early protandric phase resulted in the differentiation and enlargement of the prostate glands of the implants while the oviducal and albumen glands remained small and undifferentiated. Transplantation of an undifferentiated tract into animals of later male stages resulted in the enlargement of the whole implanted tract. However, after transplantation of undifferentiated tracts into the haemocoel of egg-laying animals, only the female parts of the tracts enlarged considerably, the prostate glands remained undeveloped. From these results the investigators concluded that two blood-borne hormones are present, one controlling the prostate gland, the other the development of the female glands.

Castration of shell-bearing pulmonate snails is impossible in the species in which the gonad is embedded in the digestive gland. Operations on this region of the body are highly lethal. The survival times are short or the operations result in only partial castration (Filhol, 1938; Goddard, 1960); however, in planorbid snails the ovotestis is separate from the digestive gland in the apex of the shell, and Harry (1965) succeeded in the castration of juvenile specimens of Biomphalaria glabrata (Say). After complete castration the whole reproductive tract failed to develop, whereas after partial castration a normal development occurred. Also these data suggest, that the hormone(s) involved with growth and differentiation of the reproductive tract in pulmonate gastropods originate from the gonad.

Geraerts and Algera (1972) studied the endocrine control of the growth and differentiation of the reproductive tract of Lymnaea stagnalis. They combined the techniques of implantation of the genital tract with the anlage of the accessory glands into the haemocoel with cauterization of endocrine centres in the cerebral ganglia (dorsal bodies and light green and caudo-dorsal cells). They concluded from their results that the tracts develop normally if implanted into the haemocoel of intact adult snails, and furthermore that the growth and differentiation of the female accessory glands is dependent on the

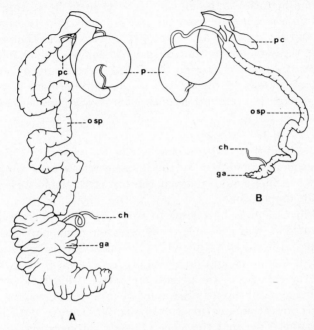

Fig. 26. Effect of castration of adult *Limax maximus* on the genital tract. A, control; B, castrated specimen. ch, hermaphrodite duct; ga, albumen gland; o sp, spermoviduct; p, penis; pc, bursa copulatrix. (After Laviolette, 1954a).

presence of the dorsal bodies of the host, whereas the development of the male part is under control of neurosecretory cells. These results are in agreement with those of Joosse and Geraerts (1969) and confirm the work on *Agriolimax* (Runham *et al.*, 1973) from which it was concluded that the male and the female part of the reproductive tract are under separate hormonal control (see above). Apparently the source of these two hormones is the dorsal bodies and the neurosecretory cells in the cerebral ganglia (Geraerts and Algera, 1972). The hormones may influence the reproductive tract via the gonad. This hypothesis is sustained by the work of Bailey (1973) who obtained good survival and growth of the prostate gland of *Agriolimax reticulatus* in an organ culture medium only if gonadial tissue, cerebral ganglia and optic tentacles were present. If, on the other hand, gonadial tissue or the cerebral ganglion–optic tentacle complex were added separately then prostate gland material did not survive.

All reports mentioned suggest that the growth and differentiation of the reproductive tract of pulmonate gastropods is under the control of endocrine factors. At the moment it is most likely to suppose that these hormones originate from the gonad, and that two factors are involved, one for the male and one for the female part. The production of each of these factors is controlled by hormones from centres in or near the central nervous system: the dorsal body hormone controls the female factor produced by the gonad, a neurohormone from neurosecretory cells in the cerebral ganglia controls the male factor.

5. Chemistry of the hormones

From a comparative endocrinological point of view it can be expected that all hormones produced by the neurosecretory cells are peptides. Up to now nothing is known about the size of the molecules and their amino acid composition.

Several investigators have tried to study the effect of the well known steroid gonadial hormones of vertebrates on pulmonate gastropods. Among these, the work of Aubry (1962) is most extensive. She injected oestradiol, testosterone and progesterone into *Lymnaea* and *Helix* and described the changes in the gonad. Similarly, Bridgeford and Pelluet (1952) injected a mixture of FSH and LH from the vertebrate anterior hypophysis into the limacid *Agriolimax reticulatus*. With Durchon (1967) it is suggested that this type of study on invertebrates has not as a rule given useful information.

Up to now it is also difficult to evaluate the experiments of Gottfried *et al.* (1967), Gottfried and Lūsis (1966), and Gottfried and Dorfman (1970b). They studied the synthesis of steroids from labelled cholesterol *in vivo* and/or *in vitro* in *Arion ater rufus* and *Ariolimax californicus*. Various steroids appeared to be synthetized. However, it is not yet clear whether these compounds exert hormonal effects. Gottfried and Dorfman (1970c) studied also the modifying effect of steroids on the influence of optic tentacle removal in *Ariolimax*. Further investigations are needed to elucidate the data obtained so far.

C. Endocrine control of osmotic equilibrium and ion transport

The first study in which the presence of osmotically active factors of supposed neuroendocrine nature has been demonstrated in pulmonate gastropods is that of Hekstra and Lever (1960) on *L. stagnalis*. The authors found an increase of the body weight, due to water uptake, after extirpation of the pleural ganglia. Extirpation of the parietal ganglia had a moderate effect, as had sectioning of the pleural connectives. In further experiments (Lever *et al.*, 1961a) the implantation of pleural ganglia after pleuralectomy proved to annihilate the swelling. Injection of homogenates of the pleural ganglia into intact animals resulted in a decrease of the body weight during the first hours after

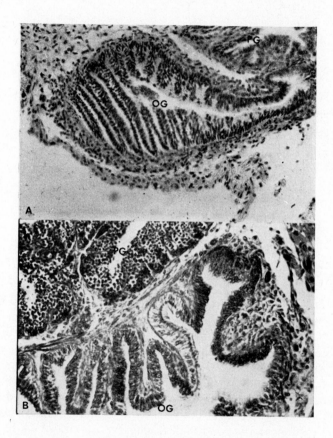

Fig. 27. Differentiation of the Anlage of the reproductive tract of *L. stagnalis* after 4 weeks of implantation in adult hosts. A, no differentiation of the oothecal (OG) and prostate gland (PG) occurs in a host lacking the dorsal bodies and the light green and caudo-dorsal neurosecretory cells (× 200). B, if only the dorsal bodies are removed no differentiation of the female gland occurs. The male gland on the other hand shows clear secretion phenomena. × 200. (Courtesy Geraerts and Algera.)

injection. The authors concluded that the pleural ganglia produce a
factor which exerts a diuretic influence. Since this factor is apparently
able to act via the blood stream, it was assumed to be an endocrine
substance.

These results have been substantiated by Wendelaar Bonga (1970a,
1971b, 1972) in some extensive electron microscopical studies. In these

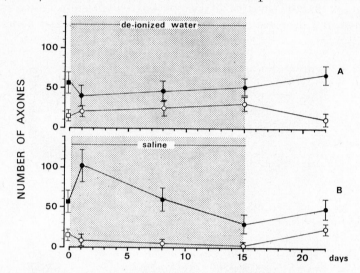

Fig. 28. *L. stagnalis.* Quantitative data concerning the release of neurosecretory
material of the DGC in the neurohaemal zone after exposure of animals to
deionized water (A) or to 0·1 M NaCl (B). (From Wendelaar Bonga, 1971b.)

studies it was demonstrated that in *Lymnaea stagnalis* two types of
neurosecretory cells are involved in the control of the water and ion
transport, viz. the dark green neurosecretory cells and the yellow
neurosecretory cells (for the location of these cells and their neurohaemal areas, see Fig. 4).

The experiments of Wendelaar Bonga were based on the studies of
Van Aardt (1968) and Greenaway (1970) which were concerned with the
osmoregulatory capacities of *Lymnaea stagnalis.* Like freshwater
molluscs in general, the pond snail *L. stagnalis* maintains a high osmotic
gradient between blood and environment. This gradient causes an
inward flow of water and an outward diffusion of solutes. To maintain
a steady state the production of large volumes of a hypotonic urine
and active uptake of ions from the environment may be involved. It was
demonstrated that in deionized water the water elimination and the
ion-uptake mechanisms are activated, whereas in saline these processes

are suppressed. As a consequence, the presumed neuroendocrine systems engaged in the control mechanisms should be stimulated in deionized water and inactivated in saline.

Since the pleural ganglia are involved in osmoregulation as appeared from the extirpation experiments (see above) and since the dark green cells are the only neurosecretory cells in these ganglia, it can be assumed that these cells produce the diuretic factor. Thus, Wendelaar Bonga (1970a, 1971b) primarily studied these cells with quantitative electron microscopy.

For comparison some other neurosecretory cell types were also investigated. Snails were exposed for two weeks either to a medium of deionized water or to a 0.1 M saline solution. It was observed that the dark green cells were activated in deionized water: a statistically significant increase was noted in the volume of the cells, in the extent of the granular endoplasmic reticulum, in the activity of the Golgi complex, and in the release activity in the neurohemal zones (Fig. 28). After exposure to a saline solution the secretory processes were suppressed and the cells showed a general cytoplasmic regression. So the hypothesis that the dark green cells produce a diuretic factor is confirmed by these experiments.

However, a second cell type, the yellow cells of the parietal ganglia and the visceral ganglion, also showed inactivation in saline and activation in deionized water (Wendelaar Bonga, 1970a, 1972). The meaning of this second factor can tentatively be explained from the work of Chaisemartin (1968) on *Lymnaea limosa* (L.). This investigator studied the effect of cauterization of various ganglia on the osmotic equilibrium and ion transport by measuring urine formation and sodium transport. Cauterization of the pleural ganglia caused an increase in body weight and a reduction in urine formation, but did not affect the sodium fluxes. However, after cauterization of the parietal ganglia a decreased turnover of Na^+ ions was demonstrated. Therefore, Chaisemartin suggested that two neuroendocrine substances are involved in osmoregulation: a substance from the pleural ganglia with a diuretic effect, and a substance from the parietal ganglia that stimulates the ion transport mechanisms located in the body wall and the ureter (Little, 1965).

So, interpretation of the results of Chaisemartin (1968) and of Wendelaar Bonga (1972) leads to the conclusion that in *Lymnaea stagnalis* the dark green cells activate water elimination, while the yellow cells stimulate the ion-uptake mechanisms. This latter conclusion is sustained by the fact that the yellow cells show peripheral neurosecretion in the region of the ureter (see above), where active uptake of ions from the prourine takes place (Van Aardt, 1968; Wendelaar Bonga and Boer, 1969). Comparable studies on other pulmonate snails are as yet not available.

The presence of an antidiuretic factor in *Lymnaea stagnalis* was suggested by Lever and Joosse (1961). Some neurosecretory cells in the lateral lobes of the cerebral ganglia were depleted of secretory material in snails exposed to a saline solution. However, these experiments have been repeated several times with negative results. Moreover, no clear physiological indication of antidiuresis has been presented.

D. Endrocine control of growth

The experimental work on the endocrine control of growth in Pulmonata is extremely limited. The only report is that of Joosse and Geraerts (1969) on *L. stagnalis*. In preliminary experiments they found that shell growth ceased in animals in which the dorsal bodies and light green neurosecretory cells in the cerebral ganglia had been cauterized. If each of these centres were removed separately the growth of the shells decreased. Since these structures are morphologically closely related, cauterization of one of them may affect the activity of the other one. Therefore it is as yet not clear whether both centres or only one of them are important for the control of shell growth in this species. Investigations on this subject are in progress in this laboratory.

IV. References

Aardt, W. J. van (1968). *Neth. J. Zool.* **18**, 253–312.

Abeloos, M. (1943). *C.r. hebd. Séanc. Acad. Sci. Paris* **216**, 90–91.

Altmann, G. and Kuhnen-Clausen, D. (1959) *Ann. Univ. Saraviensis Scientia* **8**, 135–140.

Aubry, R. (1962). *Archs Anat. microsc. Morph. exp.*, (*Suppl.*) **50**, 521–602.

Badino, G. (1967). *Arch. Zool. Ital.* **52**, 271–275.

Bailey, T. G. (1969). *Experientia* **25**, 1225.

Bailey, T. G. (1973). *Neth. J. Zool.* **23**, 72–85.

Barth, R. and Jansen, G. (1961). *Mem. Inst. Oswaldo Cruz* **59**, 83–114.

Bekius, R. (1972). *Neth. J. Zool.* **22**, 1–58.

Bern, H. A. (1962). *Gen. comp. Endocrinol.*, suppl. **1**, 117–132.

Berry, A. C. and Chan, L. C. (1968). *Aust. J. Zool.* **16**, 849–855.

Bierbauer, J. (1967). *Gen. comp. Endocrinol.* **9**, 433–434.

Bierbauer, J. and Török, L. J. (1968). *Acta biol. Acad. Sci. Hung.* **19**, 133–143.

Bierbauer, J. and Vigh-Teichmann, I. (1970). *Acta biol. Acad. Sci. Hung.* **21**, 11–24.

Bierbauer, J., Kiss, J. and Vigh, B. (1968). *In* "Symposium on Neurobiology of Invertebrates" (J. Salánki Ed.) pp. 135–142. Plenum Press, New York.

Bierbauer, J., Török, L. J. and Teichmann, I. (1965). *Zool. Jb.* (*Zool. Physiol.*) **71**, 545–551.

Boer, H. H. (1963). *Gen. comp. Endocrinol.* **3**, 687–688.

Boer, H. H. (1965). *Arch. néerl. Zool.* **16**, 313–386.

Boer, H. H. and Lever, J. (1959). *Proc. Kon. Ned. Akad. Wet.* C, **62**, 76–83.

Boer, H. H., Douma, E. and Koksma, J. M. A. (1968a). *Symp. zool. Soc. Lond.* **22**, 237–256.

Boer, H. H., Slot, J. W. and Van Andel, J. (1968b). *Z. Zellforsch. mikrosk. Anat.* **87**, 435–450.

Böhmig, L. (1931). *Sber. Akad. Wiss. Wien, math.-naturw. Kl., Abt. I* **140**, 319–335.

Bridgeford, H. B. and Pelluet, D. (1952). *Can. J. Zool.* **30**, 323–337.

Brink, M. and Boer, H. H. (1967). *Z. Zellforsch. mikrosk. Anat.* **79**, 239–243.

Chaisemartin, C. (1968). *C.r. Séanc. Soc. Biol.* **162**, 1994–1998.

Charniaux-Cotton, H. (1965). *In* "Organogenesis" (R. de Haan and H. Ursprung, Eds) pp. 701–740. Holt, Rinehart and Winston, New York.

Chétail, M. (1963). *Archs Anat. microsc. Morph. exp.* (*Suppl.*) **52**, 129–203.

Choquet, M. (1969). "Contribution à l'étude du cycle biologique et de l'inversion du sexe chez *Patella vulgata* (L.) (Mollusque Gastéropode Prosobranche)." Thesis, University of Lille.

Choquet, M. (1971). *Gen. comp. Endocrinol.* **16**, 59–73.

Cook, H. (1966). *Arch. néerl. Zool.* **17**, 1–72.

Cottrell, G. A. and Osborne, N. (1969). *Comp. Biochem. Physiol.* **28**, 1455–1459.

Drawert, J. (1968). *Zool. Jb.* (*Zool. Physiol.*) **74**, 292–318.

Durchon, M. (1967). "L'endocrinologie des Vers et des Mollusques." Paris, Masson et Cie.

Filhol, J. (1938). *Arch. Anat. micr.* **34**, 155–439.

Foh, E. and Schlote, F. W. (1965). *Kurznachr. Akad. Wiss. Gottingen* **7**, 33–38.

Gabe, M. (1954). *Année biol.*, Ser 3, **30**, 6–62.

Gabe, M. (1966). "Neurosecretion." Pergamon Press, Oxford.

Geraerts, W. P. M. and Algera, L. H. (1972). *Gen. comp. Endocrinol.* **18**, 592.

Goddard, C. K. (1960). *Aust. J. biol. Sci.* **13**, 378–386.

Gomot, L. (1971). *In* "Invertebrate Tissue Culture" (C. Vago Ed.) Academic Press, London and New York.

Gomot, L. and Guyard, A. (1964). *C.r. hebd. Séanc. Acad. Sci. Paris* **258**, 2902–2905.

Gomot, L. and Guyard, A. (1968). *Proc. IInd Int. Coll. Invert. Cult.*, Milan, 22–31.

Gottfried, H. and Dorfman, R. I. (1970a). *Gen. comp. Endocrinol.* **15**, 101–119.

Gottfried, H. and Dorfman, R. I. (1970b). *Gen. comp. Endocrinol.* **15**, 120–138.

Gottfried, H. and Dorfman, R. I. (1970c). *Gen. comp. Endocrinol.* **15**, 139–142.

Gottfried, H. and Lūsis, O. (1966). *Nature, Lond.* **212**, 1488–1489.

Gottfried, H., Dorfman, R. I. and Wall, P. E. (1967). *Nature, Lond.* **215**, 409–410.

Greenaway, P. (1970). *J. exp. Biol.* **53**, 147–163.

Griffond, B. (1969). *C.r. hebd. Séanc. Acad. Sci. Paris*, Ser. D, **268**, 963–965.

Guyard, A. (1967). *C.r. hebd. Séanc. Acad. Sci. Paris* Ser. D, **265**, 147–149.

Guyard, A. (1969a). *C.r. hebd. Séanc. Acad. Sci. Paris*, Ser. D, **268**, 162–164.

Guyard, A. (1969b). *C.r. hebd. Séance. Sci., Paris*, Ser. D, **268**, 966–969.

Guyard, A. (1970). *Annls. Biol.* **9**, 401–408.

Guyard, A. (1971). "Étude de la différenciation de l'ovotestis et des facteurs controlant l'orientation sexuelle des gonocytes de l'escargot *Helix aspersa* Müller." Thesis, University of Besançon.

Guyard, A. and Gomot, L. (1964). *Bull. Soc. zool. Fr.* **89**, 48–56.

Harry, H. W. (1965). *Trans. Am. microsc. Soc.* **84**, 157.

Hekstra, G. P. and Lever, J. (1960). *Proc. Kon. Ned. Akad. Wet.* C, **63**, 271–282.

Herlant-Meewis, H. and Van Mol, J. J. (1959). *C.r. hebd. Séanc. Acad. Sci., Paris* **249**, 321–322.

Highnam, K. C. (1965). *Zool. Jb. (Zool. Physiol.)* **71**, 558–582.

Highnam, K. C. and Hill, L. (1969). "The comparative Endocrinology of the Invertebrates." Edward Arnold, London.

Jariol, M. S. and Scudder, G. G. E. (1970). *Z. Morph. Ökol. Tiere* **68**, 269–299.

Joosse, J. (1963). *Acta physiol. pharmac. Néerl.* **12**, 99–100.

Joosse, J. (1964). *Arch. néerl. Zool.* **15**, 1–103.

Joosse, J. (1972). *Gen. comp. Endocrinol.*, Suppl. **3**, 591–601.

Joosse, J. and Geraerts, W. J. (1969). *Gen. comp. Endocrinol.* **13**, 540.

Joosse, J. and Lever, J. (1959). *Proc. Kon. Ned. Akad. Wet.* C, **62**, 145–149.

Joosse, J. and Reitz, D. (1969). *Malacologia* **9**, 101–109.

Joosse, J. and Veld, C. J. (1972). *Gen. comp. Endocrinol.* **18**, 599–600.

Jungstand, W. (1962). *Zool. Jb. (Zool. Physiol.)* **70**, 1–23.

Knowles, F. G. (1965). *Arch. Anat. micr. Morph. exp.* **54**, 343–358.

Knowles, F. G. and Bern, H. A. (1966). *Nature, Lond.* **210**, 271.

Krause, E. (1960). *Z. Zellforsch. mikrosk. Anat.* **51**, 748–776.

Krkac, N. and Mestrov, M. (1970). *Bull. Soc. zool. France* **95**, 627–644.

Kuhlmann, D. (1963). *Z. Zellforsch. mikrosk. Anat.* **60**, 909–932.

Kuhlmann, D. (1966). *Z. wiss. Zool.* **173**, 218–231.

Kuhlmann, D. (1970). *Z. Zellforsch. mikrosk. Anat.* **110**, 131–152.

Kuhlmann, D. and Greven, H. (1973). *Histochemie* **34**, 177–190.

Kuhlmann, D. and Nolte, A. (1967). *Z. wiss. Zool.* A. **176**, 271–286.

Kupfermann, I. (1972). *Am. Zool.* **12**, 513–520.

Landing, B. J., Hall, H. E. and West, C. D. (1956). *Lab. Invest.* **5**, 256–266.

Lane, N. J. (1962). *Q. Jl. microsc. Sci.* **103**, 211–226.

Lane, N. J. (1964). *Q. Jl. microsc. Sci.* **105**, 35–47.

Lane, N. J. (1966). *Am. Zool.* **6**, 139–157.

Laviolette, P. (1954). *Bull. Biol. Fr. Belg.* **88**, 310–332.

Lever, J. (1957). *Proc. Kon. Ned. Akad. Wet.* C, **60**, 510–522.

Lever, J. (1958a). *Arch. néerl. Zool.* **13**, 194–201.

Lever, J. (1958b). *Proc. Kon. Ned. Akad. Wet.* C, **61**, 235–242.

Lever, J. and Bekius, R. (1965). *Experientia* **21**, 395–398.

Lever, J. and Joosse, J. (1961). *Proc. Kon. Ned. Akad. Wet.* C, **64**, 630–639.

Lever, J., Boer, H. H., Duiven, R. J. Th., Lammens, J. J. and Wattel, J. (1959). *Proc. Kon. Ned. Akad. Wet.* C, **62**, 139–144.

Lever, J., Jansen, J. and De Vlieger, T. A. (1961a). *Proc. Kon. Ned. Akad. Wet.* C, **64**, 531–542.

Lever, J., Kok, M., Meuleman, E. A. and Joosse, J. (1961b). *Proc. Kon. Ned. Akad. Wet.* C, **64**, 640–647.

Lever, J., De Vries, C. H. and Jager, J. C. (1965). *Malacologia* **2**, 219–230.

Little, C. (1965). *J. exp. Biol.* **43**, 23–37.

Lofts, B. (1972). *Gen. comp. Endocrinol.* Suppl. **3**, 636–648.

Loud, A. V., Barany, W. and Pack, W. C. (1965). *Lab. Invest.* **14**, 258–270.

Lubet, P. and Streiff, W. (1969). *In* "Cultures d'Organes d'Invertébrés" (H. Lutz Ed.), pp. 141–160. Gordon and Breach, Paris, London, New York.

Lūsis, O. (1961). *Proc. zool. Soc. Lond.* **137**, 433–468.

Maddrell, S. H. P. (1967). *In* "Insects and Physiology" (J. W. L. Beament and J. E. Treherne Eds), pp. 103–118. American Elsevier, New York.

Martoja, M. (1965a). *C.r. hebd. Séanc. Acad. Sci. Paris* **260**, 2907–2909.

Martoja, M. (1965b). *C.r. hebd. Séanc. Acad. Sci. Paris* **260**, 4615–4617.

Martoja, M. (1965c). *C.r. hebd. Séanc. Acad. Sci. Paris* **261**, 3195–3196.

Martoja, M. (1968). *In* "Traité de Zoologie" (P. Grassé Ed.) Vol. V (3), 927–986.

Martoja, M. (1972). *In* "Chemical Zoology" (M. Florkin and B. T. Scheer, Eds) **7**, pp. 349–391. Academic Press, London and New York.

Mol, J.-J. van (1960a). *C.r. hebd. Séanc. Acad. Sci. Paris* **250**, 2280–2281.

Mol, J.-J. van (1960b). *Ann. Soc. r. Belg.* **91**, 45–55.

Mol, J.-J. van (1967). *Mém. Acad. r. Belg. Cl. Sci.* 8° **37**, (5), 1–168.

Mordue, W. (1967). *Gen. comp. Endocrinol.* **9**, 406–415.

Nabias, B. de (1899). *Trav. Lab. Soc. Sci. Stat. Zool. Arcachon* **4**, 1–26.

Nishioka, R. S., Simpson, L. and Bern, H. A. (1964). *Veliger* **7**, 1–4.

Nolte, A. (1963). *Gen. comp. Endocrinol.* **3**, 721–722.

Nolte, A. (1964). *Naturwiss.* **51**, 148.

Nolte, A. (1965). *Zool. Jb.* (*Anat. Physiol.*) **82**, 365–380.

Nolte, A. (1966). *Z. Zellforsch. mikrosk. Anat.* **75**, 120–128.

Nolte, A. and Kuhlmann, D. (1964). *Z. Zellforsch. mikrosk. Anat.* **63**, 550–567.

Nolte, A., Breucker, H. and Kuhlmann, D. (1965). *Z. Zellforsch. mikrosk. Anat.* **68**, 1–27.

Normann, T. C. (1969). *Exp. Cell Res.* **55**, 285–287.

Novikoff, A. B. and Holtzman, E. (1970). "Cells and organelles". Holt, Rinehart and Winston, Inc., New York, London.

Olivereau, J. M. (1970). *Z. Zellforsch. mikrosk. Anat.* **105**, 430–441.

Osborne, N. N. and Cottrell, G. A. (1971). *Z. Zellforsch. mikrosk. Anat.* **112**, 15–30.

Pelluet, D. (1964). *Can. J. Zool.* **42**, 195–199.

Pelluet, D. and Lane, N. J. (1961). *Can. J. Zool.* **39**, 691–805.

Pentreath, V. W. and Cottrell, G. A. (1970). *Z. Zellforsch. mikrosk. Anat.* **111**, 160–178.

Peute, J. and Van de Kamer, J. C. (1967). *Z. Zellforsch. mikrosk. Anat.* **83**, 441–448.

Quattrini, D. (1962). *Monit. zool. ital.* **70**, 56–96.

Quattrini, D. (1963). *Monit. zool. ital.* **72**, 3–12.

Quattrini, D. and Lanza, B. (1965). *Monit. zool. ital.* **73**, 1–60.

Renzoni, A. (1969). *Veliger* **12**, 176–181.

Rogers, D. C. (1969). *Z. Zellforsch. mikrosk. Anat.* **102**, 113–128.

Röhnisch, S. (1964). *Z. Zellforsch. mikrosk. Anat.* **63**, 767–798.

Röhlich, P. and Bierbauer, J. (1966). *Acta biol. Acad. Sci. Hung.* **17**, 359–373.

Roubos, E. W. (1973). *Z. Zellforsch. mikzosk. Anat.* **146**, 177–205.

Runham, N. W. and Laryea, A. A. (1968). *Malacologia.* **7**, 193–108.

Runham, N. W., Bailey, T. G. and Laryea, A. A. (1973). *Malacologia* **17**, 135–172.

Sakharov, D. A. (1970). *Ann. Rev. Pharm.* **10**, 335.

Sanchez, S. and Bord, C. (1958). *C.r. hebd. Séanc. Acad. Sci. Paris* **246**, 845–847.

Sanchez, S. and Sablier, H. (1962). *Bull. Soc. zool. Fr.* **87**, 319–330.

Sanchiz, C. A. and Zambrano, D. (1969). *Z. Zellforsch. mikrosk. Anat.* **94**, 62–71.

Scharrer, B. (1968). *Z. Zellforsch. mikrosk. Anat.* **89**, 1–16.

Scharrer, B. (1969). *J. neuro-visceral. Relat.* Suppl. **9**, 1–20.

Scharrer, B. and Weitzmann, M. (1970). *In* "Aspects of Neuroendrocrinology" (W. Bargmann and B. Scharrer, Eds) pp. 1–23. Springer Verlag, Berlin, Heidelberg, New York.

Scharrer, E. (1928). *Z. vergl. Physiol.* **7**, 1–38.

Schooneveld, H. (1970). *Neth. J. Zool.* **20**, 151–237.

Simpson, L. (1969). *Z. Zellforsch. mikrosk. Anat.* **102**, 570–593.

Simpson, L., Bern, H. A. and Nishioka, R. S. (1966a). *Gen. comp. Endocrinol.* **7**, 525–548.

Simpson, L., Bern, H. A. and Nishioka, R. S. (1966b). *Am. Zool.* **6**, 123–138.

Sminia, T. (1972). *Z. Zellforsch. mikrosk. Anat.* **130**, 497–526.

Smith, B. J. (1966). *J. comp. Neurol.* **126**, 437–452.

Starke, F. J. (1971). *Z. Zellforsch. mikrosk. Anat.* **119**, 483–514.

Sterba, G. (1963). *Gen. comp. Endocrinol.* **3**, 733.

Streiff, W. (1967) "Recherches cytologiques et endocrinologiques sur le cycle sexuel de *Calyptraea sinensis* L. (Mollusque Prosobranche hermaphrodite protandre)". Thesis, University of Toulouse.

Tuzet, O., Sanchez, S. and Pavans de Ceccatty, M. (1957). *C.r. hebd. Séanc. Acad. Sci. Paris* **244**, 2962–2964.

Wautier, J., Pavans de Ceccatty, M., Richardot, M., Buisson, B. and Hernandez, M. L. (1961). *Bull. mens. Soc. Linn. Lyon* **30**, 79–87.

Wendelaar Bonga, S. E. (1970a). *In* "Aspects of Neuroendocrinology" (W. Bargmann and B. Scharrer, Eds) pp. 44–46. Springer Verlag, Berlin, Heidelberg, New York.

Wendelaar Bonga, S. E. (1970b). *Z. Zellforsch. mikrosk. Anat.* **108**, 190–224.

Wendelaar Bonga, S. E. (1971a). *Z. Zellforsch. mikrosk. Anat.* **113**, 490–517.

Wendelaar Bonga, S. E. (1971b). *Neth. J. Zool.* **21**, 127–158.

Wendelaar Bonga, S. E. (1972). *Gen. comp. Endocrinol.* suppl. **3**, 308–316.

Wendelaar Bonga, S. E. and Boer, H. H. (1969). *Z. Zellforsch. mikrosk. Anat.* **94**, 513–529.

Woerdeman, M. W. and Raven, C. P. (1946). "Experimental Embryology in the Netherlands 1940–1945." Elsevier, New York.

Zylstra, U. (1972). *Z. Zellforsch. mikrosk. Anat.* **130**, 93–134.

Chapter 7

Reproduction

C. J. DUNCAN

Department of Zoology, University of Liverpool, England

I. Introduction

Anyone who has dissected a garden snail cannot fail to have been impressed by the complexity of the reproductive organs. Not only is the organization of this hermaphrodite system as complex as any in the animal kingdom, but the various ducts are convoluted and follow tortuous paths, so adding to the difficulty of elucidating the functional

309

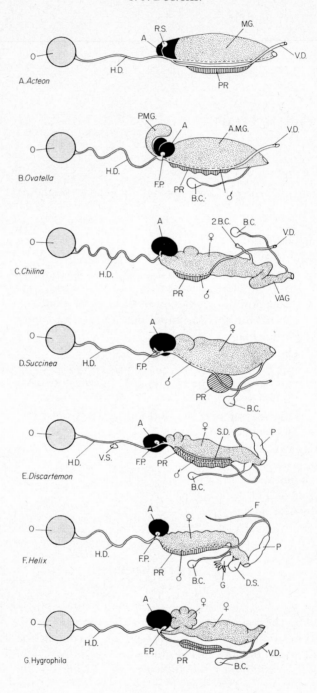

morphology. This complexity, and the numerous variants associated with it, is necessarily reflected in the many published accounts of pulmonate genital systems. The reproductive organs of all pulmonates conform to a common basic plan. This has similarities with the organization found in *Acteon*, but in this primitive opisthobranch the genital duct is associated with the body wall, as in prosobranchs, and has a straight course from gonad to genital aperture; also there is a large uninvaginable penis. In higher opisthobranchs and in pulmonates the greater complexity of the ducts is associated with the fact that they are free to coil in the haemocoel. This plan, together with the various modifications found in different families, is shown schematically in Fig. 1, where it can be seen that the reproductive system is formed of hermaphrodite, male and female sections.

Spermatozoa and ova are produced in an ovotestis and, when released, both pass anteriorly along the little hermaphrodite duct. Thereafter, male and female gametes follow separate paths; the male and female paths may be only functionally separate channels within a common duct (the spermoviduct), or they may diverge into two completely separate ducts (Fig 2). At the junction (the carrefour) of the hermaphrodite, male and female ducts, the albumen gland opens adding its nutritive secretion to the fertilized eggs. The fertilization pocket is also associated with this region; it is either a small diverticulum, or a more elaborately specialized area of the lower end of the hermaphrodite duct (Fig 2). It is in the region of the carrefour, at the head of the male and female ducts that it is particularly difficult to determine details of morphological organization and of functional roles.

The posterior part of the male duct has secretory tissue which comprises the prostate gland and anterior to this it becomes the muscular vas deferens which passes through the body wall and then re-enters the haemocoel before opening into the penis. The female duct is large and

Fig. 1. Diagrams of the organization of the reproductive systems of *Acteon* and selected pulmonates. A, *Acteon tornatilis* (L.) (Opisthobranchia) based on the description of Fretter and Graham (1954), B, *Ovatella* (Ellobiidae) modified after Morton (1955b), C, *Chilina* (Chilinidae) based on descriptions by Harry (1964) and Duncan (1960a), D, *Succinea pfeifferi* Ross, (Succineidae) E, *Discartemon* (Stylommatophora, Streptaxidae) after Berry (1965) F, *Helix pomatia* (Stylommatophora) after Fretter (1946), G, generalized plan of the reproductive systems of freshwater Basommatophora. A, albumen gland; A.M.G., anterior mucous gland; B.C., bursa copulatrix; D.S., dart sac; F, flagellum; F.P., fertilization pocket; G, accessory mucous gland; H.D., hermaphrodite duct; M.G., mucous gland; O, ovotestis; P, penis; P.M.G., posterior mucous gland; PR, prostate; R.S., receptaculum seminis; S.D., seminal duct; V.D., vas deferens; VS, seminal vesicle. ♂, male channel; 2B.C., secondary bursa copulatrix.

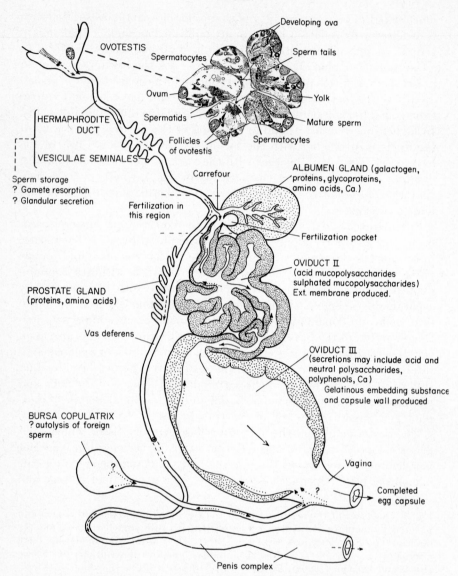

Fig. 2. Diagram to illustrate the functional anatomy of the pulmonate reproductive system. Organization based on *Physa* (Duncan, 1960b), but details of probable physiological roles based on studies of many of the higher pulmonates and described in the text. ---→ passage of home sperm; ····→ passage of foreign sperm; ——→ passage of ova. The ova receive nutritive secretions from the albumen gland and the investments of the egg capsule from the oviduct. It seems probable that the foreign sperm that effect fertilization never enter the bursa copulatrix, (see discussion on pp. 334–8).

glandular, its role being to receive sperm from a copulating partner, provide a site for fertilization and to form the egg capsule. The bursa copulatrix is a spheroidal sac with a long narrow duct opening into the vagina.

There is sometimes confusion of terminology in the literature on the morphology of the pulmonate reproductive tract. Difficulties have arisen because, 1. words originally used for vertebrates have been adopted, 2. the different portions of the female duct in particular have been independently named by different workers, and 3. although as far as possible, organs have (rightly) been named on the basis of their function, in some cases the role has been only surmised and not proven. The terminology used in the present review attempts to follow a common plan for the pulmonates, indicating the probable homologies both within the order and where possible with other gastropod groups and retains existing terminology where reasonably possible, especially where this indicates the known function of an organ (Fig 1). Thus, there is little difficulty with such organs as the ovotestis and albumen gland. However, the diverticula at the distal end of the female tract and at the bifurcation of the little hermaphrodite duct have caused considerable confusion since their precise function in copulation and fertilization is not yet certain and they may have different roles in different pulmonates. The terminology adopted by Fretter (1946) and Fretter and Graham (1962) in a study of the evolution of prosobranch genital ducts has been followed; the sac which opens into the vagina will be referred to as the bursa copulatrix, and the diverticulum (or diverticula) opening into the proximal part of the oviduct as the fertilization pockets (in prosobranchs and *Acteon* this is the site of the receptaculum). The problem of whether or not fertilization occurs in this pocket is discussed later.

II. Organization of the pulmonate genital system

A. Ellobiidae

Amongst the lower Basommatophora, the genitalia of the Ellobiidae almost certainly illustrate the most primitive condition, and the morphology of several genera is known in detail (Kowslowsky, 1933; Morton 1955a,b; Berry *et al.*, 1967). Fretter (1946) and Morton (1955a,b) have discussed the evolution of the pulmonate genital ducts from the ancestral prosobranch condition and Morton (1955a) has also shown their similarity with the tectibranch *Acteon*; he believes that the Ellobiidae stand at the point close to where the prosobranchs, opisthobranchs and pulmonates diverged. Morton (1955b) points out that the main trends in which the reproductive system of the lowest pulmonates is different from, or advanced upon, that of the prosobranchs, include the following:

1. universal hermaphroditism, 2. universal separation of the albumen gland from the reproductive tract to form a distinct outgrowth, 3. the development of the bursa copulatrix with a long duct, and 4. the retraction of the penis into an invaginable praeputium.

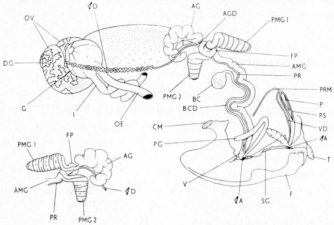

Fig. 3. Semi-diagrammatic dissection of the reproductive system of *Pythia scarabaeus* (L.) (Ellobiidae) from the right-hand side (from Berry *et al.*, 1967). Inset shows details of the glandular portions from the left. ♂A, male aperture; ♀A, hermaphrodite aperture; AG, albumen gland; AGD, albumen gland duct; AMG, anterior mucous gland; BC, bursa copulatrix; BCD, bursa copulatrix duct; CM, columellar muscle; DG, digestive gland; ♀D, little hermaphrodite duct; F, foot; FP, fertilization pouch; G, gizzard; I, intestine; OE, oesophagus; OV, ovotestis; P, penis; PG, pallial gland; PMG 1, PMG 2, upper and lower lobes of posterior mucous gland; PR, prostate; PRM, penis retractor muscle; PS, penis sac; SG, seminal groove; T, tentacle; V, vagina; VD, vas deferens.

In all ellobiids the basic pulmonate pattern is clearly established, but *Pythia*, alone among the pulmonates (Morton, 1955a; Berry *et al.*, 1967), has the primitive arrangement in which sperms leaving the common genital opening pass forward along an open, ciliated seminal groove which corresponds with that found in many tectibranchs. They then enter the true vas deferens near the right tentacle and pass to the penis (Fig.3). A completely closed, tubular vas deferens is found in other ellobiids (Figs. 1 and 5).

In the Ellobiidae, as in the Otinidae and Chilinidae, the glandular female tract is clearly demarcated into an upper region with diverticula and a lower, straight duct; these areas are generally termed the posterior and anterior mucous glands respectively.

In *Pythia* and the more primitive ellobiid genera the male and female tracts, although functionally separate, remain as confluent channels of the spermoviduct. In *Leucophytia* (Fig. 4), *Melampus* (Morton, 1955a)

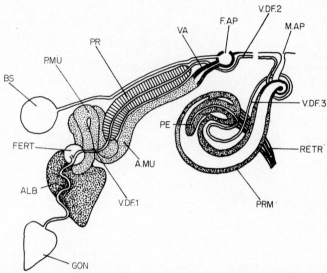

Fig. 4. Diagram of the reproductive system of *Leucophytia bidentata* (Montagu) (Ellobiidae), from Morton (1955a). ALB, albumen gland; AMU, anterior mucous gland; BS, bursa copulatrix; FAP, female aperture; FERT, fertilization pouch; GON, gonad; M.AP, male aperture; PE, penis; P.MU, posterior mucous gland; PR, prostate; PRM, praeputium; RETR, retractor muscle of the penis; VA, vagina; V.DF. 1, V.DF. 2, V.DF. 3 three portions of the vas deferens.

and *Auricula* (Berry *et al.*, 1967), however, these glandular ducts are separate, resembling the condition found in the higher, freshwater Basommatophora. A muscular genital vestibule, the pallial gland, opens close to the female aperture in *Pythia*, *Carychium* and *Cassidula* (Morton, 1955a, Berry *et al.*, 1967). The pallial gland is particularly large and muscular in *Pythia* (Fig. 3) and presumably plays a role in reproduction, since spermatozoa have now been found in it (Berry *et al.*, 1967). The pallial gland of *Carychium*, however, is glandular, with cilia in isolated tufts and shows secretory activity not only during oviposition, but throughout the year (Morton, 1955b).

Thus, Morton and Berry *et al.* agree in considering *Pythia* as having the least specialized reproductive system in the Ellobiidae and, with *Ovatella*, *Ophicardelus* and *Cassidula*, forming a basal stock from which *Leucophytia*, *Auricula* and *Melampus* are derived rather close together, but separate from the Pedipedinae and *Carychium*. The evolutionary changes in the organization of the reproductive system are summarized in Fig. 5. Specializations that have occurred in different genera are the

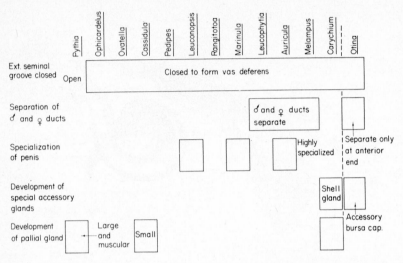

Fig. 5. Table summarizing the specializations of the reproductive system found in various genera of the Ellobiidae and in *Otina*. Based on Morton (1955a) with additional information from Morton (1955b, c) and Berry *et al.* (1967). The rectangles represent the specialized condition indicated.

shell gland of *Carychium* and the complex penis found in some genera. The shell gland is a long, glandular diverticulum at the posterior end of the mucous gland probably associated with the production of large single eggs with leathery capsules, (Morton 1955b). The penis is particularly specialized in *Ellobium aurisjudae* (L.) where both it and the praeputium are long, muscular and coiled (Fig. 15).

B. Otinidae

The reproductive system of the tiny, limpet-like *Otina otis* (Turton) closely resembles that of the ellobiids and is shown in Fig. 6 and described in detail by Morton (1955c). Open communication between the male and female tracts of the spermoviduct is retained for three-quarters of their common length. The posterior mucous gland is rather complex, being composed of three pouches, each of which is further subdivided into two (Fig. 6). It is believed that the eggs pass through each of the pouches in succession, receiving thin mucoid capsules before entering the anterior mucous gland. Another unusual feature is a second (accessory) bursa copulatrix, some 200 μm in diameter, the duct of which opens into the hermaphrodite portion of the spermoviduct. The duct of the other, larger bursa opens into the vagina.

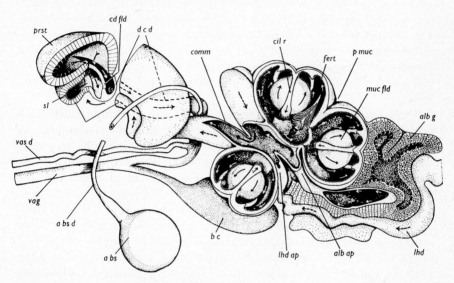

Fig. 6. *Otina otis*. Stereogram of the glandular regions of the genital tract, viewed from the median aspect and dissected to show the internal structure and the course of the ciliary currents. The anterior – most part of the albumen gland is represented in longitudinal section; the roof of the posterior mucous gland has been removed; and the prostate with the anterior mucous gland and distal common duct has been cut across transversely and the two halves separated to show the relations of the three ducts. The anterior parts of the vagina and vas deferens, with the penis, have been omitted. a bs, accessory bursa copulatrix; a bs d, duct of the accessory bursa copulatrix; alb ap, opening of the albumen gland into the posterior mucous gland; alb g, albumen gland; comm, communication between the anterior and posterior mucous glands; bc, bursa copulatrix; cd fld, longitudinal fold dividing the cavity of the distal common duct; cil r, ciliated ridge running round a fold of the posterior mucous gland; d c d, distal common duct; fert, fertilization pouch; lhd, little hermaphrodite duct; lhd ap, opening of the little hermaphrodite duct into the posterior mucous gland; muc fld, glandular fold of the posterior mucous gland; p muc, posterior mucous gland; prst, prostate; sl, longitudinal slit by which the prostate and anterior mucous gland communicate; vag, vagina; vas d, vas deferens. (From Morton 1955c).

C. Other families of the lower Basommatophora

The reproductive system of the other, lower families of the Basommatophora are known in less detail and few histological studies have been made. The morphological studies have been summarized by Simroth and Hoffmann (1928), and in all genera the basic basommatophoran plan can be discerned. In the Chilinidae (see Haeckel, 1911; Harry, 1964) the male and female ducts are incompletely separated, with an upper spermoviduct (with male and female gutters in a common duct) and a lower separation into an oviduct and a long, coiled vas deferens. The oviduct has a diverticulum, perhaps a secondary bursa copulatrix, although there are no histological details (Fig. 1). The muscular vagina secretes calcareous granules, and a spermatophore, filled with spermatozoa and bounded by a cuticle, has been found in the oviduct (Harry, 1964).

The common spermoviduct is also short in the Gadiniidae. It has a glandular diverticulum to the female tract (probably composed of columnar, mucous cells and ciliated cells), and prostatic cells open into the male tract (Schumann, 1913). Thereafter, the oviduct and vas deferens separate completely. The bursa copulatrix which opens into the lower oviduct is large.

The comparative morphology of the Siphonariidae has been described in detail by Hubendick (1945); the male and female ducts are confluent until they open into the genital atrium, close to the opening of the duct of the bursa copulatrix. In some species there is an elaboration of the penis by the addition of glandular organs and flagellum, together with a muscular portion to the sperm duct prior to the penis, termed an epiphallus. Spermatophores have been described by Abe (1940) and Marcus and Marcus (1960).

D. The freshwater Basommatophora (Hygrophila)

Details of the morphology of the reproductive systems of many species of the Hygrophila are available, although histological investigations are less numerous. These studies (many with extensive bibliographies) include:

Lymnaeidae: Crabb (1927a, b); Holm, (1946); Bretschneider (1948a, b); McCraw (1957); Walter (1969).

Physidae: Slugocka (1913); Duncan (1958).

Planorbidae: Baker (1945); Abdel-Malek (1954a, b) Hubendick (1955) Wright (1956); Pan (1958); Alaphilippe (1959); Schutte and van Eeden (1959); Stiglingh et al. (1962); Walter (1968).

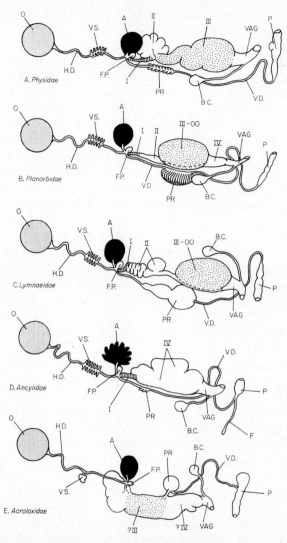

Fig. 7. Diagram to illustrate the organization of the reproductive systems of representatives of the families of the freshwater Basommatophora. Homologous organs, together with the suggested nomenclature for the various parts of the female tract are shown. A–D based on Duncan (1960b) and E on the description by Hubendick (1962). (A) *Physa fontinalis* (L.); (B) *Lymnaea peregra* (Müll.); (C) *Planorbarius corneus* (L.); (D) *Ancylus fluviatilis* Müll. (E) *Acroloxus lacustris* (L.). A, albumen gland; B.C., bursa copulatrix; F, flagellum; F.P., fertilization pocket; H.D., hermaphrodite duct; O, ovotestis; III-OO, oothecal gland, oviduct III; P, penial complex; PR, prostate gland; VAG, vagina; V.D., vas deferens; V.S., vesicula seminales; I, II, III and IV, oviducts.

Ancylidae: Lacaze-Duthiers (1899); Hoff (1940); Basch (1959a).
Acroloxidae: Hubendick (1962).

With the exception of the Acroloxidae, the Hygrophila are characterized by the complete separation of the male and female ducts from their points of departure from the carrefour. The vas deferens has the glandular prostate, the prostate of the Lymnaeidae being fundamentally different in histology from that of the other families (Hubendick, 1947; Duncan 1960b). The oviduct is more complex than in other pulmonates, and may be subdivided into clearly demarcated regions, probably associated with the production of the more elaborate egg capsules appropriate to fresh water. Considerable confusion exists between the nomenclature ascribed to these regions in different, independent descriptions, and an attempt has been made to provide a standard terminology which recognizes homologous parts of the oviduct in different families (Duncan, 1960b). This is based on a study of only a few species, but recognizes topological and histological similarities; future investigations will reveal whether these are related to common functions in the production of the different parts of the egg capsule. The nomenclature is summarized in Fig. 7, the oviduct being subdivided into regions, I, II, III, and IV.

E. Succineidae

The comparative morphology of the succineid reproductive system has been studied by Quick (1933, 1936, 1939a, b, c, 1957), Pilsbury (1948), Odhner (1950) and Patterson (1968) and histological details have been provided by Hanna (1940), Duncan (1961) and Rigby (1965). Again, the general pattern of organization of the pulmonate genital system is apparent (Fig. 1). The fertilization pocket complex consists of a pair of blind pockets opening from a diverticulum at the base of the hermaphrodite duct, although small differences of arrangement may occur in the same species (Franzen, 1963). The spermoviduct is incompletely divided into male and female sections; the vas deferens separates completely about one-third of the way along the female duct and the prostate is large and spheroidal. The female tract is simple and there is no clear differentiation into separate regions.

F. The higher Stylommatophora

In most other families of the Stylommatophora the reproductive system is less specialized (Fig. 1); there is a common spermoviduct, the female tract is not clearly differentiated into separate regions, and prostatic follicles open into the male tract (Mead 1950; Berry, 1963a, b,

1965; Rigby, 1963; Ghose, 1963). This is also true of many of the slugs (Quick, 1960; Bayne, 1967; Runham and Hunter, 1970). In some slugs the penis and dart sac are lost although the sac is retained in *Philomycus* (Kugler, 1965). *Agriolimax reticulatus* (Müll.) has "trifid" appendages to the penis (Bayne, 1967) and these are evertile (Runham and Hunter, 1970). Their function is unknown, but the cells apparently become necrotic and breakdown products appear in the lumen (Runham and Laryea, 1968). The basically unspecialized pattern is retained in the helicids, but a number of accessory structures have been developed. These may include one or more digitate mucous glands and the dart sac opening into the vagina, a flagellum (a blind diverticulum opening into the vas deferens, just proximal to its junction with the penis) and a diverticulum of the duct of the bursa copulatrix (Fig. 1); *Helix aspersa* Müll. and *H. pomatia* L. differ in some of these details (see Meisenheimer, 1907; Baecker, 1932 for morphological and histological accounts).

III. Functional anatomy and physiology of the reproductive system

A. The hermaphrodite system

1. Ovotestis and hermaphrodite duct

The ovotestis consists of a variable number of follicles, the lobes of which are embedded in the digestive gland. It is bounded externally by squamous epithelium and between its lobes is a loosely-woven network of connective tissue (Archie, 1942). It is frequently pigmented; in *Arion ater* (L.) the pigment is a lipofuscin (Lūsis, 1962), and the degree of pigmentation is correlated with the development of the gland (Lūsis, 1961).

The histology of the pulmonate ovotestis has been studied by many workers, and details are available for a variety of species (Fig.2). From these descriptions it is apparent that there is considerable uniformity in the structure throughout the group (see Gatenby 1917, 1918, 1919a, b, 1920; Hickman, 1931; Mahoney 1940; Archie, 1942; Abdel-Malek, 1954a, b; Duncan 1958). Ancel (1902), continuing the earlier work of Blomfield (1881), recognized three developmental stages in the ovotestis of *Helix pomatia:* the formation of progerminative male gametes from undifferentiated sex cells; the development of nurse cells which in turn organize further undifferentiated cells to develop into potential female gametes. The details of this problem of the organization and determination of two types of sex cells within the same organ has not yet been satisfactorily resolved (see Lūsis, 1961); it may be under hormonal control (Pelluet and Lane, 1961; Pelluet, 1964) and the role of the endo-crine system is discussed in detail in chapter 6. Gametogenesis can be affected by light (Henderson and Pelluet, 1960), low temperature (Brid-geford and Pelluet, 1952) and starvation (Joosse et al., 1968) and the proportion of the two gametes is also modified by environmental con-ditions. High temperature, low humidity (Lūsis, 1966; Smith, 1966) orγ-irradiation can all differentially affect gametogenesis (Joosse et al., 1968). 6°C favours oogenesis in Cepaea nemoralis (L.), spermatogenesis being suppressed, whereas spermatogenesis is favoured at 23°C (Bouillon, 1956).

The ellobiids and Siphonaria (Marcus and Marcus, 1960) exhibit a protandrous sexual cycle; in Carychium (Morton, 1954) and Leucophytia (Morton, 1955b) the ovotestis shows a phase of sperm production which is gradually replaced by a phase of oogenesis. In the majority of pul-monates, however, spermatogenesis and oogenesis can occur simultan-eously in the same follicle; the ova remain attached to the germinal epithelium until completely mature, while the sperms generally fall into the lumen at an early stage and continue development there (Duncan, 1958). Gametogenesis follows a similar pattern in Arion ater (Lūsis, 1961), in Helix (Ancel, 1902; Demoll, 1912; Gatenby, 1917; Lee, 1897, 1904), where spematogenesis continues through the summer (Lind, 1973), and in Lymnaea (Gatenby, 1919; Archie, 1942) and the following developmental history of the ovotestis is taken from the detailed account given for L. stagnalis (L.) by Archie (1942).

The ovotestis first appears as a solid cord of germinal epithelium and, when the snail is 5mm long, this has developed into a long sac of tissue with a small lumen. Spermatogonia appear some two weeks before the oogonia, so that although, when the animal is finally mature, eggs and sperm are present together, the early developmental phases reflect the protandric sequence of the primitive basommatophorans (Duncan, 1960a). When the spermatogonia appear, groups of them are attached to common cytoplasmic stalks extending into the lumen from the ovo-testis wall, there being up to 20 spermatogonia attached to each of these Sertoli cells. In 3–4 mm snails several nuclei resembling those of spermatogonia are found, but all are included in a single cell. A little later oval masses of cytoplasm, each possessing a large Sertoli cell nucleus and 2 spermatogonia nuclei, are found. Archie believes that during the early multiplication of the spermatogonial nuclei, one nucleus ceases to divide and becomes differentiated as the Sertoli cell nucleus. During subsequent spermatogenesis in Lymnaea up to 285 spermatids may become attached to each Sertoli cell (Fig. 2). In the helicids, the differentiation of these Sertoli cells is distinguished by a series of endo-mitotic chromosomal replications, the nucleus in *Cepaea nemoralis* becoming 32–ploid. During this polyhaploid state the cells function as secretory cells and a pairwise fusion, resulting in the polydiploid state, is required before the cells become functional as sperm nurse cells (Serra and Koshman, 1967; Koshman and Serra, 1967).

Each ovum is surrounded by a layer of 8-10 nurse (or follicle) cells and, as the ovum grows older, each nurse cell appears as a thin line with an externally-facing bulge which indicates the position of the nucleus. An active alkaline phosphatase has been reported in the nurse cells of *Biomphalaria* (Muller, 1965) and *Arion* (Lūsis, 1961). When the ovum approaches maturity the nurse cells are sloughed off and fall into the lumen of the ovotestis. Further details of oogenesis in *Lymnaea* are given by Raven (1948, 1958) and Aubry (1954a–d), and in the Planorbidae by Albanese and Bolognari (1961) and Pan (1958). There is a similar sequence in *Agriolimax* and *Arion* (see Lusis, 1961; Smith, 1966; Runham and Laryea, 1968, for detailed accounts), but the time-scale is expanded and development is markedly protandric in *Arion*, with succeeding male, hermaphrodite and female stages.

There are a number of studies on the detailed structure and histochem-istry of pulmonate sperm (Grassé *et al.*, 1956; Rebhun, 1959; Gatenby, 1960; Barth and Oliveira, 1964; Tahmisian, 1964; Galangau and Tuzet, 1966; Tuzet and Galangau, 1967; Anderson and Personne, 1967, 1969a, b, 1970; Personne and Anderson, 1969, 1970; Bayne, 1970; Starke and Nolte, 1970); spermatozoan morphology is relatively uniform in the Basommatophora, whereas considerable heterogeneity is evident in the Stylommatophora (see Thompson, 1973, for a comprehensive review). In *Planorbarius* the sperm nucleus has a spiral twist and bears super-ficial helical ridges and grooves. The axoneme has the basic pattern of 9 + 2 elements (9 + 9 + 2 just behind the head) and is ensheathed by a layer of granular material through which spiral numerous organelles. Spermatozoa of *Hedleyella falconeri* (Reeve) are 1140–1400 μm in length, the longest recorded in any mollusc to date (Thompson 1973). Sperm-atozoan ultrastructure and histochemistry are summarized in Figs 8, 9 and 10.

The gametes are liberated into the hermaphrodite duct, the walls of which are generally formed of glandular cells and ciliated cells. Their arrangement varies in different genera. For example, in *Physa* it is completely ciliated at its upper and lower ends, but the central section of the duct is largely glandular, with ciliated cells confined to a lateral tract (Duncan, 1958). Additional, glandular diverticula, the seminal vesicles, are found, and these are apparently aggregations of glandular

M

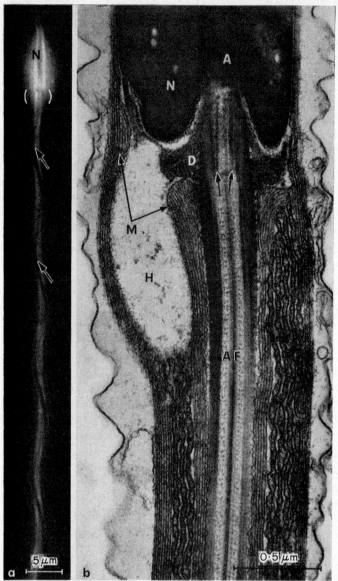

Fig. 8. a, The anterior region of a spermatozoon of *Helix aspersa* is illustrated in this phase contrast micrograph. The neck segment (in brackets) lies at the base of the elongate nucleus (N) and within the helical mitochondrial derivative (arrows). b, A longitudinal section through the neck and middle piece of a speratmozoon is illustrated in this electron micrograph. The cross-banded coarse fibres of the middle piece and the flagellar tubules (AF) terminate in an indentation (A) in the base of the head (N). In the neck region, two dense masses are seen on either side of the central flagellar tubules (arrows). Granular material (D) surrounds the connecting piece and occupies the anterior portion of the major helix (H). The paracrystalline portion of the mitochondrial derivative (M) is juxtaposed to the axial filament. (From Anderson and Personne, 1967.)

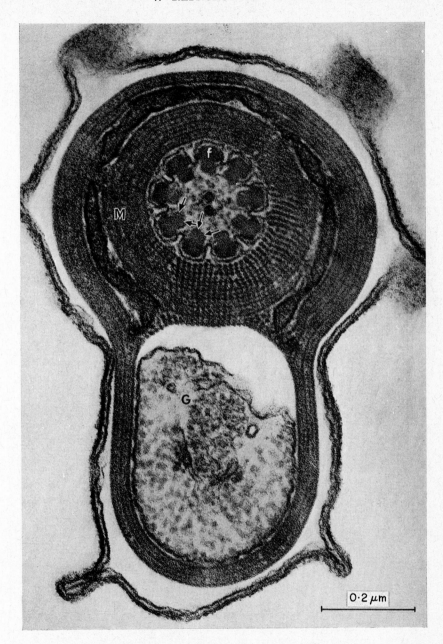

Fig. 9. A transverse section through the middle piece of a mature spermatozoon is illustrated in this electron micrograph. The mitochondrial derivative (M) containing glycogen (G) in the major helix surrounds the axial filament. In this region the peripheral flagellar tubules (arrows) exist as dense structures at the inner region of the coarse fibres (f). The coarse fibres are all alike and are arranged in radial symmetry. (From Anderson and Personne, 1967.)

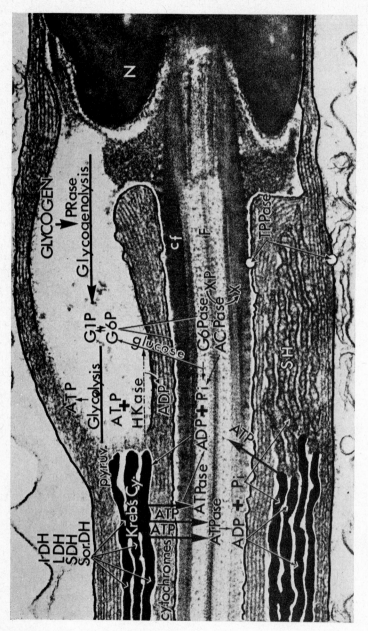

Fig. 10. Diagrammatic representation showing the enzyme localizations and possible inter-relationships in the middle piece of the mature sperm of *Helix*. ACPase, acid photophatase; ADP, adenosine diphosphate; ATP, adenosine triphosphate; ATPase, adenosine triphosphatase; G1P, glucose-1-phosphate; G6P, glucose-6-phosphate; G6Pase, glucose-6-phosphatase; HKase, hexokinase; P_i, inorganic phosphate; PRase, phosphorylase; TPPase, thiamine pyrophosphatase; XP, unspecific phosphate esters; pyruv, pyruvate; IDH, isocitric dehydrogenase; LDH, lactic dehydrogenase; SDH, succinic dehydrogenase; Sor. DH, sorbital dehydrogenase; cf, coarse fibre; F, flagellum; N, nucleus; SH, secondary helix. (From Anderson *et al.*, 1968).

cells. The seminal vesicles have a clear lumen in *Lymnaea* and the remainder of the duct is ciliated (Holm, 1946; Aubry, 1956). When the animal is mature, the hermaphrodite duct of most pulmonates is filled with sperm (Duncan, 1958; Lind, 1973) and it is probable that the whole duct, and not merely the seminal vesicles, have a role in sperm storage. Some of the sperm (in *Lymnaea*) are resorbed by the gland cells and resorption of gametes may be common in pulmonates; ripe oocytes of *Lymnaea* and *Agriolimax* are also resorbed by their nurse cells (Joosse *et al.*, 1968; Runham and Hunter, 1970).

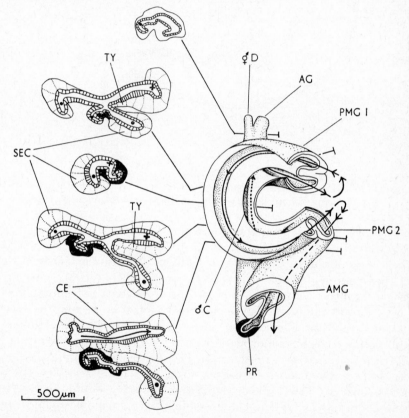

Fig. 11. Details of the region of the fertilization pouch of *Pythia scarabaeus* (L.) (Ellobiidae), from above, showing ciliary tracts with sections at levels shown on the left. In sections, black = prostate tissue, + = ciliary current passing backward or apically, ● = ciliary current passing forward or abapically. (From Berry *et al.*, 1967.) AG, albumen gland; AMG, anterior mucous gland; CE, ciliated epithelium; PMG 1 & 2, upper and lower lobes of posterior mucous gland; PR, prostate tissue; SEC, secretory cells of mucous glands; TY, typhlosole; ♂C, male channel; ♀D, little hermaphrodite duct.

2. The carrefour

The junction of the hermaphrodite duct with the male and female tracts is termed the carrefour (Baker, 1945) and the duct from the albumen gland also opens at this point. The detailed arrangement varies in different genera (Figs. 11 and 12), but the action of the ciliated tracts,

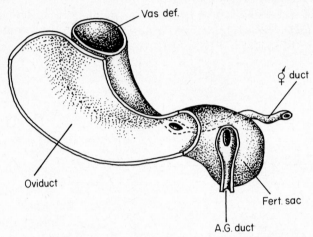

Fig. 12. Diagrammatic reconstruction of the carrefour and fertilization sac of *Planorbarius corneus*. (After Alaphilippe and Regondaud, 1959.)

muscle sheaths and sphincters must serve to direct the outgoing sperm into the male groove or duct (Fig. 2). After fertilization the ova receive their coats of albumen and must then be passed into the female section of the spermoviduct (or oviduct), to receive their tertiary envelopes which form the egg capsule. Foreign sperm received in copulation will also arrive at the carrefour, when cross-fertilization occurs, the sperm having ascended the female tract. In almost all descriptions of pulmonate genital systems, one or more diverticula are found in the region of the carrefour; they may open directly into the carrefour or may be blind caeca at the distal end of the hermaphrodite duct (e.g. *Succinea*, Rigby, 1965). They are here termed fertilization pockets (= talon in Stylommatophora; also = receptaculum seminis, receptaculum complex and spermathecal sacs); fertilization may or may not take place here, but certainly occurs in this general vicinity, before the eggs receive their albumen. The arrangement of the fertilization pockets in different genera is given in many of the anatomical descriptions cited above, but detailed descriptions are given, for example by Alaphilippe and Regondaud (1959, Hygrophila), Rigby (1965, *Succinea*), Berry *et al.* (1967, Ellobiidae) and Lind (1973, *Helix pomatia*), see Figs 11 and 12. Storage of foreign sperm, if it occurs, is almost certainly in these fertilization pockets.

B. The male system

1. Vas deferens and prostate

When the vas deferens is a separate duct (as it is in the Hygrophila) it is filiform, muscular and lined by a ciliated epithelium. Otherwise, it is a ciliated gutter along the spermoviduct and probably functionally separate from the female tract. Sperm are said to be confined to the male tract of *Arion* by inwardly-beating cilia, although eggs have been found in both sections of the spermoviduct (Lūsis, 1961). Lind (1973) has emphasized that this connection between male and female channels is probably important in *Helix*; sperm are expelled from the hermaphrodite duct at times other than at copulation and it is believed that these pass into the female tract and thence to the bursa where they are destroyed. Lind believes that the functional separation of the two tracts by longitudinal folds is produced only during copulation. Both types of vas deferens serve to convey the sperm to the penis at copulation, the ciliary beat being in this direction (Bayne, 1967), and secretions are added by the prostate gland (Fig. 2). Hubendick (1948) defines the pulmonate prostate as being a glandular structure developed from the vas deferens, proximal to its passage through the body wall. The prostate may be in the form of glandular cells or follicles opening into the male tract of the spermoviduct, (Stylommatophora, many ellobiids) or as numerous blind finger-like diverticula (*Physa*, Duncan, 1958), a small number of follicles (*Ancylus, Laevapex*) or as a separate organ complete with collecting ducts (*Succinea, Acroloxus*). In *Planorbarius* and *Biomphalaria* the vas deferens enters the prostate and histologically this section resembles one of the prostate follicles, being lined by columnar secretory cells; the prostate follicles open into a series of collecting ducts and are enclosed in a common envelope (Pan, 1958; Duncan, 1960a, b).

Hubendick (1948a) has shown that there are two fundamentally different types of prostate in the Hygrophila; firstly, in the Lymnaeidae, the surface enlargement has been achieved by the widening of the vas deferens, and the secretory cells, tall and columnar, are packed with granules and have a characteristic histology (see Duncan, 1960), and secondly in the Physidae, Planorbidae and Ancylidae the prostatic cells are large and cuboidal with large secretory globules. Although it is difficult to decide on the basis of the histological observations available, it seems probable that the prostatic cells of the majority of pulmonate

genera are of this second type (exceptions may be *Acroloxus*, see Hubendick, 1962, and some ellobiids, see Berry *et al.*, 1967). However, this
may be an oversimplified view of pulmonate prostatic histology. There
are descriptions of more than one type of secretory cell in some species
(which may represent developmental stages in the secretory cycle) and
Runham and Laryea (1968) have identified three different types of
glandular cells in the prostatic epithelium of *Agriolimax*. Additionally, a
second, separate gland opening into the male duct (the "flask gland")
has been described in *Arion ater* (L.) (Smith, 1965). Quattrini (1966a, b,
1967) has produced ultrastructural studies of the prostates of *Vaginula
Laevicaulis* and *Milax*. The glandular cells are filled with granules
contained in vacuoles (1-2 μm diameter) which correspond with
the secretory material seen by light microscopy, and the free borders
of the cells bear numerous microvilli. Most descriptions show that
ciliated cells are wedged between the apices of the glandular cells and
ciliary currents can be detected in living preparations (Duncan, 1958).

The functions of the prostatic fluid may well be different in different
families (Bayne, 1967). It contributes to the sperm mass transferred at
copulation and probably has a role, together with the flask gland, in
the formation of the spermatophore of *Arion*. The histochemical reactions of the prostatic secretion of this species have been studied in detail (Smith, 1965); the flask glands produce a weakly acidic polysaccharide, sometimes in association with Ca^{2+}, whilst the secretion from
prostatic cells is mainly phospholipo-protein. The spermatophores of
slugs can have a very complex shape which is characteristic of the
species (Quick, 1960) and which probably mirrors the internal form
of the epiphallus, the enlarged distal end of the vas deferens (Runham
and Hunter, 1970). Bayne (1967) has provided a detailed analysis of
the free amino acids and the amino acids of macromolecular components of the prostatic secretions of *Agriolimax*. Proteins and amino
acids may well be the major constituents of the pulmonate prostatic
secretion.

In most species, the vas deferens finally enters the musculature of the
body wall (although not in some Stylommatophora) before re-emerging
and joining the proximal end of the penial complex. The exception is
the primitive condition found in *Pythia*, in which the sperms pass forward along the open seminal groove. (see p. 314)

2. Penial complex

The sperm mass is transferred to the copulatory partner by the int-
romittent, muscular penis. Since the penis morphology constitutes a

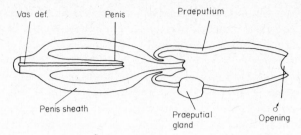

Fig. 13. Diagrammatic sagittal section of the penial complex of *Physa*.

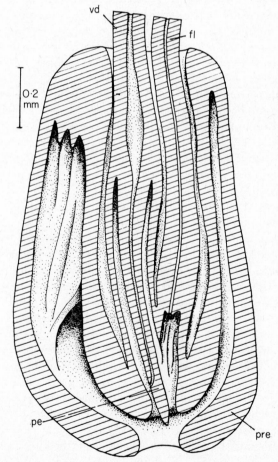

Fig. 14. Diagrammatic sagittal section of the penial complex of *Ancylus fluvia-tilis*. (From Hubendick, 1964.) fl, flagellum; pe, penis; pre, praeputium; vd, vas deferens.

valuable aid to taxonomic studies, its structure in differing species has been described in detail (see references under the descriptions of the different families above, especially Hubendick, 1947). However, in spite of considerable variation, once again a basic pulmonate pattern of organization can be discerned. Thus, in the higher limnic Basommatophora the penis is enclosed within a penis sheath and is everted through the praeputium (Figs 13, 14). The whole penial complex (a term used by Baker, 1945, to include the penis and its sheath, praeputium and the attached muscles) is muscular, but in *Lymnaea*, for example, the sheath has a thinner wall than has the praeputium in which there is an outer layer of longitudinal muscle (particularly well developed in two longitudinal wedges) and an underlying layer of circular muscle (Holm, 1946). In other families this arrangement is reversed (Rigby, 1965). The penial complex is supplied by retractor and protractor muscles (see description by Carriker, 1946, which also includes details of penial innervation).

The penis is particularly simple in the Otinidae (Morton, 1955c), being a short muscular tube perforated at its tip by the vas deferens. In the primitive Ellobiidae, however, the basic pulmonate plan is evident (Fig. 4). In *Pythia* (Fig. 3) (Berry *et al.*, 1967) and *Ovatella* (Morton, 1955b) the penis is enclosed in a penis sac and opens via an indistinct praeputtium. *Ellobium aurisjudae*, on the other hand, has an extremely specialized penial apparatus; the praeputium is very long and thrown into a strong spiral (Fig. 15). The same trend is found in *Leuconopsis* and *Marinula* (for details see Morton 1955a, b; Berry *et al.*, 1967). Specialization to form an "ultra-penis" in some planorbids and in *Laevapex* has been discussed in detail by Hubendick (1948b, 1955) and Basch (1959a).

A praeputial gland, the funtion of which is uncertain, is present in the Physidae (Duncan, 1958) and a praeputial organ is found in the Planorbidae; this organ is muscular and, in *Helisoma*, can be extended during copulation, when its role may be to act as a holdfast (Abdel-Malek, 1952). The praeputial organ in the *Helisoma* tribe has probably evolved as a specialization of the muscle pillars of the praeputium of the Basommatophora (Hubendick, 1963). A "flagellum" (a long narrow diverticulum) is present in *Ancylus* and *Helix*; in the former it is glandular, with an histology similar to that of its prostate follicles and vas deferens (Duncan, 1960). Sperm are transferred in a spermatophore in several stylommatophoran species.

Penes armed with barbs or stylets are described in a number of genera (e.g. *Siphonaria*, some planorbids, *Oophana*). In contrast the penis is absent in the Limacidae and Arionidae and in some species of the Siphonariidae and Ancylidae. Aphallism has also been described in a con-

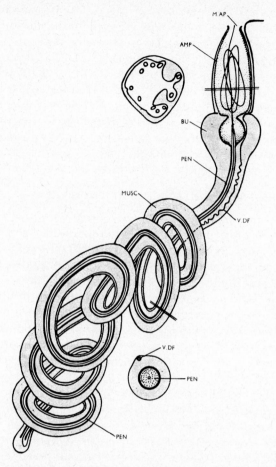

Fig. 15. Diagrammatic view of the penis of *Ellobium aurisjudae*, showing internal structure as seen in a transparent preparation. Two transverse sections are at the levels indicated. (From Morton, 1955a.) AMP, thin-walled terminal ampulla; BU, muscular bulbous portion of praeputium or penis sheath; M.AP, male opening; MUSC, muscular wall of penis sheath; PEN, filamentar portion of penis traversed by vas deferens; V.DF, inward running vas deferens.

siderable proportion of individuals of *Agriolimax laevis*, *Physa cubensis* Pfr. *Drepanotrema anatinum* (Orbigny) (= *D. yzabalensis*), *Ferrissia irrorata* (Guilding) (see Hyman, 1967) and *Bulinus contortus* (Mich.) (Larambergue, 1939).

The isolated penial complex of *Lymnaea* shows spontaneous, rhythmic contractions which persist for many hours (Duncan, 1964; Pecsi *et al.*, 1967). This preparation is sensitive to both 5-hydroxytryptamine and acetylcholine, but interpretation of the results is complicated by the possible action of these agents both directly on the muscle of the penial complex and on the associated nerves. However, the results with antagonists are comparable with those obtained by Jaeger (1962, 1963) on the penis retractor muscle of *Strophocheilus*, and Pecsi *et al.* (1967) have emphasized the similarities between these and other muscle preparations from pulmonates (e.g. *Helix* stomach, Gryglewski and Supnieswski, 1963). The isolated penis apparatus of *Helix* also shows rhythmic activity, but this is short-lived and believed to be dependent on nervous connection with the CNS (Goddard, 1962).

C. Fertilization and the bursa copulatrix

At copulation the sperms are transferred into the female orifice, but may be enclosed in either a spermatophore or jelly mass in some genera, (e.g. slugs, Ikeda, 1929; Quick, 1960; *Oxychilus*, Rigby, 1963; *Helix*, Lind, 1973). The fate of the sperm and the details of fertilization in these hermaphrodite animals are by no means clearly established. It is known that fertile eggs can be produced both as a result of self-fertilization (by autosperms) and cross-fertilization (by allosperms). Parthenogenesis (Colton, 1918; Crabb, 1927b; Ikeda, 1937) and self-copulation (Boray, 1964) are either very rare or absent, although an initial development following parthenogenesis has been reported in *Lymnaea stagnalis* (Bretschneider, 1961). An associated problem is the role of the bursa copulatrix, the duct of which usually joins the vagina close to the female opening. There are two views of fertilization (see Duncan, 1958 and Fig. 2), both of which could operate in the same snail, thus –

1. The ova are released from the ovotestis and pass down the hermaphrodite duct which is filled with autosperm and these effect fertilization by the time the ova reach the carrefour. On this view, copulation (if it occurs) serves merely as a stimulus for the release of the ova, and the foreign allosperms are received in the bursa copulatrix where they are autolysed. Mated *Lymnaea* oviposit much earlier in the breeding season than virgin individuals (McDonald, 1969), and the specific stimulus appears to be the presence of sperm in the bursa copulatrix (Horstmann, 1955).

2. The foreign sperm ascend the female tract and fertilize the ova in the region of the carrefour, perhaps in the "fertilization pockets" (Fig. 2). On this view, the bursa serves either in the autolysis of *excess* sperm as well as prostatic fluid, or all the sperm enter the bursa where they are stored or receive additional secretions or undergo physiological maturation. The bursa copulatrix could also have a role in releasing sperms

from a spermatophore (e. g. *Arion*, Smith, 1965). In such a situation the autosperms, which have not undergone "maturation" (by contact with prostatic and bursal secretions), would be less readily able to effect the fertilization of the ova. The foreign sperm would have to pass back, down the duct of the bursa before beginning the ascent of the female tract.

The problem has been studied in a number of ways: (a) raising snails in isolation, (b) investigating the nature of the secretions of the bursa copulatrix, (c) use of genetic markers, and (d) tracing the path of sperms after fertilization. From such studies it is becoming clear that there is no one common system among the pulmonates and, further, that different fertilization mechanisms may even operate in the same species under different conditions.

Very little information is available concerning fertilization in the families of the lower Basommatophora. There are a number of reports of living sperm being found in the bursa of ellobiids after copulation and, subsequently, histological evidence of a disintegrating sperm mass (Morton, 1954, 1955a, b). A similar account is given for *Otina*, although Morton (1955c) believes that copulation occurs several months before fertilization and that living sperm do not remain long in the bursa, but are transferred and stored in the accessory bursa.

The majority of studies on fertilization mechanisms have been conducted with the freshwater basommatophorans, especially *Lymnaea stagnalis*, and from these studies it seems clear that:

1. Foreign sperms appear in the bursa after copulation and at least some subsequently undergo disintegration, following the pattern of the Ellobiidae; the loss of the sperm tail has been described in *Philomycus* (Ikeda, 1929). This disintegration may be actively assisted by secretions of the bursa (as in *Lymnaea*, Horstmann, 1955).

2. Many pulmonates are capable of producing fertile eggs when bred in isolation. There are examples in the Lymnaeidae (Colton 1918; Walton and Jones, 1926; Crabb, 1927b; Boycott *et al.*, 1931; Colton and Penny-packer, 1934; Cain, 1956; DeWitt and Sloan, 1958; Boray, 1964); Planorbidae (Chadwick, 1903; Brumpt, 1928; Fraga de Azevedo *et al.*, 1959; Allen, 1935; Abdel-Malek, 1954a, b; Richards and Ferguson, 1962; Richards, 1969), although several species in this family fail to self-fertilize: Physidae (Crabb, 1927b; Baker, 1933; Brown, 1937; DeWitt, 1954b; Duncan 1958; DeWitt and Sloan, 1959); Arionidae (Williamson, 1959), *Philomycus*, (Ikeda, 1937), *Limax cinereoniger* Wolf. (Oldham, 1942) *Milax gagates* (Drap.) (Karlin and Bacon, 1961); *Archachatina* (Larambergue and Alaphilippe, 1959), *Achatina* and *Macrochla-*

mys (Ghose, 1959). Furthermore, cultures of *Lymnaea peregra* (Müll.) and *L. columella* Say have been maintained for many generations by self-fertilization, as have some species of *Agriolimax* (Maury and Reygro-bellet, 1963), and *Vaginula borelliana* Colosi and *Laevicaulis alte* (Fér.) (Lanza and Quattrini, 1964), and Aubry (1955) has even suggested that fertilization is possible in *L. stagnalis* whilst the eggs are still within the ovotestis. On the other hand, self-fertilization has apparently not been found in some genera, e.g. *Cepaea nemoralis* (although the sperm transferred from a single cross-fertilization may last for four years, A. J. Cain and P. M. Sheppard, personal communication). Self-fertilization is rare in the Ancylidae, but is reported in *Ferrissia shimekii* (Pilsbry) (Basch, 1959b). Self-fertilization is also reported as absent or very rare in *Helix aspersa* (Lang, 1911), and in helicids generally (Frömming, 1954), *Bradybaena similaris* (Fér.) (Ikeda and Emura, 1934) and *Agriolimax reticulatus* (Luther, 1915; Runham and Hunter, 1970; Bayne, 1970).

4. Studies on *Lymnaea*, using genetic markers (albinism), have shown that both self- and cross-fertilization can occur. In *L. peregra*, the first egg capsules laid after copulation were cross-fertilized, whereas later eggs were self-fertilized, perhaps when the foreign sperm were exhausted (Boycott *et al.*, 1931). Foreign sperm were also more effective in *L. stag-nalis* and remained viable in the recipient snail for as long as 116 days, after which time it was either exhausted or non-viable, and self-fertilization took place (Horstmann, 1955; Cain, 1956). *Lymnaea tomentosa* (Pfr.) is also preferentially cross-fertilizing (Boray, 1964), and fertilization is exclusively by foreign sperm in *Biomphalaria glabrata* (Say) for 20–60 days after copulation, but thereafter self-fertilization operates (Paraense, 1959). Mixed broods, utilizing both home and foreign sperm, are found in *Arion* (Williamson, 1959). Barbosa *et al.* (1958) have described a technique of artificial insemination of *Biomphalaria* which may prove of value in studies with genetic markers.

In summary, therefore, great variability is found in the pulmonates, depending both on the species and on environmental conditions (see Paraense, 1959). Fertilization occurs in the region of the carrefour, perhaps in the fertilization pockets, using either home or foreign sperm (Figs 2, 11, 12). The latter ascend the female tract (using a ciliated sperm groove which occurs in the oviduct of some families, e. g. *Lymnaea*, Horstmann, 1955), but probably do not penetrate into the herma-phrodite duct and seminal vesicles (Cain, 1956). Foreign sperm can often be identified in the fertilization pockets because it is oriented (Rigby, 1963, 1965; Lind, 1973). Foreign sperm are generally more effective in fertilization, either because of some biochemical barrier to self-fertilization or by virtue of their exposure to prostatic secretions. However, self-fertilization is certainly possible in many species.

A recent study by Lind (1973) has clarified the details of fertilization in *Helix pomatia*; the penis of one snail is inserted into the duct of the bursa of its partner and the spermatophore is slowly introduced by peristalsis. The spermatophore head reaches the bursa in 3–4 h and its tail is within the bursa duct 4–5 h after the start of copulation. The spermatozoa escape via the tail canal of the spermatophore into the bursa duct and ascend the oviduct to the fertilization pockets where they orient with their heads against the epithelial cells. Only foreign sperm are stored in the fertilization pockets, which in *H. pomatia* are 3–5 blind sacs termed spermathecal sacs by Lind. The majority of sperm do not escape in this way, but pass into the bursa and are destroyed. (See also Németh and Kovács, 1972.)

At fertilization, the sperm moves through the whole thickness of the egg cytoplasm and comes to rest near the egg cortex. Raven (1958) describes how the tail then breaks away in *Lymnaea*, *Limax* and *Bulinus*, and the head changes shape and rounds off. The sperm tail is resorbed in the egg cytoplasm, either during the first maturation division (*Bulinus*), or between the first and second divisions (*Lymnaea*), or after the second division (*Physa*, *Limax*, *Helix*), or as late as the first cleavage (*Arion*). Polyspermy has been found in some genera, but the supernumerary sperms soon disintegrate (Raven, 1958).

The bursa copulatrix is a spheroidal organ the duct of which opens into the distal end of the female tract. It is lined by cuboidal or columnar, secretory epithelium, and in some species the duct is ciliated. A red pigment (of unknown function) is found in the bursa of many pulmonates at the time of copulation. Smith (1965) has described the results of histochemical studies on both the contents and epithelium of the bursa of *Arion*, and studies by Horstmann (1955, Basommatophora) Németh and Kovács (1972, Stylommatophora) and Lind (1973, Stylommatophora) strongly suggest that its role is the digestion of excess sperm. Since the foreign sperm received in copulation undergo disintegration in the bursae of *Lymnaea* (Horstmann, 1955), *Agriolimax* (Bayne, 1970), and *Helix* (Lind, 1973) it seems probable that, if these gametes are to effect fertilization, they cannot remain there long, but must be stored elsewhere during the period between copulation and fertilization, which may be several weeks (Bayne, 1970) or even years. The evidence suggests that this storage site is the region of the carrefour and, in particular, the fertilization pockets.

Future studies on fertilization could include the use of radioactively labelled sperm to follow their path after copulation and a detailed investigation of the secretions of the bursa, particularly the enzyme activity, using microanalytical techniques, following the histochemical studies already published.

D. The female system

1. The albumen gland

The fertilized egg is surrounded by the nutritive secretion of the albumen gland. Judging from the published morphological and histological descriptions, it is probable that the albumen gland has a standard structure and physiological role throughout the pulmonates, although biochemical information is lacking for the lower Basommatophora. Its function is the production of albumen, a major constituent of which is galactogen. It is generally spheroidal with a duct opening at the carrefour, but is bilobed in the ellobiids *Ovatella* and *Carychium* (Morton, 1955b), formed of a series of lobules in *Otina* (Morton, 1955c) and *Ellobium* (Berry *et al.*, 1967) and ribbon-like in *Laevapex* (Basch, 1959a).

Histological studies of the albumen gland of a wide variety of pulmonates have been presented by many authors. It is a compound, tubular structure with a central duct lined by ciliated epithelium. The many secretory follicles are circular when viewed in cross section, and each has a small central duct around which a single layer of cells is arranged. The glandular cells usually have a basal nucleus and are tall and columnar in the Ellobiidae and Otinidae (Morton, 1955b, c) and more nearly cuboidal in higher pulmonates. Accounts differ in describing one or two types of cells in the acini; this may reflect genuine specific differences or, more probably, the difficulty of sectioning this organ. Two types of cell have been described in the Ellobiidae (Morton, 1955a), *Gadinia* (Schumann, 1913), *Helix* (Yung, 1911; Baecker, 1932), *Lymnaea* (Holm, 1946) *Biomphalaria* (Pan, 1958) and *Planorbis* (Alaphilippe, 1959); the larger cells are secretory and the small ones are probably the ciliated (or centrotubular) cells. The presence of two types of cells has been clearly established in the albumen gland of *Helix* by electron microscopy (Nieland and Goudsmit, 1969). The apical portion of the secretory cells contains PAS-positive globules, where the galactogen is located, (Bauer, 1933; Von Brand and Files, 1947; Grainger and Shilitoe, 1952; Fantin and Vigo, 1968). Their development and size depends on seasonal factors, and the histological picture presented by the gland after trichrome staining reveals a clear developmental cycle as the reproductive season approaches.

In electron micrographs of the albumen gland of *Helix pomatia*, the nuclei of the secretory cells are large and vesicular (Figs 16, 17). The

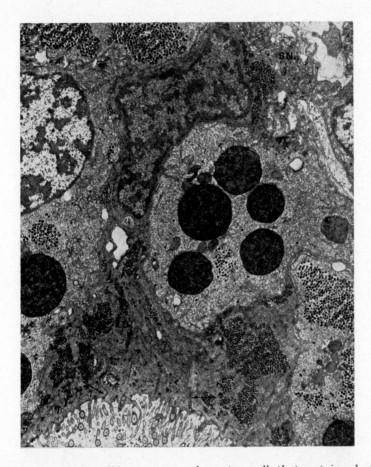

Fig. 16. Centrotubular cell between several secretory cells that contain galactogen globules. The centrotubular cell is distinguished by its smaller and denser nucleus (N), the greater density of its cytoplasm (CT), the cilia present at the luminal surface, and the absence of secretory globules. The location of desmosomes between adjacent centrotubular cells is indicated by arrows. A cell, possibly neurosecretory in function, containing many small dense bodies lies in the interstitial space (NS). × 6660. (From Nieland and Goudsmit, 1969.)

spherical, apical globules have a granular appearance, and intracytoplasmic clusters of rosette-like particles are also prominent in many areas, both supra and infranuclear. Golgi membranes are particularly prominent in the secretory cells and are frequently found adjacent to secretory globules that contain enormous numbers of small particles (approx. 20 nm) embedded in a homogeneous matrix. The secretory

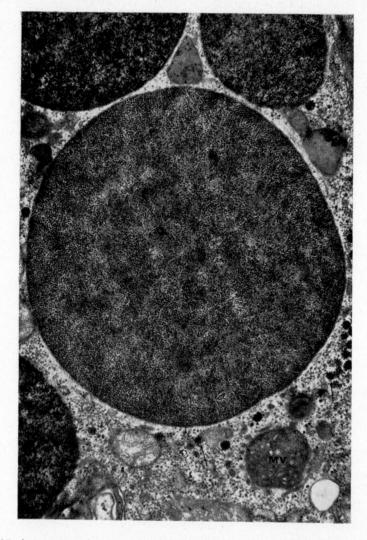

Fig. 17. A secretory globule composed of 20nm granules embedded in a homogeneous background matrix. The surrounding cytoplasm contains rosette-like particles (arrows), mitochondria, and multivesicular bodies (MV) that contain a few small granules. × 28 260. (From Nieland and Goudsmit, 1969.)

globules range in size from small droplets to globules larger than the cell nucleus. Nieland and Goudsmit (1969) believe, therefore, that the secretory droplets originate in the Golgi zone, and they have identified the 20 nm particles as galactogen. The second type of cell has been termed the centrotubular cell (Nieland and Goudsmit, 1969). They are located closer to the lumen, but rest on the basement membrane, although connected at times by a thin strand of cytoplasm only. Their nuclei are smaller and denser than those of the secretory cells. The free borders of both cells possess many microvilli, but only the centrotubular cells have cilia with a 9 + 2 arrangement of axial elements. Thus, the role of the centrotubular cells is apparently to carry the secretory products down to the main collecting ducts (Figs 16 and 17).

Galactogen is a galactose polymer which has now been reported in many pulmonates (May, 1931; Holm, 1946; Horstmann, 1956; McMahon et al., 1957; Duncan, 1958; Ghose, 1963; Rangarao, 1963; Bayne, 1967; Fantin and Vigo, 1968). Hammarsten (1885) first described "sinistrin" in extracts of the albumen gland, a substance which he reported to be glycoproteid in nature, but similar to glycogen. Subsequently (May, 1931; 1932a, b; 1934a, b, c; May and Kordowich, 1933) "sinistrin" was shown to be a polysaccharide composed of galactose residues. The polymer has been shown to consist of both D- and L- galactose in the albumen gland of *Helix pomatia* (Bell and Baldwin, 1940) in the approximate ratio of 6:1 (Bell and Baldwin, 1941). In their earlier studies Baldwin and Bell (1938) had shown that methylation produces equimolar proportions of 2, 3, 4, 6-tetramethyl- and 2, 4-dimethyl-D-galactose and suggested that galactogen is formed of either a β-1→3 or a β-1→6 linear chain, in which each galactopyranose unit carries a terminal galactose constituent on C-6 or C-3 respectively. Bell and Baldwin (1941) and May and Weinland (1953, 1954) believe that the L-galactose is confined to the side chain in the galactogen in *Helix*. Only D-galactose was found in homogenized eggs of *Lymnaea stagnalis* (Fleitz and Horstmann, 1967, L-galactose <0.4%). The average molecular weight of galactogen in *Helix* is given as 4×10^6 (Horstmann, 1964) and that of *Lymnaea* as 2.2×10^6 (Fleitz and Horstmann, 1967). The suggestion has also been advanced that the galactogen consists of highly branched chains (O'Colla, 1953; also Corrêa et al., 1967, *Biomphalaria*).

Most of the detailed biochemical studies on the synthesis of galactogen have been performed with albumen glands from *Helix pomatia*. The synthetic mechanism is similar to the path by which glycogen is manufactured and in which uridine diphosphate (UDP) sugars have a key role. The reaction sequence for the formation of D-galactose and its incorporation into galactogen can be summarized (Sawicka and Chojnacki, 1968):

Glucose-6-phosphate $\xrightarrow{(1)}$ glucose-1-phosphate $\xrightarrow{(2)}$ UDP glucose $\xrightarrow{(3)}$ UDP galactose $\xrightarrow{(4)}$ galactogen.

Albumen gland extracts are able to catalyse this sequence of reactions and Meenakshi and Scheer (1968, 1969) have shown that injection of ^{14}C glucose into the slug *Ariolimax* results in the labelling of the galactogen.

Thus, the formation of UDP-D-galactose from glucose-1-phosphate (reactions 2-3) was shown by Sawicka and Chojnacki (1968); the process depends on the presence of UTP, and the conversion of UDP glucose to UDP galactose (reaction 3) is believed to be the rate-limiting step. They also demonstrated that UDP-galactose could be formed from galactose-1-phosphate in the presence of a uridyl donor (e. g. UTP or UDP-glucose).

Goudsmit and Ashwell (1965) showed that UDP-D-galactose-^{14}C was used in the formation of a radioactive galactose polymer (reaction 4) which was tentatively identified as galactogen. Galactogen prepared from *Helix pomatia* was required as a primer and reaction 4 can probably be summarized:

$$\text{UDP-D-galactose} + (C_6H_{10}O_5)_n \rightarrow (C_6H_{10}O_5)_{n+1} + \text{UDP}$$
$$\text{galactogen} \qquad\qquad \text{galactogen}$$

UDP-D-glucose (reaction 3) and TDP-D-galactose were also incorporated into galactogen and, when UDP-D-glucose was used as the substrate, only galactose was produced on the acid hydrolysis of the galactogen formed. Little or no galactogen synthesis was found, on the other hand, with D-galactose, ADP-D-glucose, or CDP-D-glucose.

The reactions described above relate to the incorporation of the D-galactose molecule into galactogen. Goudsmit and Neufeld (1966) have shown that UDP-L-galactose also can be isolated from the albumen gland of *H. pomatia* ($0 \cdot 16$ μmol from 150 g of glands), so that L-galactose is similarly built into the galactogen molecule (reaction 4). However, UDP-L-galactose synthesis follows a different pathway, being formed from GDP-D-mannose in the presence of cell-free albumen gland extracts. D-mannose and L-galactose differ in their configurations about C3 and C5 (Goudsmit and Neufeld, 1967):

GDP- D - mannose   GDP-L-galactose

Galactogen is apparently confined in the pulmonates to the secretions of the albumen gland and, of course, to the albumen surrounding the fertilized eggs.

Freshly hatched *Lymnaea* also still contained large amounts of galactogen from the albumen which is ingested by the developing embryo (McMahon *et al.*, 1957). Electrophoretic studies reveal differences in detail between the galactogen from stylommatophorans and higher basommatophorans (McMahon *et al.*, 1957) and correspondingly, when galactogen from *Lymnaea stagnalis* replaced that from *Helix pomatia* as a primer, the formation of galactogen from UDP-D-galactose by albumen gland extracts of *Helix* was greatly reduced (Goudsmit and Ashwell, 1965). Geldmacher-Mallinckrodt and his fellow workers (1957–1963) have shown that *Helix* galactogen will separate into three bands in the ultracentrifuge, although the galactogens from *Helix*, *Lymnaea* (Horstmann, 1964) and *Agriolimax* (Bayne, 1966) are each believed to be homogeneous, the appearance of bands in the ultracentrifuge being ascribed by Horstmann (1964) to pretreatment with alkali.

Meenakshi (1954) has identified galactogen in the albumen gland of the prosobranch *Viviparus*, but McMahon *et al.* (1957) found that in *Lanistes boltenianus* (Roth) (Prosobranchia) true galactogen was absent and was replaced by a galactose-fucose polysaccharide which was confined to females, the males containing glycogen only. Bayne (1966) suggests that the galactose reported in prosobranchs and that in the female tract (excluding the albumen gland) of the slug *Agriolimax* is not present in the form of galactogen.

Glycogen in addition to galactogen has been detected in the albumen glands of several pulmonates; in conflict with some of the earlier studies McMahon *et al.* (1957) and Fantin and Vigo (1968) agree that it is absent from the albumen gland of *Lymnaea*. It is present in much smaller concentrations than is galactogen, and is reported in *Helix* (Grainger and Shillitoe, 1952), *Philomycus* (epithelium of the albumen gland canal, Kugler, 1965), *Ariolimax* (Meenakshi and Scheer, 1968) and *Biomphalaria* (although not in the eggs, McMahon *et al.*, 1957). The failure to identify glucose in hydrolysed extracts, as Bayne (1967), working with *Agriolimax* points out, may well be due to the much larger quantities of galactose in the hydrolysate. The possible identification of glycogen and its relation to electron micrographs has been discussed in detail by Nieland and Goudsmit (1969). The galactogen content of the albumen gland shows seasonal variation in *Ariolimax*, being highest in the breeding season, and Meenakshi and Scheer (1969) believe that growth of the gland is under hormonal control from the centre in the optic tentacle, the endocrine effects on the synthesis of galactogen being a consequence of the effect on growth.

It is now evident that the albumen surrounding the fertilized egg is a complex fluid containing much more than galactogen. Proteins, glyco-protein, some free amino acids (probably in low concentration, Bayne, 1966) and calcium have been identified (Morrill, 1963; Morrill *et al.*, 1964; Bayne, 1966). It has not been established with certainty whether all these substances are produced by the albumen gland. Studies on the latter have been conducted with homogenized glands (with the con-sequent possibility of contamination by intracellular substances which are not secreted, Bayne, 1967) or with albumen from the egg capsules. However, all of the 18 amino acids and glucosamine identified in the perivitelline fluid of *Agriolimax* have been found in an albumen gland hydrolysate (Bayne, 1967) and histochemical evidence that protein is associated with the galactogen globules has been obtained by Grainger and Shillitoe (1952, *Helix*), Fantin and Vigo (1968, *Lymnaea*) and Rizzotti (1963, *Helix*). The other nutritive source available to the developing embryo is the yolk, which consists partly of proteins and partly of fatty substances (see Raven, 1958 for details).

The mode of secretion by the albumen gland cells is believed to vary with the degree of distension by the secretory globules. Small membrane-limited vesicles are seen below the apical membrane in cells with a low secretory content and appear to discharge their contents into the lumen. Cells distended by large globules apparently rupture at their luminal surface and discharge the contents of one or more globules (Nieland and Goudsmit, 1969). The alkaline phosphatase activity detected histo-chemically in the cell membranes (Kugler, 1965) and surrounding the secretory globules (Muller, 1965) may be associated with the secretory process.

2. The female tract

After being fertilized and receiving the secretions of the albumen gland, the eggs pass down the female duct (Fig. 2) and are invested with the envelopes that form the egg mass or capsule (Fig 18). The female tract appears to be relatively simple in pulmonates possessing a common spermoviduct, although there is a clear distinction into the anterior and posterior mucous glands in the Ellobiidae and Otinidae. In the fresh-water Basommatophora, an increase in complexity of the female duct, by the formation of additional glandular areas (Duncan, 1960b), is re-flected in the complexity of the egg capsule. The epithelium of the ovi-duct of these families is lined by a basic cell type, a columnar glandular cell, and differences in the parts of the female duct relate to the additions

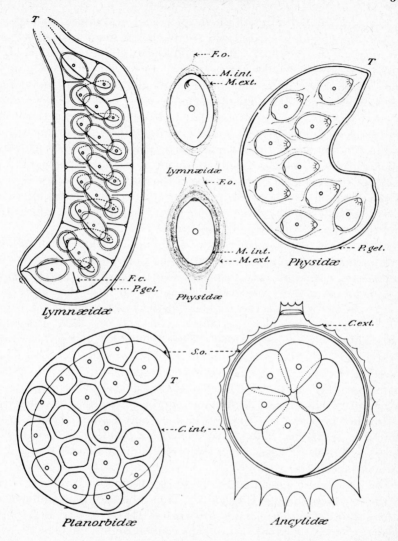

Fig. 18. Diagram showing the four main types of hygrophilan capsules with "torsion". Lymnaeidae: anti-clockwise "torsion". The other three: clockwise "torsion". In *Ancylus fluviatilis* Müll. (bottom, right): + quaternary enveloping layer. In *Acroloxus lacustris* (L.): +quaternary enveloping layer, but no "torsion" at all. Between the two upper figures the two types of eggs with external membranes are shown. C. int, internal capsules (tertiary enveloping layer of the 3rd order); C.ext., external capsules (quaternary enveloping layer of foot secretion); F.c., capsular strings; F.o., egg strings; M.int., internal membrane; M.ext, external membrane; P.g., gelatinous coating; S.o., operculate suture; T, terminal tail. The corkscrew line at top, left, indicates the position of the eggs in the cylindrical capsule. (From Bondesen, 1950.)

of tracts of ciliated cells and the development of complex foldings (see p. 320; Fig. 7). Small, ciliated cells are frequently found wedged between the apices of the glandular cells. The glandular cells of the oviduct of the basommatophorans undergo a clear developmental cycle in early spring and should prove an excellent tissue for an ultrastructural study of the development and physiology of mucus-secreting cells. The Golgi apparatus is probably intimately associated with this secretory activity, and this material might be ideal for an investigation of this organelle.

The functioning of the different parts of the female duct has been elucidated by histological and histochemical studies of the oviduct at different stages during the formation of the egg capsule, and by analysis of the secretions of the female tract and of the components of the completed capsule. Studies of the capsule have yielded detailed information and the results are presented in section IV.

Each zygote is surrounded by perivitelline fluid and then enclosed within the perivitelline (or internal) membrane. Further tertiary and quaternary envelopes may then be added (see p. 348) by the female tract, one or more zygotes being assembled to form the egg capsule characteristic of the species. Studies on the functioning of the female tract of higher pulmonates (Duncan, 1958; Ghose, 1960; Rangarao, 1963; Smith, 1965; Kugler, 1965; Bayne 1966; 1967; Berry *et al.*, 1967; Fantin and Vigo, 1968) agree that the nutritive perivitelline fluid is produced by the albumen gland (section III. D. 1., p. 344). The perivitelline membrane of *Agriolimax* is formed of the same biochemical constituents as the perivitelline fluid. Electron micrographs reveal that it is composed of two layers; there is a thin, outer layer and a much thicker, inner, fibrous layer (Bayne, 1966). Bayne (1967) suggests that the perivitelline layer of *Agriolimax* forms when the perivitelline fluid comes in contact with the surrounding jelly.

The reports also agree that the successive protecting layers are added as the eggs pass through the oviduct. Thus, the eggs pass through the convolutions of oviduct II and into oviduct III in *Physa* and *Lymnaea* (Duncan, 1958, 1960a, b), or from one convolution to the next in *Arion* (Smith, 1965). A similar sequence is described in *Otina* (Morton, 1955). These envelopes and jelly-like substances which constitute the egg capsule serve for the maintenance of water balance, mechanical support and protection from predators and parasites (Bayne, 1966). These components of the capsule fall within the general classes of mucopolysaccharides, polysaccharide–protein complexes and mucoproteins. In some species histological or, more usually, histochemical differences can be detected between the glandular cells of the different sections of the female duct. Thus, in *Lymnaea stagnalis* the large glandular cells of oviduct II secrete acid mucopolysaccharides whilst other smaller cells produce sulphated mucopolysaccharides, and two zones are distinguishable in oviduct III, on the basis of the properties of the mucopolysaccharides they produce (Fantin and Vigo, 1968). Acid mucopolysaccharides are also the characteristic secretion of the female part of the spermoviduct of *Nanina* (Rangarao, 1963), and Kugler (1965) reports both neutral and acid mucopolysaccharides in *Philomycus*. The histochemistry of the spermoviduct of *Arion* has been studied in detail by Smith (1965). The secretions of the female tract are complex, showing a large number of staining reactions. They have the properties of a strongly acidic polysaccharide, and markedly positive histochemical reactions are also given for polyphenols, phenylamines and calcium. These reactions are also shown by the inner shell membranes of the completed egg mass, whilst the outer layer is heavily calcified. The calcium is associated with a lipid coating, forming an unusual coat for a pulmonate egg mass (Bayne 1968a). The Ca^{2+} is believed to be derived from stores in the digestive gland, and the active alkaline phosphatase in the female tract might be associated with its secretion, but Meenakshi (1955) has shown that carbonic anhydrase may well fulfil this role in the prosobranch *Pila*. The genital atrium of *Arion* is surrounded by the spongy gland, the secretion from which is almost pure lipid (Smith, 1965). Its role could well be to contribute to the outermost layer of the specialized egg mass of this species.

These histochemical observations have been supplemented by analyses of the amino acids and sugars of extracts of the female tract of *Agriolimax*; those from the upper portion of the duct correspond with analyses of the jelly of the egg mass, confirming that it is secreted there. The presence of additional macromolecular material in the extracts indicated contamination during extraction. Fucose, absent from the egg jelly, was scarce in extracts from the upper part of the female tract, whereas it was common in analyses of both the lower section and the egg shell. Thus, the results confirm that the shell of the capsule is secreted in this lower portion of the duct (Bayne, 1967; 1968a), and that the acidic mucopolysaccharides of the jelly and shell layers are the products of the essentially acidic mucopolysaccharide secretions of the female tract. Hydrolysis of this polysaccharide from *Ariolimax columbianus* (Gould) yielded galactose, fucose and glucosamine (Meenakshi and Scheer, 1968).

The specializations in the oviduct found in the different families of the freshwater Basommatophora (Fig. 7) can be correlated with the more complex egg capsules that they produce (section IV). Only in the Physidae and Lymnaeidae are the zygotes enclosed within a second individual envelope (the external membrane). In both *Physa* and *Lymnaea* oviduct II is greatly developed and it is probably in this region that the external membrane is produced (Bretschneider, 1948b; Duncan, 1958). This region of the oviduct is absent from *Ancylus fluviatilis* Müll. and exists merely as a straight tube in *Planorbarius*, and in these genera only the internal (= perivitelline) membrane is found. The viscous substance in which the "eggs" are embedded, together with the capsular membrane, are secreted by oviduct III in *Physa* and *Lymnaea* (Duncan, 1960b). The completed egg capsule is finally extruded through the muscular vagina.

IV. Egg capsules

The pulmonate egg capsule is a necessary development in a group of molluscs adapted for terrestrial and freshwater life. Throughout the order, the zygote is surrounded by perivitelline fluid (the albumen) which is enclosed in the perivitelline (or internal) membrane. This "egg" is then either surrounded by further protective layers and deposited singly, or a number are packaged together, embedded in jelly and enclosed in a common egg capsule (as in the freshwater Basommatophora). The eggs of many ellobiids (Morton, 1954, 1955a, b; Berry *et al.*, 1967) and of *Otina* (Morton, 1955c) are also deposited in batches, the less formally-organized egg masses. The eggs are laid singly in *Carychium*, in batches of 8–30 in other species, and hundreds are contained in a long string in the egg mass of *Ellobium*.

The morphology of the more complex egg capsules of the freshwater Basommatophora (Fig. 18). have been described in detail by Bondesen (1950) and reference should be made to this work for a very extensive bibliography A primary (vitelline) membrane formed from the ovum itself has been demonstrated in *Lymnaea*, but a secondary envelope (chorion) has not been described in the freshwater pulmonates (Raven, 1958). There are a number of tertiary envelopes, however. In all families each zygote and the albumen in which it is embedded are enclosed in an internal (perivitelline) membrane. In the Lymnaeidae and Physidae there is a second individual membrane, the external membrane, which has a fine lamellar structure and is continued in the egg strings which are found only in the egg capsules of these two families. The capsules of the Physidae and Lymnaeidae are also distinguished by having convex egg capsules whereas those of the other freshwater families are flat. In all families of the Hygrophila each of these "eggs" is then embedded in a common gelatinous substance, enclosed by the capsule wall. Quaternary envelopes, formed by secretions from the foot, have been demonstrated in the Ancylidae and Acroloxidae (Bondesen, 1950). A terminal tail is found at the end of the capsule, which is prolonged into a spout or tube, and during oviposition the snail tends to turn away from the capsule, which may become curved or even spiral. Dextral and sinistral families therefore have their capsules curved in opposite directions (Bondesen, 1950).

The egg masses of the Stylommatophora have been studied in particular by Jura and George (1958), George and Jura (1958) and Bayne (1966; 1968a, b) The eggs are laid one at a time in a selected (usually damp) site, forming a heap. The number in each egg mass varies both with the species and the individual; for example 40–65 in *Helix pomatia*, 12–80 in *Arion hortensis* Fér., 100–368 in *Limax maximus* L. (see Hyman, 1967). The zygote is, as in all pulmonates, enclosed in albumen and the perivitelline membrane, and then each such packet is individually surrounded by jelly and shell layers. Details for several species are given by Bayne (1968a) and the biochemistry of the different components agree with the studies on the female tract. The albumen is composed of neutral polysaccharide (galactogen) and proteins, and a more detailed study of *Agriolimax reticulatus* (Müll.) also revealed glycoprotein, water, free amino acids, calcium and probably other inorganic ions (Bayne, 1966). Of the outer coverings the shell is probably of greatest importance in providing mechanical support, while the jelly provides a short term buffer against changes in the water relations of the egg, although measurements of desiccation rates show that there is little resistance to water loss (Bayne, 1966; 1968b; Chernin and Adler, 1967); nevertheless, the embryos are able to survive considerable dehydration (Bayne, 1969). Protein is sparse in these outer coverings, which are primarily formed of acid mucopolysaccharides. These layers frequently contain calcium, which is deposited as crystals of calcium carbonate in *Helix aspersa*, *Cepaea nemoralis* and *Arion ater* (Bayne, 1968a).

V. Reproductive cycles

There is a wealth of information on the reproductive cycles of pulmonates. In particular, details are known for many species of the Basommatophora. The sequence found in the primitive *Leucophytia*, *Carychium* and *Otina* (Morton, 1954, 1955a, b, c) is clearly protandric (Fig. 19), with a shorter phase of sperm production preceding the longer phase of egg maturation. The development of the female duct, however, long precedes the ripening of the eggs. In *Carychium tridentatum* (Risso) the penis and vas deferens are present only in specimens collected in July and August during the male phase. In *Leucophytia* and *Otina* the cessation of sperm production in the ovotestis is accompanied by a great reduction of the glandular section of the male duct. The ovotestis of *Cassidula*

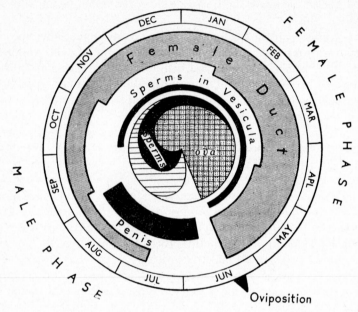

Fig. 19. *Carychium tridentatum*. Diagram illustrating the course of the reproductive cycle of specimens at Box Hill, England, over a single season. The area at the centre represents the changes taking place in the gonad, and the outer part of the diagram shows the state of development of the accessory genital glands and ducts. (From Morton, 1954.)

aurisfelis (Brug.) from Malaya shows almost continuous spermatogenesis, superimposed on which are intermittent accumulations of oocytes, this abundance perhaps being related to tidal wetting (Berry, 1968); *Melampus bidentatus* Say is also a simultaneous hermaphrodite (Apley, 1970). Copulation and spawning in *Siphonaria* occur during May–July at Asamushi, Japan, and Abe (1940) reports that these processes have a lunar periodicity. Egg-laying, hatching and larval settlement of populations of *Melampus* living in the higher levels of salt marshes are each confined to cycles of about 4 days in phase with spring tides. These snails produce free-swimming veligers, so that synchrony of hatching with spring high tides is of particular importance (Russell-Hunter *et al.*, 1970,

1972), and a semi lunar periodicity is found in *Melampus bidentatus*; this species has a 3–4 year life cycle (Apley, 1970).

The condition in the freshwater Basommatophora probably represents a condensation of the protandric sequence of many members of the Ellobiidae. In *Physa fontinalis* (L.) (Fig. 20) the spermatogonia appear before the oogonia, but at maturity both types of gamete are pre sent together. Snails hatching in the spring in England have mature

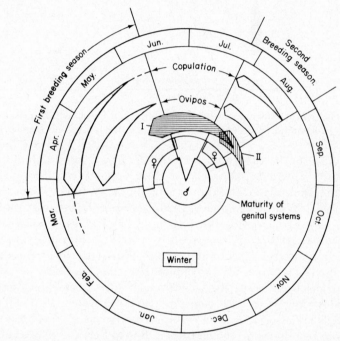

Fig. 20. *Physa fontinalis.* Diagram to illustrate the life cycle. The shaded areas indicate the occurrence of young snails (generations I and II). The areas in the centre show the maturity of the male and female systems, whilst the periphery indicates the times of copulation and oviposition. This is an example of a year where a complete second breeding season is found; for further details see text.

systems by September; the hermaphrodite duct is filled with sperm and the whole of the male system is fully developed, and histological studies suggest that it is capable of active secretion. The female system, including the albumen gland and bursa, on the other hand is rudimentary at this time of year, and does not begin a rapid maturation until the following March (Duncan, 1959). The animals die after copulation and oviposition. This simple annual cycle is typical of other freshwater Basommatophora, e.g. *Physa acuta* Drap. (Duncan, 1959), *Physa gyrina* Say

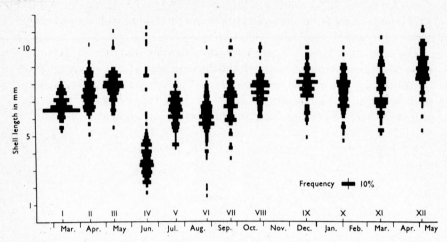

Fig. 21. Histogram showing the percentage distribution of shell length of *Physa fontinalis* in a pond in Middlesex, England during 1954–5. (From Duncan, 1959.)

(DeWitt, 1954a, b, 1955; Clampitt, 1970), *Physa integra* Hald. (Clampitt, 1970), *Aplexa hypnorum* (L.) (Hartog and Wolf, 1962), *Ancylus fluviatilis* (Hunter, 1953; Geldiay, 1956; Maitland, 1965), *Lymnaea truncatula* (Müll.) (Walton and Jones, 1926), *Lymnaea humilis* Say (McCraw, 1961), *Lymnaea peregra* (Müll.) (Hunter, 1961a, b), *Planorbis albus* Müll. (Hunter, 1961b), *Planorbarius corneus* (L.) (Berrie, 1963).

There is a number of variants on this basic pattern of reproduction (in which the snail lives for 12–16 months); thus, *Lymnaea stagnalis* in Scotland has been shown to be biennial, with most snails breeding during May–August in both years (Berrie, 1965, 1966). Probably other large snails are also biennial in certain situations (e.g. *Planorbarius*, see discussions in Bondesen, 1950; Hunter, 1961b, Berrie, 1963) and the reproductive cycle is greatly distorted when the animals are kept in captivity. Longevity may be increased (see review by Comfort, 1957) and the animals may breed at a much earlier age. Such actuarial studies of laboratory populations have been given by Crabb (1929), Baily (1931), Winsor and Winsor (1935), Noland and Carriker (1946), Barlow and Muench (1951), Kendall (1953), DeWitt (1954b), DeWitt and Sloan (1958) and Ritchie *et al.* (1966).

In some smaller species (*Physa fontinalis*, DeWit, 1955; Duncan, 1959; *F. gyrina, P. integra*, Clampitt, 1970; *Lymnaea truncatula*, Walton and Jones, 1926; *L. humilis*, Van Cleave, 1935; McCraw, 1961) a second generation may be produced in the late summer under favourable circumstances (Figs 20 and 21). The young snails mature rapidly and oviposition begins in early August, the spring-born generation being completely replaced by an overwintering generation. This second generation is not found in populations living in the colder parts of the animal's range and water temperature must be an important, although not the only, factor determining maturation rate (Vaughn, 1953; Duncan 1959; Berrie, 1966). Growth of *Biomphalaria* – a tropical snail – is most rapid at 30°C (following acclimation at 25°C), but reproduction is suppressed and pathological changes are found in the genital organs; the relatively low temperature of 15°C inhibits both growth and reproduction, although the effect is readily reversible (Michelson, 1961). The severity of the winter, amount of spring sunshine and low rainfall in late summer may all affect population densities of freshwater snails (Hunter, 1961a). Hunter (1961b) also describes variations in the reproductive cycle: 1. a delay in the time of breeding, with further post-winter growth (e.g. *Lymnaea palustris* (Müll.) in Ontario, McCraw, 1970) and 2. a second generation produced in the late summer which does not replace the animals hatching in the spring, the autumn breeding season providing only a small fraction of the overwintering stock. Where the temperature of the environment is artificially elevated, the snails develop rapidly and breed continuously (Agersborg, 1932; Kevan, 1943).

Stylommatophorans often live longer, longevities of 3 years being recorded in the wild and up to 15 years in captivity (see survey by Comfort, 1957), although only 15–18 months is recorded in *Succinea* (Rigby, 1965) and some slugs (Comfort, 1957). The smaller helicids mature more quickly and the total duration of life (12–18 months) is less than in larger species (Chatfield, 1968). The most detailed information on reproductive cycles is available for slugs (Carrick, 1938; Barnes and Weil, 1944, 1945; Quick, 1960; Bett, 1960; Galangau, 1964; Smith, 1966; Hunter, 1968; Runham and Laryea, 1968) and differences in detail are reported between populations in southern (Bett, 1960) and northern England (Hunter, 1968). *Agriolimax reticulatus* has two main breeding seasons each year: the spring generation hatches about May and the slugs complete their life-cycle in 5 months whilst the autumn generation hatches in late September and takes 7 months to mature. However, breeding in Britain probably continues throughout the year since eggs may be found in all monthly samples (Bett, 1960; Hunter 1968; Runham and Laryea, 1968). Most *Arion hortensis* in Northern England have a simple annual cycle, with hatching in July, but a few animals take nearly 2 years to complete the life cycle (Hunter, 1968). Few *Arion ater* succeed in overwintering in Wales after breeding, the majority survive as eggs or very young animals. Copulation occurs from July–September and is followed by oviposition, after which most of the animals die (Smith, 1966). *Milax budapestensis* (Hazay) hatches during May–August in northern Britain

and matures during its second autumn and winter, a biennial life-cycle.
A few hatch in April and lay eggs during their first winter (Hunter,
1968). *M. budapestensis* hatches during the autumn and winter in
southern England and is sexually mature by the following October
(Bett, 1960), whilst *M. gagates* (Drap.) has two main hatching periods in
Mediterranean areas, spring and autumn, and sexual activity decreases
during the intervening dry season (Galangau, 1964). The zonitid *Oxychilus cellarius* (Müll.) lays eggs from January–August, with a peak of
activity in April–May; growth continues throughout the year and duration of life is about 12–18 months in England and, whilst the breeding
season is long and flexible, it is at a peak in the autumn when the ground
is moist (Chatfield, 1968). Maturation and growth of stylommatophorans
are again very dependent on temperature (Bouillon, 1956; Herzberg
and Herzberg, 1960; Arias and Crowell, 1963; Wolda, 1967) and in tropical regions growth, maturation and spermatogenesis are largely continuous, although oviposition may still be seasonal, probably related to
rainfall (Ghose, 1959, 1960; Berry, 1963a, b). Even under these conditions, spermatogenesis precedes oogenesis, and the fluctuations in the
activity of the female system are also indicative of a condensed protandric sequence (see also Mead, 1950, and Galangau, 1964 for further
details of protandric development in stylommatophorans). Oviposition
frequency in *Cepaea* also is affected by drought (Wolda, 1965) and maturation of *Arion ater* can be retarded by a dry summer, delaying oviposition in many individuals until the following spring (Laviolette,
1950). This sensitivity to drought may well be true for many stylommatophorans.

VI. Copulation

Copulation can be readily induced in freshwater snails during the
breeding season by bringing a number into the laboratory, when they
apparently copulate with the first individual encountered (Duncan,
1959). Isolated individuals also copulate readily when brought together;
Noland and Carriker (1946) describe an active search, but the encounter
is probably haphazard (Barraud, 1957). Behavioural details have been
given for *Lymnaea stagnalis* (Noland and Carriker, 1946; Barraud, 1957)
and *Physa fontinalis* (Duncan, 1959). The "male" usually mounts the
shell of the "female" with both snails facing the same way and the everted
penis is then inserted into the female aperture. At the moment of penetration the *Lymnaea* acting as a female retracts the whole forepart of
the body and may lose its foothold, so that both animals float to the

surface (Barraud, 1957). Barraud believes that penetration in this species is seldom achieved, the penis merely being inserted beneath the shell.
Coupling may last from a few minutes to 12 h but is usually from 30–90
min. Reciprocal copulation, in which the animals remount with the
roles reversed, has been described in a number of planorbids (Precht,
1936) and in *Lymnaea stagnalis* (Barraud, 1957), but simultaneous reciprocal copulation in *Helisoma trivolvis* (Say), lasting 4·5 h, has also
been described by Abdel-Malek(1952). Copulatory chains have been described in *Physa fontinalis* (Duncan, 1959) and *P. heterostropha* Say (Bondesen, 1950), several lymnaeids (Bondesen, 1950) and *Ancylus fluviatilis*
(Geldiay, 1956); each individual in a chain functions as a female to the
snail above and as a male to the one below. The maximum number of
Physa fontinalis observed participating in a chain was 5, but 3 or 4 was
more usual. Formation of copulatory chains is probably largely a laboratory phenomenon, favoured by the sudden concentration of snails in
breeding condition, but Geldiay (1956) concludes that chain copulation
is probably the rule in *Ancylus* in Lake Windermere. Whilst temperature
and maturity of the genital system are fundamental factors, copulation
of the freshwater snails in the laboratory can be produced most readily
by sudden environmental changes and in the physids congregation of
snails previously kept in isolation is usually the most effective stimulus.

Copulation in the Stylommatophora, on the other hand, is frequently
preceded by an elaborate courtship behaviour and often involves reciprocal exchange of sperm. Copulation is usually (but not invariably,
Ingram, 1946) nocturnal and takes place less readily in some slugs in
captivity than it does in freshwater snails, although copulation apparently occurs readily in captive *Helix aspersa* (Herzberg and Herzberg,
1962) and *Cepaea*. Detailed descriptions of copulatory behaviour in individual species of slugs have been given by Gerhardt (1933–1937),
Quick (1960) and Karlin and Bacon (1961) and have been summarized by
Runham and Hunter (1970). In *Agriolimax reticulatus*, for example,
the initial positional movements cover a circular area 5–8 cm in diameter, the slugs crawling around the periphery of the circle and facing
opposite directions. This circular crawling pattern is common in slug
courtship and is often accompanied by oral stimulation and feeding on
the partner's slime. In *Agriolimax* each individual periodically thrusts
forward its head and strikes the other slug and, as the diameter of the
circle is decreased, and their right sides come into contact, each animal
everts its sarcobelum and strokes its partner. Courtship lasts 30–90 min,
but sperm transfer is effected in less than one minute. The sarcobela
have previously become swollen and interlaced and are violently rotated
back and forth. Large quantities of mucus are produced and sperm

N

transfer is accomplished in small slime balls, since true spermatophores are not produced in this species (Karlin and Bacon, 1961). Courtship is aerial in *Limax maximus*; initial excitatory movements are on a vertical surface, where the slugs intertwine their bodies. They can then fall off, suspended by a thick slime thread, with penes extruded and intertwined. The animals remain suspended for 10–15 min and copious mucus production accompanies the rotation of the penes. The animals climb up the mucous thread or fall to the ground after copulation is finished. Most slugs lack a true penis and the genital apertures are simply apposed at copulation. Transfer of sperm is not invariably reciprocal and one individual may act as the male and the other the female, although the roles may be reversed in subsequent copulations.

The mating behaviour of helicids is also complex and has been summarized by Hyman (1967); when *Helix pomatia* meet they erect the anterior part of the body, pressing the foot soles together and reciprocally stimulating with the tentacles and oral lappets. Preliminary manoeuvring may take 15–30 min before ejection of the darts. Eventually the penis is everted and cross-copulation follows when the snails eventually negotiate their female openings into the correct position. The interval between the start of copulation and spermatophore transfer is 4·5 min and between the start of copulation and penis withdrawal is 5·6 min (mean of 17 observations). For a long time the tail of the spermatophore lies freely exposed on the ventral surface of the foot. The behaviour of the copulants is identical and nearly synchronous (Lind, 1973). In *H. aspersa*, precopulatory manoeuvring takes about 30 min and when the copulatory organs finally reach a position where copulation becomes possible, the union of the two animals occurs very rapidly and from that time the animals cannot be separated without injury. Copulation takes 4–10 h and many *H. aspersa* copulate at frequent intervals. Herzberg and Herzberg (1962) also observed the darts during and after copulation; they were generally near the snails and occasionally pierced the foot of one of them. They are apparently not essential for copulation and their role has been briefly reviewed by Goddard (1962).

VII. Oviposition

Oviposition in freshwater snails is primarily dependent on sexual maturity and water temperature, and since growth and maturation are in turn dependent on temperature this factor has a major role in controlling reproductive cycles. Probably 7–11°C is the critical range for the initiation of oviposition in *Physa fontinalis* in Britain (Duncan, 1959) and similar temperatures probably hold for other freshwater snails in temperate regions (see Bondesen, 1950; R. M. DeWitt, 1955; W. F. DeWit, 1955), whereas 15°C is the critical temperature for *Biomphalaria* in Brazil (Paulini and Camey, 1964). Light and day-length are probably unimportant in the field (DeWitt, 1967; Van der Steen, 1967); copulation may serve as an important stimulus, but is not essential in self-fertilizing species. Most freshwater pulmonates oviposit readily in the laboratory and many species lose all trace of the normal annual cycle (see discussion, Duncan, 1959). Most animals oviposit regularly every few days, but this rhythmicity can be disturbed and a particularly detailed study of the factors affecting the oviposition of *Lymnaea stagnalis* in laboratory culture has been given by Van der Steen (1967). A change of water in the aquarium causes an increased oviposition either within 3 hours or during the following night (*Lymnaea*, Linville, 1900; Van der Steen, 1967; *Physa*, Duncan, 1959), and the introduction of the plant *Hydrocharis morsus-ranae* when illuminated (perhaps due to the increased oxygen supply), or a gradual rise of water temperature (18 to 28°C) caused oviposition in *Lymnaea stagnalis* (see Raven, 1948) within a few hours. Van der Steen (1967) reports that a sudden rise in temperature also stimulates oviposition, although not immediately, whereas a sudden fall has an inhibitory effect. Oviposition is markedly lower at 15°C than at 20°C. Experiments by Timmermans (1959) also suggest that changes in water temperature can stimulate oviposition in *Planorbarius corneus*, *Planorbis planorbis* (L.) and *Physa acuta*. Paired *Lymnaea stagnalis* in captivity begin oviposition at a much earlier age than isolated snails (McDonald, 1969). The nature of the diet may be influential (Noland and Carriker, 1946; Lagrange, 1957; Frank, 1963; Van der Steen, 1967), and it is probable that a wide variety of environmental changes can serve as a stimulus to oviposition. Thus, oviposition activity is related to a change in atmospheric pressure (Van der Steen, 1967). Capsules are generally laid at night, a rhythmicity which persists even when the illumination is altered (Cole, 1925; Duncan, 1959; Paulini and Camey, 1964). For a full study of the rhythmicity of oviposition in *Lymnaea*, see Van der Steen (1967). Egg production increases when *Physa pomilia* Conrad is reared in total darkness and DeWitt (1967) suggests that light may inhibit egg production although the effect is not very marked. On the other hand, complete darkness causes gamete resorption in the slug *Arion* (Smith, 1966), a reduction in fecundity of *Bulinus contortus* (Deschiens and Bijan, 1956) and a decrease in oviposition of *Lymnaea stagnalis* (Van der Steen, 1967).

Stylommatophorans lay their eggs in humid situations (Carrick, 1938) and *Helix aspersa, H. pomatia* and *Nanina ligulata* (Fér.) dig shallow holes. Low temperature and low humidity inhibit activity in *H. aspersa* (Basinger, 1931); very wet or dry soil is unsuitable for nest building and oviposition follows copulation after 9–13 days (Herzberg and Herzberg, 1962) and takes an average of 9 h. Details of oviposition in *Anguispira alternata* (Say) in the laboratory are given by Elwell and Ulmer (1971). A number of stylommatophorans are ovoviviparous; the eggs may be held in the female tract until they have developed into young snails (Pelseneer, 1901; Berry, 1963b, 1965).

Pulmonate behaviour is both limited and slow. However, these two features do facilitate analysis of activity and it should be possible to produce a very complete mathematical description of the interacting effects of the various factors that influence a relatively simple piece of behaviour. Thus, McFarland (1965c) has provided an analysis, based on control theory and the use of flow graphs, for hunger and thirst in the Barbary dove (McFarland, 1964, 1965a, b) and the studies of Van der Steen (1967) on the oviposition of *Lymnaea* could well prove a good starting point for a comparable study.

VIII. References

Abdel-Malek, E. T. (1952). *Am. Midl. Nat.* **48**, 94–102.
Abdel-Malek, E. T. (1954a). *Trans. Am. microsc. Soc.* **73**, 103–124.
Abdel-Malek, E. T. (1954b). *Trans. Am. microsc. Soc.* **73**, 285–296.
Abe, N. (1940). *Sci. Rep. Tohoku Univ.* Ser. 4 **15**, 59–95.
Agersborg, H. P. K. (1932). *Nautilus* **45**, 121–123.
Alaphilippe, F. (1959). *Bull. biol. Fr. Belg.* **93**, 260–287.
Alaphilippe, F. and Régondaud, J. (1959). *Bull. Soc. zool. Fr.* **84**, 485–493.
Albanese, M. P. and Bolognari, A. (1961). *Caryologia* **14**, 329–347.
Allen, C. (1935). *J. Conch. Lond.* **20**, 126.
Ancel, P. (1902). *Archs Biol. Paris* **19**, 389–652.
Anderson, W. A. and Personne, P. (1967). *J. Microscopie* **6**, 1033–1042.
Anderson, W. A. and Personne, P. (1969a). *J. Microscopie* **8**, 87–96.
Anderson, W. A. and Personne, P. (1969b). *J. Microscopie* **8**, 97–102.
Anderson, W. A. and Personne, P. (1970). *J. Histochem. Cytochem.* **18**, 783–793.
Anderson, W. A., Personne, P. and André, J. (1968). *J. Microscopie* **7**, 367–390.
Apley, M. L. (1970). *Malacologia* **10**, 381–397.
Archie, V. E. (1942). Unpublished thesis, University of Wisconsin.
Arias, R. O. and Crowell, H. H. (1963). *Bull. Sth. Calif. Acad. Sci.* **62**, 83–97.
Aubry, R. (1954a). *C.r. Séanc. Soc. Biol.* **148**, 1498–1500 (b) 1626–1629 (c) 1856–1858 (d) 2075–2077.

Aubry, R. (1955). *C.r. Séanc. Soc. Biol.* **149,** 390–392.

Aubry, R. (1956). *C.r. Séanc. Soc. Biol.* **150,** 1786–1789.

Baecker, R. (1932). *Z. ges. Anat. (Ergebn. Anat. EntwGesch.)* **29,** 449–585.

Baily, J. L. Jr. (1931). *Biologia gen.* **7,** 407–428.

Baker, F. C. (1933). *Nautilus* **47,** 35.

Baker, F. C. (1945). "The Molluscan Family Planorbidae". University of Illinois Press, Urbana.

Baldwin, E. and Bell, D. J. (1938). *J. chem. Soc.* 1938, 1461–1465.

Barbosa, F. S., Barbosa, I. and Carneiro, E. (1958). *Anais Inst. Med. trop. Lisb.* **15,** 397–400.

Barlow, C. H. and Muench, H. (1951). *J. Parasit.* **37,** 165–173.

Barnes, H. F. and Weil, J. W. (1944). *J. Anim. Ecol.* **13,** 140–175.

Barnes, H. F. and Weil, J. W. (1945). *J. Anim. Ecol.* **14,** 71–105.

Barraud, E. M. (1957). *Brit. J. Anim. Behav.* **5,** 55–59.

Barth, R. and Oliveira, F. B. (1964). *Anais Acad. bras. Cienc.* **36,** 559–564.

Basch, P. F. (1959a). *Misc. Publs Mus. Zool. Univ. Mich.* No. **108,** 1–56.

Basch, P. F. (1959b). *Trans. Am. microsc. Soc.* **78,** 269–276.

Basinger, A. J. (1931). *Rep. Calif. Coll. Agric.* Bull. No. **515.**

Bauer, H. (1933). *Z. mikrosk.-anat. Forsch.* **33,** 143.

Bayne, C. J. (1966). *Comp. Biochem. Physiol.* **19,** 317–338.

Bayne, C. J. (1967). *Comp. Biochem. Physiol.* **23,** 761–773.

Bayne, C. J. (1968a). *Proc. malac. Soc. Lond.* **38,** 199–212.

Bayne, C. J. (1968b). *J. Zool., Lond.* **155,** 401–411.

Bayne, C. J. (1969). *Malacologia* **9,** 391–401.

Bayne, C. J. (1970). *Z. Zellforsch. mikrosk. Anat.* **103,** 75–89.

Bell, D. J. and Baldwin, E. (1940). *Nature, Lond.* **146,** 559–560.

Bell, D. J. and Baldwin, E. (1941). *J. chem. Soc.* 1941, 125–132·

Berrie, A. D. (1963). *Nature, Lond.* **198,** 805–806.

Berrie, A. D. (1965). *Proc. malac. Soc. Lond.* **36,** 283–295.

Berrie, A. D. (1966). *Proc. malac. Soc. Lond.* **37,** 83–92.

Berry, A. J. (1963a). *Proc. zool. Soc. Lond.* **141,** 361–369.

Berry, A. J. (1963b). *Proc. malac. Soc. Lond.* **35,** 139–150.

Berry, A. J. (1965). *Proc. malac. Soc. Lond.* **36,** 221–228.

Berry, A. J. (1968). *J. Zool., Lond.* **154,** 377–390.

Berry, A. J., Loong, S. C. and Thum, H. H. (1967). *Proc. malac. Soc. Lond.* **37,** 325–337.

Bett, J. A. (1960). *Proc. zool. Soc. Lond.* **135,** 559–568.

Blomfield, S. E. (1881). *Q. Jl microsc. Sci.* **21,** 415–431.

Bondesen, P. (1950). *Natura jutl.* **3,** 1–208.

Boray, J. C. (1964). *Aust. J. Zool.* **12,** 231–237.

Bouillon, J. (1956). *Nature, Lond.* **177,** 142–143.

Boycott, A. E., Diver, C., Garstang, S. and Turner, F. M. (1931). *Phil. Trans. R. Soc.* B **219,** 51–131.

Bretschneider, L. H. (1948a). *Proc. K. ned. Akad. Wet.* C **51,** 358–363.

Bretschneider, L. H. (1948b). *Proc. K. ned. Akad. Wet.* C **51,** 616–626.

Bretschneider, L. H. (1961). *Verh. K. ned. Akad. Wet. Afg. natuurk.* **54,** 1–39.

Bridgeford, H. B. and Pelluet, D. (1952). *Can. J. Zool.* **30**, 323–337.

Brown, T. F. (1937). *Am. Midl. Nat.* **18**, 251–259.

Brumpt, E. (1928). *C.r. hebd. Séanc. Acad. Sci. Paris* **186**, 1012–1015.

Cain, G. L. (1956). *Biol. Bull. mar. biol. Lab., Woods Hole* **111**, 45–52.

Carrick, R. (1938). *Trans. R. Soc. Edinb.* **59**, 563–597.

Carriker, M. R. (1946). *Trans. Wis. Acad. Sci. Arts Lett.* **38**, 1–88.

Chadwick, W. H. (1903). *J. Conch., Lond.* **10**, 265.

Chatfield, J. E. (1968). *Proc. malac. Soc. Lond.* **38**, 233–245.

Chernin, E. and Adler, V. L. (1967). *Ann. trop. Med. Parasit.* **61**, 11–14.

Clampitt, P. T. (1970). *Malacologia* **10**, 113–151.

Cole, W., H. (1925). *Am. Nat.* **59**, 284–286.

Colton, H. S. (1918). *Biol. Bull. mar. biol. Lab., Woods Hole* **35**, 48–49.

Colton, H. S. and Pennypacker, M. (1934). *Am. Nat.* **68**, 129–136.

Comfort, A. (1957). *Proc. malac. Soc. Lond.* **32**, 219–241.

Corrêa, J. B. C., Dmytraczenko, A. and Duarte, J. H. (1967). *Carbohydrate Res.* **3**, 445–452.

Crabb, E. D. (1927a). *Biol. Bull. mar. biol. Lab., Woods Hole,* **53**, 55–66.

Crabb, E. D. (1927b). *Biol. Bull. mar. biol. Lab., Woods Hole* **53**, 67–98.

Crabb, E. D. (1929). *Biol. Bull. mar. biol. Lab., Woods Hole* **56**, 41–63.

Deschiens, R. and Bijan, H. (1956). *Bull. Soc. Path. exot.* **49**, 658–661.

DeWit, W. F. (1955). *Basteria* **19**, 35–73.

DeWitt, R. M. (1954a). *Trans. Am. microsc. Soc.* **73**, 124–137.

DeWitt, R. M. (1954b). *Am. Nat.* **88**, 159–164.

DeWitt, R. M. (1955). *Ecology* **36**, 40–44.

DeWitt, R. M. (1967). *Malacologia* **5**, 445–453.

DeWitt, R. M. and Sloan, W. C. (1958). *Trans. Am. microsc. Soc.* **77**, 290–294.

DeWitt, R. M. and Sloan, W. C. (1959). *Anim. Behav.* **7**, 81–84.

Duncan, C. J. (1958). *Proc. zool. Soc. Lond.* **131**, 55–84.

Duncan, C. J. (1959). *J. Anim. Ecol.* **28**, 97–117.

Duncan, C. J. (1960a). *Proc. zool. Soc. Lond.* **134**, 601–609.

Duncan, C. J. (1960b). *Proc. zool. Soc. Lond.* **135**, 339–356.

Duncan, C. J. (1961). *Proc. zool. Soc. Lond.* **136**, 575–580.

Duncan, C. J. (1964). *Z. vergl. Physiol.* **48**, 295–301.

Elwell, A. S. and Ulmer, M. J. (1971). *Malacologia* **11**, 199–215.

Fantin, A. M. B. and Vigo, E. (1968). *Histochemie* **15**, 300–311.

Fleitz, H. and Horstmann, H. J. (1967). *Hoppe-Seyler's Z. physiol. Chem.* **348**, 1301–1306.

Fraga, de Azevedo, J., Costa Faro, M. M. Da and Pequito, M. M. G. (1959). *Archs. port. Sci. biol.* **12**, 35–44.

Frank, G. H. (1963). *Bull. Wld Hlth Org.* **29**, 531–537.

Franzen, D. S. (1963). *Nautilus* **76**, 82–95.

Fretter, V. (1946). *J. mar. biol. Ass. U.K.* **26**, 312–351.

Fretter, V. and Graham, A. (1954). *J. mar. biol. Ass. U.K.* **33**, 565–585.

Fretter, V. and Graham, A. (1962). British Prosobranch Molluscs. Ray Society, London.

Frömming, E. (1954). Biologie der Mitteleuropaischen Landgastropoden, Duncker und Humblot, Berlin.

Galangau, V. (1964). *Bull. Soc. zool. Fr.* **89**, 510–513.

Galangau, V. and Tuzet, O. (1966). *C.r. hebd. Séanc. Acad. Sci. Paris*, Ser. D, **262**, 2364–2366.

Gatenby, J. B. (1917). *Q. Jl microsc. Sci.* **62**, 555–611.

Gatenby, J. B. (1918). *Q. Jl microsc. Sci.* **63**, 197–258.

Gatenby, J. B. (1919a). *Q Jl microsc. Sci.* **63** 401–444.

Gatenby, J. B. (1919b). *Q. Jl microsc. Sci.* **63**, 445–492.

Gatenby, J. B. (1920). *Q. Jl microsc. Sci.* **64**, 267–303.

Gatenby, J. B. (1960). *Cellule* **60**, 287–300.

Geldiay, R. (1956). *J. Anim. Ecol.* **25**, 389–402.

Geldmacher-Mallinckrodt, M. (1957). *Hoppe-Seyler's Z. physiol. Chem.* **308**, 220–224. **309**, 190–195.

Geldmacher-Mallinckrodt, M. and May, F. (1957a). *Hoppe-Seyler's Z. physiol. Chem.* **307**, 179–190.

Geldmacher-Mallinckrodt, M. and May, F. (1957b). *Hoppe-Seyler's Z. physiol. Chem.* **307**, 191–201.

Geldmacher-Mallinckrodt, M. and Horstmann, H. J. (1963). *Hoppe-Seyler's Z. physiol. Chem.* **333**, 226–231.

Geldmacher-Mallinckrodt, M. and Träxler, G. (1960). *Hoppe-Seyler's Z. physiol. Chem.* **322**, 112–121; (1961) **325**, 116–121.

George, J. C. and Jura, C. (1958). *Proc. K. ned. Akad. Wet. C* **61**, 598–603.

Gerhardt, U. (1933). *Z. Morph. Ökol. Tiere* **27**, 401–450; (1934) **28**, 229–258; (1935), **30**, 297–332; (1936) **31**, 433–442; (1937) **32**, 518–541.

Ghose, K. C. (1959). *J. Bombay nat. Hist. Soc.* **56**, 183–187.

Ghose, K. C. (1960). *Proc. zool. Soc., Calcutta* **13**, 91–96.

Ghose, K. C. (1963). *Proc. zool. Soc. Lond.* **140**, 681–695.

Goddard, C. K. (1962). *Aust. J. biol. Sci.* **15**, 218–232.

Goudsmit, E. M. and Ashwell, G. (1965). *Biochem. biophys. Res. Commun.* **19**, 417–422.

Goudsmit, E. M. and Neufeld, E. F. (1966). *Biochim. biophys. Acta* **121**, 192–195.

Goudsmit, E. M. and Neufeld, E. F. (1967). *Biochem. biophys. Res. Commun.* **26**, 730–735.

Grainger, J. N. R. and Shillitoe, A. J. (1952). *Stain Technol.* **27**, 81–85.

Grassé, P. P., Carasso, N. and Favard, P. (1956). *Annls Sci. nat. (Zool.)* **18**, 339–380.

Gryglewski, R. and Supniewski, J. (1963). *Bull. Acad. pol. Sci. Cl. II Sér. Sci. biol.* **11**, 53–56.

Haeckel, W. (1911). *Zool. Jb.* Suppl. **13** Fauna Chilensis **4** 89–136.

Hammarsten, O. (1885). *Pflügers Arch. ges Physiol.* **36**, 373–456.

Hanna, G. D. (1940). *J. Morph.* **66**, 115–129.

Harry, H. W. (1964). *Malacologia* **1**, 355–385.

Hartog, C. den, and Wolf, L. de (1962). *Basteria* **26**, 61–72.

Henderson, N. E. and Pelluet, D. (1960). *Can. J. Zool.* **38**, 173–178.

Herzberg, F. and Herzberg, A. (1960). *J. exp. Zool.* **145**, 191–196.

Herzberg, F. and Herzberg, A. (1962). *Am. midl. Nat.* **68**, 297–306.

Hickman, C. P. (1931). *J. Morph.* **51**, 243–289.

Hoff, C. C. (1940). *Trans. Am. microsc. Soc.* **59**, 224–242.

Holm, L. W. (1946). *Trans. Am. microsc. Soc.* **65**, 45–68.

Horstmann, H.-J. (1955). *Z. Morph. Ökol. Tiere* **44**, 222–268.

Horstmann, H.-J. (1956). *Biochem. Z.* **328**, 348–351.

Horstmann, H.-J. (1964). *Biochem. Z.* **340**, 548–551.

Hubendick, B. (1945). *Zool. Bidr. Upps.* **24**, 1–216.

Hubendick, B. (1947). *Zool. Bidr. Upps.* **25**, 141–164.

Hubendick, B. (1948a). *Proc. malac. Soc. Lond.* **27**, 186–196.

Hubendick, B. (1948b). *Ark. Zool.* **40**, No. 16, 1–63.

Hubendick, B. (1955). *Trans. zool. Soc. Lond.* **28**, 453–542.

Hubendick, B. (1962). *Göteborgs K. Vetensk.-o. VitterhSamh Handl.* B **9**(2), 1–68.

Hubendick, B. (1963). *Proc. malac. Soc. Lond.* **35**, 64–66.

Hubendick, B. (1964). *Göteborgs K. Vetensk.-o. VitterhSamh. Handl.* B **9**(6), 1–72.

Hunter, P. J. (1968). *Malacologia* **6**, 379–389.

Hunter, W. R. (1953). *Proc. zool. Soc. Lond.* **123**, 623–636.

Hunter, W. R. (1961a). *Proc. zool. Soc. Lond.* **136**, 219–253.

Hunter, W. R. (1961b). *Proc. zool. Soc. Lond.* **137**, 135–171.

Hyman, L. (1967). "The Invertebrates" Vol. 6, McGraw-Hill, New York.

Ikeda, K. (1929). *Annotnes zool. jap.* **12**, 295–317.

Ikeda, K. (1937). *J. Sci. Hiroshima Univ.* Ser. B **5**, 67–123.

Ikeda, K. and Emura, S. (1934). *Venus* **4**, 208–224.

Ingram, W. M. (1946). *Bull. Sth. Calif. Acad. Sci.* **45**, 152–159.

Jaeger, C. P. (1962). *Comp. Biochem. Physiol.* **7**, 63–69.

Jaeger, C. P. (1963). *Comp. Biochem. Physiol.* **8**, 131–136.

Joosse, J., Boer, M. H. and Cornelisse, C. J. (1968). *Symp. zool. Soc. Lond.* **22**, 213–235.

Jura, C. and George, J. C. (1958). *Proc. K. ned. Akad. Wet.*, C **61**, 590–594.

Karlin, E. J. and Bacon, C. (1961). *Trans. Am. microsc. Soc.* **80**, 399–406.

Kendall, S. B. (1953). *J. Helminth.* **27**, 17–28.

Kevan, D. K. McE. (1943). *Proc. R. Soc. Edinb.* B **61**, 430–461.

Koshman, R. W. and Serra, J. A. (1967). *Can. J. Genet. Cytol.* **9**, 31–37.

Kowslowsky, F. (1933). *Jena. Z. Naturw.* **68**, 117–192.

Kugler, O. E. (1965). *J. Morph.* **116**, 117–132.

Lacaze-Duthiers, H. de (1899). *Archs. Zool. exp. gén.* **7**, 33–120.

Lagrange, E. (1957). *Bull. Soc. Path. exot.* **50**, 804–811.

Lang, A. (1911). *Z. indukt. Abstamm.-u. VererbLehre* **5**, 97–138.

Lanza, B. and Quattrini, D. (1964). *Monitore zool. ital.* **72**, 93–141.

Larambergue, M.de. (1939). *Bull. biol. Fr. Belg.* **73**, 19–231.

Larambergue, M.de and Alaphilippe, F. (1959). *C.r. Séanc. Soc. Biol.* **153**, 1443–1447.

Laviolette, P. (1950). *Bull. mens. Soc. linn. Lyon* **19**, 52–56.

Lee, A. B. (1897). *Cellule* **13**, 199–278.

Lee, A. B. (1904). *Cellule* **21**, 77–116; 399–345.

Lind, H. (1973). *J. Zool., Lond.* **169**, 39–64.

Linville, H. P. (1900). *Bull. Mus. comp. Zool. Harv.* **35**, 213–248.

Lūsis, O. (1961). *Proc. zool. Soc. Lond.* **137**, 433–468.

Lūsis, O. (1962). *Nature, Lond.* **194**, 1191–1192.

Lūsis, O. (1966). *Proc. malac. Soc. Lond.* **37**, 19–26.

Luther, A. (1915). *Acta Soc. Fauna Flora fenn.* **40**, (2) 3–42.

Mahoney, F. J. (1940). *Univ. Colo. Stud. gen.* Ser. A **26**, 81–83.

Maitland, P. S. (1965). *Proc. malac. Soc. Lond.* **36**, 339–347.

Marcus, E. and Marcus, E. (1960). *Zoologia* **23**, 107–130.

Marcus, E. and Marcus, E. (1962). *Zoologia* **24**, 217–254.

Maury, M.–F. and Reygrobellet, D. (1963). *C. r. hebd. Séanc. Acad. Sci., Paris* **257**, 276–277.

May, F. Z. (1931). *Z. Biol.* **91**, 215–220; (1932a) **92**, 319–324; (1932b) **92**, 325–330; (1934a) **95**, 277–297; (1934b) **95**, 401–430; (1934c) **95**, 614–634.

May, F. and Kordowich, F. (1933). *Z. Biol.* **93**, 233—238.

May, F. and Weinland, H. (1953). *Z. Biol.* **105**, 339–347.

May, F. and Weinland, H. (1954). *Hoppe-Seyler's Z. physiol. Chem.* **26**, 154–166.

McCraw, B. M. (1957). *Can. J. Zool.* **35**, 751–768.

McCraw, B. M. (1961). *Trans. Am. microsc. Soc.* **80**, 16–27.

McCraw, B. M. (1970). *Malacologia* **10**, 399–413.

McDonald, S. L. C. (1969). *Sterkiana* No. 36, 1–17.

McFarland, D. J. (1964). *J. comp. physiol. Psychol.* **58**, 174–179.

McFarland, D. J. (1965a). *Anim. Behav.* **13**, 286–292.

McFarland, D. J. (1965b). *Anim. Behav.* **13**, 293–300.

McFarland, D. J. (1965c). *Anim. Behav.* **13**, 478–492.

McMahon, P., Von Brand, T. and Nolan, M. O. (1957). *J. cell. comp. Physiol.* **50**, 219–240.

Mead, A. R. (1950). *Bull. Mus. comp. Zool. Harv.* **105**, 218–291

Meenakshi, V. R. (1954). *Curr. Sci.* **23**, 301–302.

Meenakshi, V. R. (1955). *Curr. Sci.* **24**, 415–416.

Meenakshi, V. R. and Scheer, B. T. (1968). *Comp. Biochem. Physiol.* **26**, 1091–1097.

Meenakshi, V. R. and Scheer, B. T. (1969). *Comp. Biochem. Physiol.* **29**, 841–845.

Meisenheimer, J. (1907). *Zool. Jb. (Syst.)* **25**, 461–502.

Michelson, E. H. (1961). *Am. J. Hyg.* **73**, 66–74.

Morrill, J. B. (1963). *Acta Embryol. Morph. exp.* **6**, 339–343.

Morrill, J. B., Norris, E. and Smith, S. D. (1964). *Acta Embryol. Morph. exp.* **7**, 155–166.

Morton, J. E. (1954). *Proc. malac. Soc. Lond.* **31**, 30–46.

Morton, J. E. (1955a). *Proc. zool. Soc. Lond.* **125**, 127–168.

Morton, J. E. (1955b). *Phil. Trans. R. Soc.* B **239**, 89–160.

Morton, J. E. (1955c). *J. mar. biol. Ass. U.K.* **34**, 113–150.

Muller, R. (1965). *Proc. zool. Soc. Lond.* **144**, 229–237.

Németh, A. and Kovács, J. (1972) *Acta biol. Acad. Sci. hung.* **23**, 299–308.

Nieland, M. L. and Goudsmit, E. M. (1969). *J. ultrastruct. Res.* **29**, 119–140.

Noland, L. E. and Carriker, M. R. (1946). *Am. midl. Nat.* **36**, 467–493.

O'Colla, P. (1953). *Proc. R. Ir. Acad.* Ser B. **55**, 165–170.

Odhner, N. H. (1950). *Proc. malac. Soc. Lond.* **28**, 200–210.

Oldham, C. (1942). *Proc. malac. Soc. Lond.* **25**, 9–10.

Pan, C. T. (1958). *Bull. Mus. comp. Zool. Harv.* **119**, 237–299.

Paraense, W. L. (1959). *Am. Nat.* **93**, 93–101.

Patterson, C. M. (1968). *Malacol. Rev.* **1**, 1–13.

Paulini, E. and Camey, T. (1964). *Revta bras. Malar. Doenç. trop.* **16**, 499–504.

Pécsi, T., Kuziemski, H. and Róza, K. S. (1967). *Annls Biol. Tihany Hung. Acad. Sci.* **34**, 41–49.

Pelluet, D. (1964). *Can J. Zool.* **42**, 195–199.

Pelluet, D. and Lane, N. J. (1961). *Can. J. Zool.* **39**, 789–805.

Pelseneer, P. (1901). *Mem. Acad. r. Sci. Lett. Belg.* **54**, (3) pp. 76.

Personne, P. and Anderson, W. A. (1969). *J. Cell Sci.* **4**, 693–708.

Personne, P. and Anderson, W. A. (1970). *J. Cell Biol.* **44**, 20–28.

Pilsbry, H. A. (1948). "Land Mollusca of America." Academy of Natural Sciences, Philadelphia.

Precht, H. (1936). *Zool. Anz.* **115**, 80–89.

Quattrini, D. (1966a). *Monitore zool. ital.* **74**, 1–30.

Quattrini, D. (1966b). *Monitore zool. ital.* **74**, 125–141.

Quattrini, D. (1967). *Monitore zool. ital.* (N.S.). **1**, 109–128.

Quick, H. E. (1933). *Proc. malac. Soc. Lond.* **20**, 295–318.

Quick, H. E. (1936). *Ann. Natal Mus.* **8**, 19–45.

Quick, H. E. (1939a). *Proc. malac. Soc. Lond.* **23**, 298–302.

Quick, H. E. (1939b). *Bull. Maurit. Inst.* **1**, 57–61.

Quick, H. E. (1939c). *Proc. malac. Soc. Lond.* **23**, 333–335.

Quick, H. E. (1957). *Proc. malac. Soc. Lond.* **32**, 203–206.

Quick, H. E. (1960). *Bull. Br. Mus. nat. Hist. Zool.* **6**, 105–226.

Rangarao, K. (1963). *J. Anim. Morph. Physiol.* **10**, 158–163.

Raven, C. P. (1948). *Biol. Rev.* **23**, 333–369.

Raven, C. P. (1958). "Morphogenesis: The Analysis of Molluscan Development." Pergamon Press, London.

Rebhun, L. P. (1959). *J. biophys. biochem. Cytol.* **3**, 509–524.

Richards, C. S. (1969). *Malacologia* **9**, 339–348.

Richards, C. S. and Ferguson, F. F. (1962). *Trans. Am. microsc. Soc.* **81**, 251–256.

Rigby, J. E. (1963). *Proc. zool. Soc. Lond.* **141**, 311–359.

Rigby, J. E. (1965). *Proc. zool. Soc. Lond.* **144**, 445–486.

Ritchie, L. S., Hernandez, A. and Rosa-Amadov, R. (1966). *Am. J. trop. Med. Hyg.* **15**, 614–617.

Rizzotti, M. (1963). *Riv. Istochim. Norm. Patolog.* **8**, 287–293.

Runham, N. W. and Hunter, P. J. (1970). "Terrestrial Slugs." Hutchinson, London.

Runham, N. W. and Laryea, A. A. (1968). *Malacologia* **7**, 93–158.

Russell-Hunter, W. D., Apley, M. L. and Hunter, R. D. (1970). *Biol. Bull. mar. biol. Lab., Woods Hole* **139**, 434.

Russell-Hunter, W. D., Apley, M. L. and Hunter, R.D. (1972). *Biol. Bull. mar. biol. Lab., Woods Hole* **143**, 623–656.

Sawicka, T. and Chojnacki, T. (1968). *Comp. Biochem. Physiol.* **26**, 707–713.

Schumann, W. (1913). *Zool. Jb.* Suppl. **13** Fauna Chilensis **4**, 1–88.

Schutte, C. H. J. and Van Eeden, J. A. (1959). *Ann. Mag. nat. Hist.* Ser 13 **2**, 136–156.

Serra, J. A. and Koshman, R. W. (1967). *Can. J. Genet. Cytol.* **9**, 23–30

Simroth, H. and Hoffmann, H. (1928). H. G. Bronns Klassen und Ordnungen des Tier-Reichs III. II. 2, 1–1354, Akademische Verlagsgesellschaft, Leipzig.

Slugocka, M. (1913). *Revue suisse Zool.* **21**, 75–109.

Smith, B. J. (1965). *Ann. N. Y. Acad. Sci.* **118**, 997–1014.

Smith, B. J. (1966). *Malacologia* **4**, 325–349.

Starke, F–J. and Nolte, A. (1970). *Z. Zellforsch. mikrosk. Anat.* **105**, 210—221.

Stiglingh, I., Eeden, J. A. van and Ryke, P. A. J. (1962). *Malacologia* **1**, 73–114.

Tahmisian, T. N. (1964). *Z. Zellforsch. mikrosk. Anat.* **64**, 25–31.

Thompson, T. E. (1973). *Malacologia* **14**, 167–206.

Timmermans, L. P. M. (1959). *Proc. K. ned. Akad. Wet.* C **62**, 363–372.

Tuzet, O. and Galangau, V. (1967). *C. r. hebd. Séanc. Acad. Sci., Paris* D **264**, 337–339.

Van Cleave, H. J. (1935). *Ecology* **16**, 101–108.

Van der Steen, W. J. (1967). *Archs néerl. Zool.* **17**, 404–468.

Vaughn, C. M. (1953). *Am. midl. Nat.* **49**, 214–228.

Von Brand, T. and Files, V. S. (1947). *J. Parasit.* **33**, 476–482.

Walter, H. J. (1968). *Malacol. Rev.* **1**, 35–89.

Walter, H. J. (1969). *Malacol. Rev.* **2**, 1–102.

Walton, C. L. and Jones, N. W. (1926). *Parasitology* **18**, 144–147.

Williamson, M. (1959). *Proc. R. phys. Soc. Edinb.* **27**, 87–93.

Winsor, C. P. and Winsor, A. A. (1935). *J. Wash. Acad. Sci.* **25**, 302–307.

Wolda, H. (1965). *Archs néerl. Zool.* **16**, 387–399.

Wolda, H. (1967). *Evolution* **21**, 117–129.

Wright, C. A. (1956). *Proc. malac. Soc. Lond.* **32**, 88–104.

Yung, E. (1911). *C. r. Séanc. Soc. Phys. Hist. nat. Genève* **28**, 53–54.

Chapter 8

Development

C. P. RAVEN

Zoological Laboratory, University of Utrecht, The Netherlands

I. Introduction

When the egg cells of pulmonates are laid, each is surrounded by a spacious egg capsule filled with albumen. In freshwater pulmonates, as a rule, a certain number of egg capsules are combined into a common mass by an outer membrane or jelly. In land pulmonates, the egg capsules are usually laid singly, each surrounded by a calcareous shell.

The egg cells are generally small (± 70–200 µm) and poor in yolk. As a rule, the eggs pass their whole embryonic development within the egg capsules, from which young snails finally hatch. However, free-living marine veliger larvae are found in *Amphibola crenata* Martyn, *Onchidium nigricans* Quoy and Gaimard and various *Siphonaria* species (Fioroni, 1971). In *Onchidella celticai* (Forbes and Hanley) a

367

well-developed veliger larva is formed, but it does not hatch at this stage and metamorphoses within the capsule to a creeping young snail (Fretter, 1943). In most Basommatophora somewhat reduced trocho- phore and veliger stages are passed within the capsules. In the Styl- ommatophora the course of development deviates still more from the typical molluscan pattern. A veliger stage can hardly be distinguished; the velum, when it is not lacking altogether, as in *Limax* and *Arion*, is strongly reduced. A swollen head vesicle and a contractile foot vesicle or podocyst are important transitory organs for respiration and blood circulation. At the time of hatching in both groups most larval organs have disappeared; externally, the hatching snails resemble the adults, but the digestive tract may still be incompletely differentiated and the gonads are largely undeveloped.

II. Oogenesis

Egg formation in the pulmonates is of the follicular type. In *Lymnaea* (Bretschneider and Raven, 1951) the egg cells, after their formation from the germinal epithelium, pass first through a period of amoeboid motility. They then become sedentary and settle in the distal part of the acini of the gonad. They begin to grow and a follicle forms around

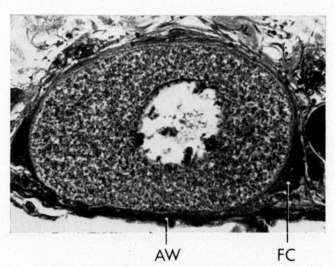

AW   FC

Fig. 1. Growing oocyte of *Lymnaea stagnalis* in the gonad. AW, connective tissue wall of acinus; FC, follicle cell. Staining: azan. × 660.

their protruding part, while the basal surface of the oocyte remains in contact with the connective tissue wall of the gonad (Fig. 1). The follicle consists more or less distinctly of two layers. The cells of the inner layer are connected with their basal surface to the gonadial wall by means of micropodia (Recourt, 1961). In *L. stagnalis* (L.), all oocytes above a certain size are surrounded by six of these inner follicle cells; in *L. peregra* (Müll.), there are 7–9 inner follicle cells. In both species the cells show a characteristic arrangement; apparently, the follicle is built according to a determinate pattern which is at the same time polar, dorsoventral and asymmetric (Raven, 1963; Ubbels *et al.*, 1969) (Fig. 2). The follicle cells are at first closely applied to the surface of the oocyte. Later a cleft appears on the apical and lateral sides between the oocyte and the follicle cells. With the approach of ovulation the follicular cavity extends around the egg cell towards its basal side; the surface of attachment is rapidly reduced and the full-grown oocyte becomes free by autolysis of the follicle cells.

In *Vaginula*, follicle cells are likewise formed from the epithelial layer of the gonad. The inner protruding part of the oocytes is completely covered by follicle cells, whereas their basal part remains in contact with the basal membrane of the acinus wall (Quattrini and Lanza, 1965).

The first part of the growth period of the oocyte (previtellogenesis) mainly consists of an increase in the amount of protoplasm. It is followed by a phase of rapid growth, in which the yolk is accumulated in the growing oocyte (vitellogenesis). The yolk consists partly of proteins and polysaccharides, partly it is of a fatty nature.

The fatty yolk consists of droplets or globules of lipid substances which can be accumulated by centrifugal force into one part of the egg. The fat content, determined by this method, is about 5% in *Lymnaea stagnalis* (Raven, 1945). The globules of the fatty yolk arise independently of visible cell structures in the cytoplasm (Parat, 1928; Hartung, 1947; Bretschneider and Raven, 1951; Ubbels, 1968).

The protein yolk consists of droplets or platelets of varying size, shape and composition. In centrifuged eggs of *Lymnaea stagnalis*, the protein yolk occupies about 50% of the egg volume (Raven, 1945; Bretschneider and Raven 1951).

The protein yolk of *Lymnaea* (Bretschneider and Raven, 1951), *Myxas* and *Succinea* (Jura, 1960) is composed of two kind of platelets, called β- and γ-granules. They differ in size, shape and probably in chemical composition (Andrew, 1959; Ubbels, 1968).

Electron microscopy has shown that the yolk granules of *Lymnaea* (Elbers, 1957; Recourt, 1961) and *Planorbis* (Favard and Carasso, 1958) contain numerous electron-dense particles, for the greater part arranged in a regular crystalline pattern (cf. Fig. 9). The particles, probably globular macromolecules of an iron-containing protein, have a diameter of about 6–7 nm.

Various cell structures may play a part in the formation of the yolk platelets. According to many authors, these platelets are formed in

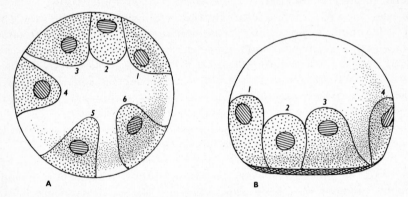

Fig. 2. Diagram of arrangement of the inner layer of follicle cells around the oocyte of *Lymnaea stagnalis*. A, from apical pole of oocyte, B, from the side. Cross-hatched, area of contact with gonadial wall; 1—6, follicle cells.

Golgi bodies (Parat, 1928; Fahmy, 1949; Bretschneider and Raven, 1951; Recourt, 1961). In *Planorbis*, however, according to Favard and Carasso (1958) the mitochondria play the main part in yolk platelet formation.

The so-called protein yolk by no means consists exclusively of proteins. The yolk platelets may also contain carbohydrates (Ubbels, 1968; Guraya, 1969) and phospholipids (Malhotra, 1960; Ubbels, 1968). Malhotra (1960) observed that the yolk granules of *Lymnaea* contain calcium, probably as calcium phosphate.

III. Egg maturation

In most pulmonates, fertilization of the egg cells takes place soon after they have left the gonad, in the first part of the spermoviduct. At the same time, egg maturation begins. The centrioles move apart and come to lie at opposite ends of the germinal vesicle. Small asters are formed around them. By a condensation of the chromatin the tetrads become visible, and orient themselves in the equator of the first maturation spindle (Fig. 4). The egg nucleolus is resorbed in the cytoplasm (Garnault, 1888; Lams, 1910; Bretschneider, 1948).

The first maturation spindle, which is provided with large asters, moves towards the animal pole, and places itself at right angles to the surface (Fig. 4). Its peripheral aster connects with the surface at the animal pole. Somewhat later, a cytoplasmic protrusion is formed at this place (Fig. 3A). The first polar body is now constricted off by an annular furrow. The polar body remains for some time connected with the egg surface by a mid-body (Raven *et al.*, 1958; Raven, 1964) (Fig. 3B).

During the first maturation division one group of dyads is expelled into the polar body; the other dyads at the end of anaphase have reached the margin of the centrosphere of the inner aster remaining in the egg (Fig. 3A). Soon thereafter the formation of the second maturation spindle begins. This process exhibits a bewildering diversity in pulmonates even among closely related species. At least four different ways in which this spindle may be formed can be distinguished:

1. The centrosphere of the inner aster of the first maturation spindle enlarges, while the surrounding astral rays break down. The centriole in this centrosphere had divided at an earlier stage; now the daughter centrioles move apart. At the same time the centrosphere elongates along the line connecting the centrioles. When the centrioles have reached the ends of the long axis of the centrosphere, the latter gets a fibrillar structure, in this way transforming into the second maturation spindle. Asters develop at both ends of the spindle, centred on the centrioles lying there. The dyads, at first situated as a compact group on one side of the developing spindle, secondarily penetrate into it and arrange themselves in its middle region. The spindle rotates into a radial position, and connects with the egg surface at the animal pole.

This mode of formation has been observed in *Physa* (Kostanecki and Wierzejski, 1896; Raven, 1959), *Arion* (Lams, 1910), *Limax* and *Deroceras* (Byrnes, 1900; Linville, 1900; Raven *et al.*, 1958), and, probably, *Helix* (Garnault, 1889); in a slightly modified form also in *Succinea* (Raven, 1959).

2. The second maturation spindle develops likewise by transformation of the centrosphere of the deep aster of the first maturation spindle. Its centriole has not divided, however, and the single centriole moves to the superficial pole of the elongating centrosphere. The dyads move along the outer surface of the developing second maturation spindle toward its equatorial region, where they penetrate into the spindle. At the pointed outer end of the spindle, an aster forms in the cytoplasm. The deep aster of the second maturation spindle is provided by the sperm aster fusing secondarily with its blunt inner end.

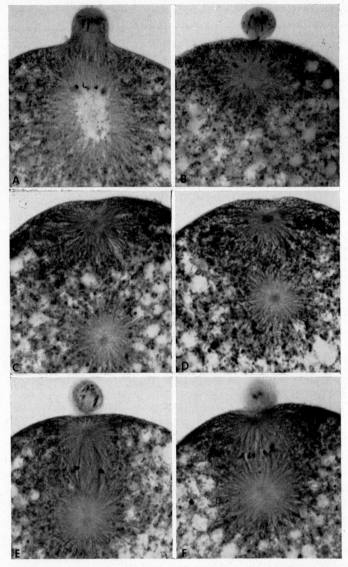

Fig. 3. Formation of the second maturation spindle in *Lymnaea peregra*. A, telo-phase of first maturation division showing beginning of first polar body and inner aster with vacuolated centrosphere. B, first polar body extruded, connected to egg by mid-body; beginning of centrifugal movement of dyads. C, sperm aster has appeared beneath maturation aster; centrifugal movement of dyads continues. D, sperm aster approaches maturation aster; dyads have reached periphery of maturation aster. E, second maturation spindle formed between both asters; dyads begin to penetrate spindle equator. F, early metaphase of second maturation division. $\times$ 800.

This manner of formation of the second maturation spindle has first been observed in *Lymnaea stagnalis* (Raven, 1949; Raven *et al.*, 1958). It occurs in a similar way in *L. palustris* (Raven, 1964).

3. The second maturation spindle does not develop from the centrosphere, but arises by "spinning" of spindle fibres between the egg centriole in the first maturation aster and a centriole in the sperm aster. The original inner aster of the first maturation spindle is taken over more or less unchanged as the outer aster of the second spindle. The latter's inner aster is provided by the sperm aster. The dyads move centrifugally along the rays of the first-mentioned aster, which then, curving inwards, carry them to the surface of the spindle (Fig. 3).

The second maturation spindle is formed in this manner in *Planorbis planorbis* (L.) and *Planorbarius corneus* (L.) (Raven, 1959) and in *Lymnaea peregra* (Raven, 1964).

4. The formation of the second maturation spindle in *Lymnaea ovata**
and *Myxas glutinosa* (Müll.) shows a great resemblance to the previous case. However, it appears that in these species the main mass of the second maturation spindle is formed by folding together of the first maturation aster rather than by "spinning" of new spindle fibres (Raven, 1964).

The formation of the second polar body closely resembles that of the first. A mid-body is also formed in this case.

Immediately after the extrusion of the second polar body the chromosomes in the egg begin to swell into little vesicles, the karyomeres. The latter then coalesce to form the female pronucleus. This remains at the animal pole where later the copulation of the two pronuclei takes place.

IV. Fertilization

In the pulmonates the fertilizing sperm may enter at any part of the egg surface. Fertilization cones have been described by Ikeda (1930) in *Eulota*. As a rule the whole spermatozoon enters the egg. The sperm traverses the egg cytoplasm and comes to rest with its head somewhere beneath the egg surface. Soon the tail breaks away from the head. It is resorbed in the egg cytoplasm; the time at which this occurs is more or less fixed, but differs in different species.

Polyspermy has been found in *Lymnaea* (Crabb, 1927; Bretschneider, 1948), *Bulinus* (De Larambergue, 1939) and *Helix* (Garnault, 1889). Only one of the sperm forms a pronucleus; the supernumerary sperm disintegrate.

*Editors' note: According to Hubendick (1951) this is a synonym of *peregra*.

Sooner or later a sperm aster becomes visible. Its origin can be traced back to the middle piece of the spermatozoon (Kostanecki and Wierzejski, 1896; Raven, 1945). The sperm aster rapidly increases in size, and a centrosphere appears in it.

The time of appearance of the sperm aster differs in different species. Its further development also varies considerably among the species. The following possibilities can be distinguished:

1. The sperm aster usually divides and forms an amphiaster with central spindle. In *Physa* this was supposed to become the cleavage spindle by Kostanecki and Wierzejski (1896). In planorbids the sperm amphiaster participates in the formation of the second maturation spindle and becomes its deep aster (Raven, 1959).

2. The sperm aster remains undivided as a rule, but a more or less distinct division may take place occasionally, leading either to the formation of an amphiaster with central spindle, of a dicentric aster or of two loose asters which soon disappear: e.g. in *Lymnaea* (Linville, 1900; De Larambergue, 1939; Raven, 1964), *Succinea* (Raven, 1959) and *Limax flavus* L. (Raven *et al.*, 1958).

3. No sperm aster has been observed in *Arion* (Lams, 1910), *Agriolimax agrestis* (L.) (Byrnes, 1900) and *Agriolimax reticulatus* (Müll.) (Raven *et al.*, 1958).

The participation of the sperm aster in the formation of the second maturation spindle in *Lymnaea* and *Planorbis* has already been mentioned (Fig. 3).

The sperm head initially retains its compact structure. Its transformation into the male pronucleus does not begin until the second polar body has been extruded. At this moment it begins to swell, synchronously with the egg chromosomes, and rapidly changes into a pronucleus. At the same time it commences its migration towards the female pronucleus at the animal pole (Fig. 5). When the two meet, they apply themselves closely against each other. No actual fusion of the pronuclei takes place, however.

Two new cytocentra appear against the closely applied pronuclei. It is likely that their centrioles are descendents of the sperm centriole (Raven, 1959). A central spindle is formed between them. The nuclear membranes of the pronuclei are dissolved and their chromosomes become arranged in the first cleavage spindle.

V. Ooplasmic segregation

Under the term "ooplasmic segregation" may be comprised a number of movements, by which various components and inclusions of the egg cytoplasm are accumulated or concentrated at certain places in the egg

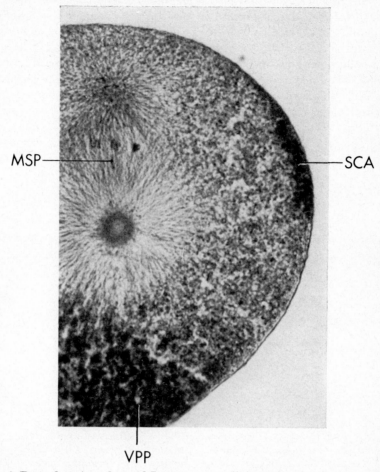

Fig. 4. Part of section of egg of *Lymnaea stagnalis* immediately after oviposition. MSP, first maturation spindle in metaphase; SCA, subcortical accumulation; VPP, vegetative pole plasm. Staining: azan. × 1250.

cell. This begins immediately after ovulation and fertilization, during the passage of the egg cells through the genital tract. It brings about a situation in which different parts of the egg differ in their cytoplasmic composition. The ooplasmic segregation continues during cleavage. The segregated egg substances are distributed unequally among the cleavage cells, which are endowed from the outset with a different chemical composition of their cytoplasm.

In the egg cells of *Lymnaea stagnalis*, a vegetative pole plasm is formed between ovulation and oviposition. It occupies a well-defined sector at the vegetative pole (Fig. 4). During maturation it spreads beneath the egg surface towards the animal side. After the extrusion of the second polar body it forms a continuous subcortical layer of uniform thickness (Raven, 1945). A similar subcortical layer of protoplasm has also been observed in *Physa fontinalis* (L.), *Planorbis planorbis* and *Succinea putris* (L.).

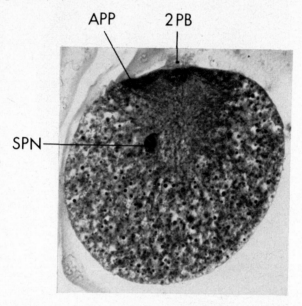

Fig. 5. Section of egg of *Lymnaea stagnalis* shortly after second maturation division showing beginning of formation of animal pole plasm (APP). 2PB, second polar body; SPN, swelling sperm nucleus on its way to animal pole. Staining: iron hematoxylin–eosin. × 540.

In addition to the vegetative pole plasm, the eggs of *L. stagnalis* also exhibit a ring of six "subcortical accumulations" (SCA) in the equatorial region at the moment of oviposition (Fig. 4). They are not evenly spaced around the equator but are arranged according to a dorsoventral and slightly asymmetric pattern, which apparently reflects the arrangement of the follicle cells during oogenesis (Raven, 1963, 1967). The SCA have also been formed by ooplasmic segregation between ovulation and oviposition. In *L. peregra*, there are 6 or 7 SCA (Ubbels *et al.*, 1969). In other *Lymnaea* species, similar SCA have been observed, their number and arrangement differing with the species.

Finally an animal pole plasm is formed in *Lymnaea, Myxas, Planorbis, Physa* and *Succinea* (Fig. 5). In *Succinea* it has appeared when the first maturation spindle connects with the surface (Jura, 1960); in the other species it becomes visible only after the first or even the second maturation division (Raven, 1964). The animal pole plasm consists of a dense matrix, which is very rich in mitochondria (Raven and Van Der Wal, 1964). According to Guerrier (1968), an animal pole plasm is likewise formed after the extrusion of the second polar body in *Limax*.

Shortly before the third cleavage the vegetative pole plasm in *Lymnaea stagnalis* concentrates markedly toward the animal side of the cells and unites with the animal pole plasm. At the subsequent divisions, this common pole plasm is differentially distributed among the cleavage cells, so that the relative amount in the cells of the blastula decreases from the animal towards the vegetative pole. Meanwhile, the SCA have been displaced toward the most vegetative part of the embryo. Their substance is distributed in a regular way among the macromeres and second and third micromeres. In the macromeres, coarse dark granules are formed in the SCA plasm at the vegetative pole. These granules move during the twenty-four-cell stage toward the inner end of the macromeres (Fig. 6), where they condense into compact dark bodies (Raven, 1945, 1946, 1963, 1967, 1970, 1974). Similar granules and dark bodies have already been observed by Wierzejski (1905) in *Physa fontinalis* (L.). Recent observations by Sathananthan (1970) in *Arion ater rufus* (L.) suggest that in this species a comparable situation is found.

VI. Cleavage and gastrulation

All pulmonates exhibit spiral cleavage. The first cleavage divides the egg equally and meridionally into the cells *AB* and *CD*. At the next division, which is also meridional, *AB* divides into *A* and *B, CD* into *C* and *D. A, B, C* and *D*, which are called the quadrants of the egg, as a rule follow upon each other in this sequence in a clockwise direction, when the egg is viewed from the animal pole. At the third division, in each quadrant an animal micromere is separated from a vegetative macromere. The cleavage spindles have an oblique position with respect to the egg axis; as a rule, they are arranged according to a right-handed spiral (dexiotropic), so that each micromere is displaced in a clockwise direction with respect to its macromere (Fig. 7A). In the next division, the spindles deviate toward the other side (laeotropic); a second quartette of micromeres is formed from the macromeres, while the first micromeres also divide laeotropically. By alternating dexiotropic and laeotropic divisions, two or three more micromere quartettes are formed from the macromeres, while the earlier formed micromeres continue to divide (Fig. 7B). Sooner or later, however, the regular character of spiral cleavage is lost, and bilaterally symmetrical divisions take place.

This description refers to the most common case, where the third cleavage is dexiotropic and the subsequent cleavages follow according

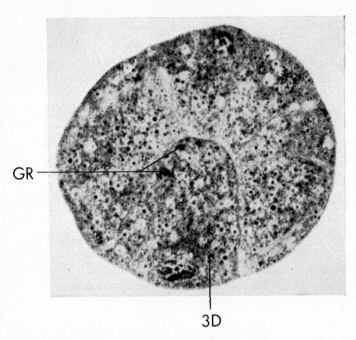

Fig. 6. Section of egg of *Lymnaea stagnalis*, 24-cell stage; cleavage cavity oblite-rated. Cell 3D projects into central part of egg, in contact with animal cells. GR, coarse SCA-granules migrating towards central ends of macromeres. Staining: iron hematoxylin–eosin. × 720.

to the rule of alternation. However, in a few cases cleavage is the mirror image of the ordinary pattern. This is linked up with the direction of coiling of the adult snails, as was first discovered by Crampton (1894): those species where the shell is coiled according to a left-handed spiral have a reversed cleavage. Occasionally, in normally dextral species, sporadic sinistral individuals or sinistral races may occur; in this case (e.g. *Lymnaea peregra*) the type of cleavage is also correlated with the direction of coiling.

Owing to the strictly determinate course of spiral cleavage, it has been possible to study the cell lineage in great detail. Such studies have been made in *Planorbis* (Holmes, 1900), *Biomphalaria* (Camey and Verdonk, 1970), *Physa* (Wierzejski, 1905), *Lymnaea* (Verdonk, 1965) and *Limax* (Kofoid, 1895; Meisenheimer, 1896). Generally the results are comparable in all species investigated. The first three quartettes of micromeres produce the ectoderm. The pretrochal part of the ectoderm,

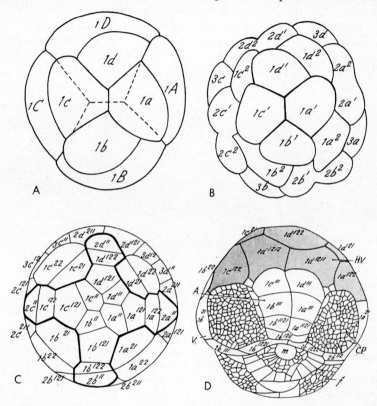

Fig. 7. Diagrams to illustrate cleavage (A, B, C,) and development of the "larval" head pattern (D) in *Lymnaea stagnalis* (after Verdonk, 1965). A, 8-cell stage; B, 24-cell stage; C, formation of the molluscan cross. D, larval head pattern at the trochophore stage. A, apical plate; CP, cephalic plate; F, foot; HV, head vesicle M, mouth; V, "velum".

which later develops into the head region, derives mainly from the first quartette; the post-trochal ectoderm comes from the second and third quartettes. The fourth micromere in the *D*-quadrant (4*d*) produces, besides a number of endodermal elements, the whole primary mesoderm. The remaining part of the endoderm comes from the cells 4*a*–4*c* and the macromeres 4*A*–4*D*.

The cells of the first quartette of micromeres ($1a$–$1d$) at their first division divide into $1a^1$–$1d^1$ and $1a^2$–$1d^2$ (Fig. 7A,B). The latter occupy the four corners of a square surrounding the animal pole. These cells, the primary trochoblasts, divide only once more. Their descendants form large parts of the velum and head vesicle of the larva. Between the trochoblasts the derivatives of $1a^1$–$1d^1$ extend in a radial direction. In this way, they form the four arms of the molluscan cross (Fig. 7C). These arms have a radial position (ventral, dorsal, left and right), whereas the trochoblasts lie interradially.

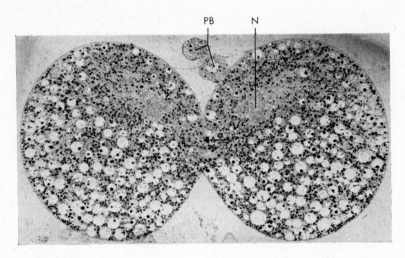

Fig. 8. First cleavage of the egg of *Lymnaea stagnalis* in section. The primary constriction delimits the two blastomeres still connected by a stalk. A fold of the cell membrane has advanced from the animal side about halfway through the stalk (arrow a). N, reconstituting nucleus of blastomere; PB, polar bodies. Electron micrograph. $\times$ 1000. (Courtesy of Dr W. Berendsen.)

At the twenty four-cell stage the macromere $3D$ partly withdraws from the surface, and bulges with the greater part of its mass into the cleavage cavity, where it touches the inner side of the animal cells. The resulting cell configuration at the vegetative side forms the first visible expression of the dorsoventrality of the embryo. At least in several species a break of some hours in the progress of cleavage occurs, during which the cleavage cavity disappears altogether (Wierzejski, 1905; Raven, 1946, 1970; Sathananthan 1970) (Fig. 6). At the next division of $3D$ its larger interior part is constricted off as the cell $4d$. At further development, $4d$ divides almost bilaterally in two equal mesendoblasts;

on further division the latter first produce one or more enteroblasts, while their remaining parts form the teloblasts giving rise to the left and right mesoderm bands.

Berendsen (1971) has made a close study of the first cleavage in *Lymnaea* by means of electron microscopy. It begins with an indentation of the egg surface at the animal pole. This constriction extends rapidly around the egg. The egg is constricted from all sides, and the two halves round off to a nearly spherical shape. After about ten minutes, from the bottom of the constriction at the animal side a narrow fold of the cell membrane invaginates into the egg (Fig. 8). The advancing tip of this furrow pushes the remnant of the spindle towards the vegetative side. Equatorial thickenings in the spindle, consisting of laterally fused spindle tubules, are compressed to a mid-body, which forms the last connection between the daughter cells. Cell division is completed by the isolation of the disk-like mid-body through ingrowth of the extracellular cavity on either side of it. The daughter cells flatten against each other into two hemispheres, separated by a flat partition wall.

Waddington, Perry and Okada (1961) describe "membrane knots", probably made up of highly folded regions of cell membranes, as intercellular connections between the blastomeres in *Lymnaea peregra*. These structures have not been found in *L. stagnalis*, but both septate desmosomes and maculae adhaerentes are present in this species (Elbers, 1959; Bluemink, 1967; Berendsen, 1971) (Fig. 9).

Cleavage in all pulmonates gives rise to a coeloblastula with wide cleavage cavity. Gastrulation takes place by invagination of the archenteron at the vegetative pole. The blastopore, which is originally situated in the centre of the vegetative hemisphere, then shifts toward the ventral side.

VII. Osmotic regulation in eggs and cleavage stages

In the eggs of freshwater pulmonates, a great discrepancy exists between the inner and outer osmotic pressure. The envelopes and capsular membranes of *Lymnaea* and *Biomphalaria* are freely permeable to water and inorganic ions (Beadle, 1969a; Raven, 1970). One may expect to find in these eggs special conditions serving to prevent excessive swelling.

In *Physa* (Clement, 1938) and *Lymnaea* (Raven, 1945), a slow swelling of the eggs occurs during the uncleaved stage. This swelling is an osmotic phenomenon; its rate is dependent on the osmotic pressure of the medium (Raven and Klomp, 1946).

The surface membrane of the egg of *Lymnaea* probably has a very low ion permeability (Elbers, 1959, 1966, 1969; Raven, 1970), while its permeability constant for water is rather low, in comparison with values found in marine eggs (Raven, 1966).

Uncleaved eggs of *Lymnaea* chilled to 3°C or treated with 10^{-3}M potassium cyanide show a slight increase in volume over control eggs (Raven *et al.*, 1953). Although these results point to the existence of an active water excretion mechanism, apparently this plays, at best, only a minor part during the uncleaved stage.

The swelling of the eggs of *Lymnaea* continues after the beginning of cleavage, at a greatly increased rate. This must be due to an increase in water permeability of the egg surface (Raven *et al.*, 1952; Raven, 1966).

Both in freshwater and land pulmonates, there is, from the two cell stage on, a wide cleavage cavity, which opens periodically to the exterior and ejects its contents to the medium. This recurrent cleavage cavity has first been described in detail by Kofoid (1895). It has since been found in various freshwater and land pulmonates (Meisenheimer, 1896; Holmes, 1900; Wierzejski, 1905; Carrick, 1939; Raven, 1946; Ghose, 1962). A special study of its physiology in various snails has been made by Comandon and De Fonbrune (1935).

The very early formation of a recurrent cleavage cavity in the pulmonates represents the precocious development of a water-excreting mechanism as a special adaptation to development in a strongly hypotonic environment. Kofoid observed that *Physa* eggs raised in salt solutions had the cleavage cavity appearing later and remaining smaller than eggs in water. In *Lymnaea* eggs, the vacuolization of the cells does not markedly increase after the cleavage cavity has appeared; at later cleavage stages, there is even a strong decrease in the size of the cytoplasmic vacuoles, which finally disappear almost entirely. When the eggs are kept in distilled water, no cleavage cavity is formed; under these circumstances the vacuolization of the cytoplasm becomes excessive, and the eggs cytolyse at the four cell stage (Raven and Klomp, 1946; Raven and Van Zeist, 1950). When one of the blastomeres of a two-cell stage of *Succinea* is killed by u.v. irradiation before the appearance of a cleavage cavity, the other blastomere invariably dies. If irradiation takes place when the cleavage cavity between the two cells is fully formed, the nonirradiated blastomere may pursue its development, produce a typical recurrent cleavage cavity during subsequent cleavages, and develop to a partial larva (Jura, 1971).

The cleavage cavity of *Biomphalaria* shows a great swelling after subjecting the eggs to anoxia, cyanide, or low temperature. Since the cleavage cells are not obviously swollen after such treatment, it is concluded that water is not actively pumped into the cavity from the cells, but the active process is concerned with removing fluid from the cavity (Beadle, 1969b).

VIII. Breakdown and utilization of the yolk

It has been mentioned above (p. 369) that the protein yolk in *Lymnaea* and some other pulmonates consists of two kinds of platelets, β- and γ-granules, differing in various aspects. During the passage of the ovulated egg through the oviduct the γ-granules of *Lymnaea* show the first indications of a peculiar swelling process (Bretschneider and Raven, 1951). This swelling becomes much more pronounced after oviposition, when clear, watery vacuoles form around the γ-granules and gradually increase in size (Raven, 1945) (Figs. 5,8).

Vacuoles surrounding γ-granules have also been observed in other *Lymnaea* species, in *Myxas glutinosa* (Müll.) and in *Succinea putris* (Jura, 1960). In other pulmonates vacuoles appear at a similar stage, though no γ-granules can be seen, e.g. in *Planorbis planorbis*, *Planorbarius corneus* and *Physa fontinalis*. According to Clement (1938), those in *Physa* are also formed from coarse yolk granules.

Though the β- and γ-granules of *Lymnaea* exhibit clear cytochemical differences (Raven, 1945; Andrew, 1959; Bluemink, 1967), their submicroscopic structure in the electron microscope shows great similarities. There is a gradual transition between the two, suggesting that γ-granules arise from β-granules by a process of swelling and disintegration (Elbers, 1959; Recourt, 1961; Berendsen, 1971; Van Der Wal, 1974).

At later cleavage stages, the vacuolization of the cytoplasm in *Lymnaea* diminishes strongly (cf. Fig. 6). The γ-granules are probably entirely broken down during early cleavage. Breakdown of the yolk continues at the blastula stage (Bluemink, 1967). Golgi bodies probably play a part in the breakdown and transformation of the yolk (Raven, 1946; Malhotra, 1960). Hydrolases (acid phosphatase) produced in Golgi cisternae are presumably transported by Golgi vesicles to the yolk granules (Bluemink, 1967).

IX. Embryonic nutrition

The egg capsule fluid or "albumen" of the pulmonates contains, as a rule, both polysaccharides and proteins. Egg clutches of *Lymnaea*, *Biomphalaria* and *Helix* contain galactogen (May and Weinland, 1953; McMahon *et al.*, 1957). The egg capsule fluid of *Lymnaea stagnalis* contains about 14·3% dry matter; half of this is galactogen, the other

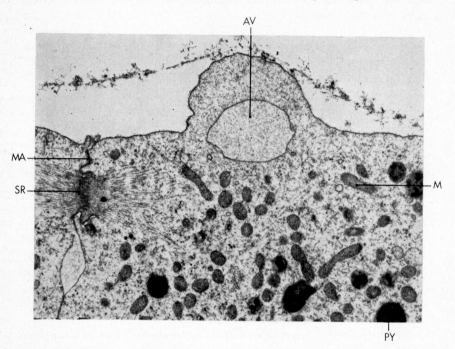

Fig. 9. Electron micrograph of a section of a superficial part of a cell from a blastula of *Lymnaea stagnalis*. AV, albumen vacuole just ingested by means of pinocytosis; M, mitochondrion; MA, macula adhaerens between adjacent cells; PY, protein yolk granule with crystalline inclusions; SR, spindle remnant. × 28 000. (Courtesy of Dr J. G. Bluemink.)

half mainly protein. According to George and Jura (1958), the egg capsule fluid of *Succinea putris* also consists of polysaccharides and proteins.

The eggs of pulmonates at cleavage stages begin to ingest albumen in all cells; this is laid down in the ectoplasmic part of the cells in special albumen vacuoles (Meisenheimer, 1896; Holmes, 1900; Raven, 1946; Jura, 1960; Sathananthan, 1968). The contents of these vacuoles exhibit strong polysaccharide reactions. The uptake of the albumen in

Lymnaea takes place partly by a peculiar process of pinocytosis in large crater-like pits at the surface of the cells (Fig. 9), partly by micropinocytosis. After their formation, the albumen vacuoles coalesce. The vacuole membrane may rupture within 15 min after vacuole formation, releasing the albumen into the cytoplasm. Coalescence of albumen vacuoles with yolk granules could also be demonstrated. Older albumen vacuoles exhibit positive reactions for acid phosphatase and for organophosphorus-resistant esterase. It is believed that the albumen vacuoles acquire these hydrolytic enzymes by coalescence with yolk granules. Calculations show that it is the capsule fluid, not the intracellular yolk, that is the main nutrient reserve for the developing embryo. The function of the yolk granules is thought to be mainly enzymatic; they supply the hydrolytic enzymes necessary for the digestion of the albumen (Bluemink, 1967).

After gastrulation the uptake of albumen is more and more restricted to the endoderm. Part of the endoderm develops into the albumen sac or larval digestive gland, which plays an important part in the uptake and digestion of the albumen. The ingested albumen is first accumulated in smaller vacuoles in the apical part of the cells, then in much bigger irregular vacuoles in the basal part (Fig. 10). A change in stainability of the albumen in the vacuoles indicates that chemical transformations are taking place (Raven, 1946; Weiss, 1968). The intracellular digestion may be preceded by an extracellular digestion in the gut lumen by means of enzymes secreted by the cells. Extrusion of secretion droplets into the lumen by the cells has sometimes been observed; in *Lymnaea*, these cells temporarily exhibit a distinct apocrine secretion (Raven, 1952).

The dry weight of *Lymnaea* embryos remains practically the same throughout cleavage. Apparently, the uptake of albumen in this period just compensates for the weight lost by respiration. After gastrulation, a rapid increase in weight takes place. The embryos ingest a great deal of the galactogen present in the egg capsule fluid. Galactogen breakdown can be demonstrated after blood circulation has begun. A great part of the ingested galactogen is, moreover, utilized in fat synthesis (Horstmann, 1956a, 1956b, 1958, 1964). In *Helix*, galactogen from the egg capsule fluid is taken up in the stomach, but its breakdown occurs only after hatching (Hunger and Horstmann, 1968).

The proteins of the egg capsule fluid are likewise used for the growth of the embryo. In *Lymnaea*, the protein content of the embryo begins to rise after gastrulation, with a concomitant decrease in protein of the egg capsule fluid (Morrill, 1964). A protein hydrolyzate of the egg capsule fluid of *Lymnaea* contains eighteen amino acids. The same eighteen amino acids, in similar proportions, are found in free form in the 7-day old embryo; it seems probable that they are derived from ingested albumen (Morrill, 1963).

In the Stylommatophora the albumen sac, and in later stages the paired primordia of the digestive gland play a part in the digestion of the albumen. Shortly before hatching the remaining albumen in the capsule is ingested. In the Limacidae it is mainly stored in the greatly dilated cavities of stomach and digestive gland; but in the Arionidae, intracellularly in the digestive gland (Weiss, 1968).

X. "Larval" development

As stated in the introduction, most pulmonates pass their development within the egg capsules, from which they hatch as crawling snails. Only in some Onchidiidae, Amphibolidae and Siphonariidae are there free living veliger larvae. In the other groups the development of some typical larval organs is more or less suppressed, but in most Basommatophora, stages somewhat comparable to the typical molluscan trochophore and veliger can still be distinguished; in Stylommatophora development generally shows greater deviations from the typical course.

In *Onchidella celtica* first a somewhat top-shaped trochophore stage is formed, with a prototroch surrounding the whole embryo; it has a mouth and shell gland in the post-trochal region, but lacks an apical tuft (Joyeux-Laffuie, 1882). At further development, the prototroch grows out into a bilobate velum. The Basommatophora as a rule have a rather well-developed prototroch. However, only its ventral half is developed as such, from the ventral primary trochoblasts and some adjacent cells; the dorsal half is incorporated into the head vesicle. This type of development also occurs in *Vaginula*. In *Helix* the prototroch is greatly reduced and remains only for a short time, while in *Arion* and *Limax* no prototroch develops at all.

In the pretrochal region the following parts may be distinguished: 1, the dorsally situated head vesicle, 2, the median apical plate, and 3, the ventrolateral cephalic plates (Fig. 7D).

The cells of the dorsal part of the pretrochal region soon begin to flatten after gastrulation. The ectoderm is lifted from the endoderm, so that a large vesicle, covered by a very thin epithelium, is formed. Mesoderm cells penetrate into this space, forming a loose network.

The head vesicle reaches its greatest development in the Stylommatophora. It swells to such an extent that it temporarily becomes much bigger than the rest of the body. Later it lags behind in growth and is gradually taken up into the body. As a rule the vesicle possesses no active contractility, but is passively distended by the contractions of the podocyst. However, according to Brisson (1968) the head vesicle of *Archachatina* possesses muscle cells and shows active contractions at late stages of development.

In the Basommatophora the head vesicle remains smaller. In *Lymnaea* and *Planorbis* it is covered by 12 large flat cells, which do not bear cilia (Fig. 7D).

The apical plate is most conspicuous in the Basommatophora. It extends from the head vesicle to the ventromedian part of the prototroch and consists of six or seven cells (Fig. 7D), which are densely covered with cilia.

On either side of the apical plate, a pair of small-celled areas, the cephalic plates, are formed by rapid cell multiplication (Fig. 7D). From these plates the cerebral ganglia, tentacles and eyes develop. The cerebral ganglia may either be formed by cell proliferation or by paired invaginations, the inner ends of which detach themselves from the epithelium (e.g. *Limax*). The tentacles develop as outgrowths of the ventrolateral parts of the cephalic plates, while the eyes are formed laterally, mostly as small invaginations, which are constricted off from the surface.

In the post-trochal region, on the dorsal side of the embryo, the shell gland develops at an early stage. First an ectodermal thickening appears by elongation of the cells, and in its centre an invagination takes place (Fig. 10). After some time, the shell gland begins to roll out again onto the surface.

In *Lymnaea* the shell gland primordium has from its first appearance a close contact with the region of small-celled endoderm of the midgut. This remains during the whole invagination period (Fig. 10). A similar close contact between shell gland primordium and small-celled endoderm exists in *Planorbis* (Rabl, 1879), *Physa* (Wierzejski, 1905), *Arion* (Sathananthan, 1970), *Limax* (Meisenheimer, 1896) and *Agriolimax* (Carrick, 1939). According to Ghose (1963a), it does not occur in *Achatina*.

During its evagination the shell gland spreads on the surface in all directions, the cells in its central part flattening greatly. The peripheral cells remain highly columnar, and form an annular fold surrounding the thin centre. This mantle fold now extends over the visceral hump.

A transparent secretion may already be extruded into the lumen of the invaginated shell gland. During evagination this spreads as a thin pellicle over the thinned-out middle part of the shell gland area, and forms in this way the larval shell (Fig. 10). Peripherally, it is bordered by the mantle fold, where further growth of the shell takes place by

o

secretion from the mantle fold. At first the shell is cup-shaped, but spiral coiling begins to appear at an early stage by a more rapid growth on one side.

In *Onchidella celtica* a larval shell develops in the veliger stage, but it is shed at metamorphosis (Joyeux-Laffuie, 1882; Fretter, 1943). In

Fig. 10 Section to show shell gland (SG) of *Lymnaea stagnalis* at beginning of evagination and first rudiment of larval shell (LS). AV, giant albumen vacuoles in cells of larval digestive gland; MG, albumen in lumen of midgut; SE, small-celled endoderm. × 870. (Courtesy of Dr L. P. M. Timmermans.)

Arion and *Limax* the shell gland develops into a closed sac, in which a rudimentary shell is secreted. In *Agriolimax* there is no true invagination, but a local proliferation of ectodermal cells, in the centre of which a lumen appears later and granules of $CaCO_3$ are secreted (Carrick, 1939).

The mantle fold, while growing forwards along the body, is elevated from the body wall, arching over the anal region and in this way forming the mantle cavity. This may be preceded by the formation of one or two special grooves in the body wall, which are secondarily overgrown by the mantle (Heyder, 1909; Carrick, 1939).

The foot originates as an ectodermal thickening on the ventral side behind the mouth. Later an outgrowth appears in this region, which is filled with mesenchyme (Fig. 12). As a rule, the foot rudiment is from an early stage covered with cilia. In Basommatophora particularly long cilia are formed on a median row of large clear elongated cells behind the mouth.

At the posterior extremity of the foot in Stylommatophora (except Succinea), a foot vesicle or podocyst with a flattened epithelium is formed at an early stage. It contains muscular elements originating from the mesenchyme. The podocyst shows powerful contractions, driving the blood into the body. In older embryos it may become very large; e.g., in *Archachatina* it lines nearly the whole inner surface of the egg capsule and surrounds the albumen (Brisson, 1968). Towards the end of the embryonic period it decreases in size, and finally is thrown off.

In *Onchidella celtica* the foot of the embryonic veliger bears a large operculum, which is later lost (Fretter, 1943). At metamorphosis the foot broadens greatly, and the small visceral sac closely connects with its upper side. Likewise, in the slugs (e.g. *Arion, Limax*) the visceral sac lies originally wholly outside the foot, and is only later partly surrounded by it.

The stomodaeum originates as an invagination on the ventral side immediately behind the prototroch at the point where the blastopore eventually closed, or in direct connection with the remaining blastoporal opening. The invagination is at first funnel-shaped. Its inner end gives rise to the oesophagus; it elongates, becomes tubular and opens into the midgut. The postero-ventral wall of the stomodaeum gives rise to the radular sac. A posteriorly directed outgrowth is formed, which soon develops a lumen. Mesoderm cells then surround the rudiment of the radular sac. The remaining outer part of the stomodaeum gives rise to the buccal cavity (Fig. 12).

In those embryos, which develop inside their capsules, albumen is conveyed into the midgut by ciliary movement. Long cilia which are found on a dorsomedian row of ciliary cells in the stomodaeum here play an important part. This row is either single or double and consists of cells with pale, sometimes strongly vacuolated cytoplasm (Fig. 12).

Immediately after gastrulation, the wall of the archenteron consists of more or less equivalent cells. Soon, however, a separation takes place into two regions, large-celled and small-celled, respectively; this is brought about by a strong swelling of the former cells, associated with the formation of large vacuoles, in a certain part of the gut wall (Fig. 10).

The small-celled endoderm lies at first more or less opposite the blastopore at the blind end of the archenteron. It then extends mainly along the dorsal and ventral walls of the midgut in the direction of the oesophagus. Later it also encroaches upon the lateral walls, and in this way grows around the lumen of the midgut, which is transformed into the stomach.

The hindgut arises as a posteriorly directed outgrowth from the small-celled part of the midgut. In the Basommatophora the hindgut primordium is at first a solid cell mass, in which a lumen appears later. According to Wierzejski (1905) it is derived in *Physa* from the entero-blasts, which are descendants of the cell *4d*. The hindgut connects with the ectoderm between shell gland and foot anlage.

The hindgut primordium lies at first medioventrally, but it is soon displaced towards the right side (the left in sinistral species) in con-sequence of torsion. At the same time the anus shifts forwards, so that the hindgut forms a loop towards the right.

The large-celled endoderm is more or less displaced from the gut lumen during further development; and, bulging out, it forms the albumen sac or larval digestive gland, which plays an important part in the digestion of the albumen. In Stylommatophora the albumen sac lies in the strongly distended head vesicle.

Bloch (1938) showed that the larval digestive gland in Basommato-phora is a typical larval organ, the cells of which have lost the capacity to divide. At the end of the embryonic period it disappears, and is very rapidly replaced by the definitive gland, which arises by proliferation from the wall of the stomach. In Stylommatophora, on the contrary, the albumen sac is incorporated at later embryonic stages into the left lobe of the definitive digestive gland of which it forms a part (Weiss, 1968). Only in *Achatina* does it separate from the midgut after which it atrophies (Ghose, 1963b).

The larval kidneys of the pulmonates are paired protonephridia. They are of two kinds. In the Basommatophora, they are looped or bent structures, consisting of four cells, with an intracellular canaliculus and a terminal flame cell (Meisenheimer, 1899). One end of the loop is directed forwards toward the cephalic plate, the other ventrally, both limbs meeting in the cell body of a huge nephroblast, situated beneath the ectoderm at the level of the lateral edges of the prototroch. The four component cells are thus – a terminal, a canalicular, a giant excretory, and an apertural cell (Fig. 11).

The basommatophoran protonephridium is derived from the mesoderm. According to Rabl (1879) and Wierzejski (1905), the nephroblasts originate on either side as large cells in the anterior half of the mesoderm bands. By division they give rise to the cells of the protonephridium.

In the Stylommatophora, the protonephridia have a very different structure. They consist of a large number of cells, bordering an intercellular nephric channel, which begins with several terminal cells

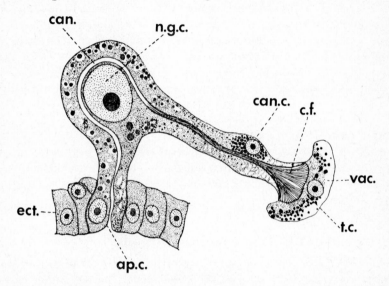

Fig. 11. Longitudinal section of protonephridium of the Basommatophora. ap. c., apertural cell; can., canaliculus of protonephridium; can. c., canalicular cell; c. f., ciliary flame; ect., ectoderm; n.g.c., nucleus of giant excretory cell; t.c., terminal cell; vac., terminal vacuole. After Meisenheimer, 1899.

(Meisenheimer, 1898, 1899). The terminal cells generally resemble those of the Basommatophora. In *Succinea* there are two terminal cells to each protonephridium, in other species there are more. The tubule opens in the lateral body wall. According to Carrick (1939) and Ghose (1963b) the distal part of the tubule is derived from the ectoderm; the inner, secretory part with the terminal cells is of mesodermal origin.

In *Onchidella celtica* there is an unpaired "larval" kidney in the wall of the mantle cavity, consisting of three cells, and resembling the secondary kidney of the opisthobranchs (Fretter, 1943).

Finally, among the "larval" differentiations of the pulmonates the nuchal cells must be mentioned, which are found in the Basommatophora and, moreover, in *Achatina* (Ghose, 1963b) and *Archachatina* (Brisson, 1968). They are very large cells, derived from the mesoderm, lying above the oesophagus in the dorsal part of the nuchal region (Fig. 12). Granules are formed in their cytoplasm, which transform into highly refringent, rod-shaped, crystal-like, often yellow-coloured concretions. Probably they have an excretory function. Towards the end of the larval period they degenerate and are extruded through gaps between the ectoderm cells in the nuchal region (Bloch, 1938; Minganti, 1950).

A few words must be said about the development of asymmetry. Rotation of the visceral sac occurs at an early stage in development. As a rule it occurs in a counterclockwise direction, when seen from the dorsal side, but in sinistral species the rotation is clockwise. The developing mantle cavity moves forwards from its original posterior position along the right side (the left side in sinistrals). The rotation is less than the 180° which is characteristic of prosobranchs. In pulmonates generally, rotation is a slow process; it is not caused by muscle contraction, but rather by differential growth of the two sides. In *Onchidella celtica* the visceral sac and mantle rotate back at later stages, bringing the mantle cavity and anus from the right side back to a median position at the hind end of the visceral sac (Joyeux-Laffuie, 1882; Fretter, 1943).

XI. Organogenesis

The thin pellicle, forming the first rudiment of the larval shell, persists as the periostracum of the adult shell. The remaining part of the shell is formed by secretion of calcium carbonate in an organic matrix beneath the periostracum. At first this secretion takes place both on the inner side of the shell and at its margin. Later the main growth takes place in the peripheral parts of the mantle.

In *Lymnaea* there is a groove on the outer surface of the mantle running parallel to its edge. It is followed by a belt of high columnar cells, which is continuous with the low columnar epithelium that covers the remaining part of the outer surface of the mantle. It has been shown by histochemical methods that there is a clear zonation in this region. The cells of the groove adjacent to the belt and the adjacent part of the belt are rich in RNA and peroxidase, and are probably involved in the synthesis, secretion, and tanning of the proteins of the periostracum. In the remaining part of the belt, the cells are also rich in RNA, but peroxidase is lacking; presumably this zone produces the inner part of the periostracum or the proteinaceous matrix of the calcareous shell layers. It is followed by a region of the low epithelium, where the cells are rich in glycogen, alkaline phosphatase, carbonic anhydrase, ATPase, cytochrome oxidase, and the enzymes of the tricarboxylic acid cycle. Apparently, these are the cells providing for the accumulation, transfer, and secretion of the calcium and carbonate ions for building the calcareous substance of the shell (Timmermans, 1969). A similar zonation has been found in *Helisoma* (Kapur and Gibson, 1968).

The development of the lung in pulmonates has been studied in *Lymnaea* (Régondaud, 1964), *Achatina* (Ghose, 1963a), *Archachatina* (Brisson, 1968), *Arion* (Heyder, 1909) and *Limax* (Meisenheimer, 1898). The lung primordium arises very early as an ectodermal invagination. At its first appearance it lies either posteriorly, and is then displaced to the right side with the rotation of the visceral sac (*Lymnaea, Limax*), or it appears immediately on the right side. The mantle cavity is formed slightly later and comes to surround the invagination which has, meanwhile, deepened; the lung now opens into the mantle cavity by a restricted opening on the right, the pneumostome. In the wall of the pulmonary cavity folds appear, enclosing blood vessels; in this way the vascular reticulum of the lung is formed.

The ganglia arise by proliferation and later delamination from the ectoderm. The commissures and connectives are formed by secondary outgrowth from the ganglia. The cerebral ganglia originate from the cephalic plates. As a rule the main body of the ganglia arises by proliferation. When it has already become partly separated from the ectoderm, a local invagination occurs at the base of each tentacle, which becomes a tubular ingrowth, the cerebral tube. The tips of these invaginations become closely applied to the cerebral ganglia and are constricted off, giving the lobus accessorius of the ganglion. In some species, as in *Arion*, the deepest part of the cerebral tube remains as a closed vesicle, the so-called cephalic gland (Chétail, 1963). The cerebral ganglia of either side become connected by the commissura cerebralis (Fig. 12).

The pedal ganglia arise from paired cell proliferations laterally or ventrally on the foot (Figs 12, 13). They connect by means of a commissure. The pleural ganglia arise from cell proliferations immediately behind the velar area and near the pleural groove (Fig. 13). The parietal ganglia are formed behind the pleural ganglia from or immediately above the pleural groove between foot and visceral sac. The abdominal (= visceral) ganglion arises from an unpaired ectodermal thickening in the floor at the posterior end of the mantle cavity (Fig. 12). According to Régondaud (1964), the visceral loop in *Lymnaea* is formed before torsion begins, and originally extends well into the visceral sac. When torsion begins, the loop becomes tilted, then relatively shorter and comes to lie anterior to the region of torsion. At first, before the right and left parietal ganglia develop, the abdominal ganglion is flanked by a supraintestinal and subintestinal ganglion. Later the right parietal fuses with the supraintestinal ganglion, and the abdominal with the subintestinal ganglion, while the left parietal ganglion remains separate.

The buccal ganglia are formed by delamination from the wall of the stomodaeum, notably from the hind wall of the suboesophageal clefts

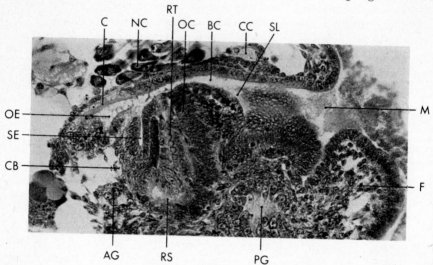

Fig. 12. Mediosagittal section through foregut of an 8-day embryo of *Lymnaea stagnalis*. AG, abdominal ganglion; BC, buccal cavity; C, ciliated cells; CB, buccal commissure; CC, cerebral commissure; F, foot; M, mouth; NC, nuchal cells; OC, odontophoral cartilage; OE, oesophagus; PG, pedal ganglion; RS, radular sac; RT, radular tooth; SE, suboesophageal cleft; SL. sublingual cleft.

(Fig. 13). They are soon united by a commissure, and connect with the cerebral ganglia by connectives (Figs 12, 13).

The eyes arise from the dorsolateral parts of the cephalic plates as small invaginations, which constrict off from the surface into vesicles with a one-layered wall. Their inner wall thickens by elongation of the cells, and becomes the retina. The outer wall of the vesicle becomes the inner part of the cornea. Its cells and those of the overlying ectoderm remain transparent. In the lumen a spherical lens is formed. Eakin and Brandenburger (1967) have made a detailed study of eye differentiation in *Helix*.

In the Stylommatophora the eyes are carried upwards with the growth of the tentacles. Members of the Onchidiacea possess accessory eyes, which are formed on the dorsal tubercles of the mantle. According to Labbé (1933) they are developed from cells derived from the epidermis (optoblasts). The lens cells of these eyes secrete silica, so that the eyes have a glass lens.

The statocysts originate on the sides of the foot primordium, near the boundary between foot and body, either as ectodermal invaginations or by delamination of a solid cluster from the ectoderm, in which

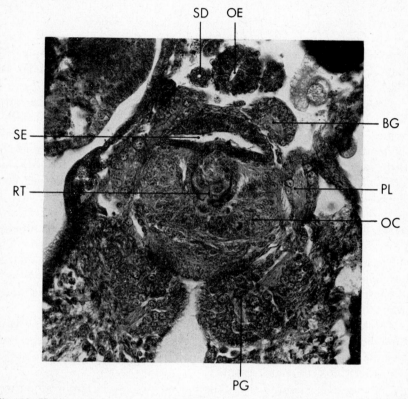

Fig. 13. Horizontal section through foregut and adjacent structures of a 9-day embryo of *Lymnaea stagnalis*. BG, buccal ganglion; OC, odontophoral cartilage; OE, oesophagus; PG, pedal ganglion; PL, pleural ganglion; RT, radular teeth; SD, salivary duct; SE, suboesophageal cleft.

a lumen appears secondarily. The cavity, which is at first very narrow, soon widens, and fine sensory hairs or cilia appear on the cells. At an early stage a large spherical statolith is formed from substances secreted by the cells. At first immobile, it later becomes free in the lumen, and begins to vibrate, set in motion by the cilia. Later a number of small accessory statoliths are often formed around the big central one.

In the walls of the stomodaeal cavity near the mouth, jaws are formed by rapid secretion of a cuticular substance by elongated cells. The stomodaeum is surrounded by mesodermal cells at an early stage. This mesodermal envelope thickens considerably at the ventral side immediately in front of the radular sac. Here the primordium of the tongue (odontophore) is formed. It is bounded anteriorly by paired sublingual clefts. In the mesoderm of the tongue a pair of odontophoral cartilages are formed (Figs 12, 13). The salivary glands appear rather late as paired diverticula in the dorsolateral angles of the buccal cavity. They grow backwards along the oesophagus (Fig. 13). The suboesophageal clefts delimit the oesophagus from the posterior side of the radular sac (Figs 12, 13).

As the radular sac grows out, it flattens dorsoventrally and becomes crescent-shaped in cross section (Fig. 13). The formation of the radula begins at an early stage with the secretion of a thin cuticular lamella on the ventral wall of the radular sac. Tooth formation takes place at the posterior end of the radular sac, where special cells (odontoblasts) are differentiated. They are large clear cells, arranged in regular cell configurations. Each group of odontoblasts secretes consecutively all the teeth of a longitudinal row, while the radula as a whole moves forwards (Schnabel, 1903; Beck, 1912; Eckardt, 1914; Spek, 1921; Hoffmann, 1932; Märkel, 1958; Runham, 1963; Keith and Krause, 1969). The chitin of the teeth which is produced by the odontoblasts is secondarily hardened by impregnation with substances secreted by the dorsal wall of the radular sac, especially proteins and in some species calcium or iron salts.

The longitudinal rows of radular teeth do not all appear at the same time, but in a certain sequence depending upon the species. The dental formula of the adult is only gradually reached (Joyeux-Laffuie, 1882; Schnabel, 1903; Carriker, 1943).

The digestive gland of the adult appears as two lateral evaginations of the stomach. Secondary evaginations give rise to separate lobules, and progressive subdivision leads to the formation of numerous follicles. The originally broad connections of the gland with the stomach later became narrower and form the collecting ducts.

The heart, pericardium and kidney arise from a common mesodermal anlage. This divides into two parts. The posterior and more laterally situated part gives rise to the kidney, the anteromedial portion to heart and pericardium, a small pericardial cavity being formed by a split in the mesoderm. The heart arises either as separate atrial and ventricular invaginations of the wall of the pericardial cavity, which unite to form a tube (Pötzsch, 1904; Heyder, 1909; Ghose, 1963c), or by splitting in a solid invagination (Meisenheimer, 1898). The blood vessels originate as clefts in the mesoderm of the primary body cavity, and connect secondarily with the ends of the heart tube. In *Achatina* and *Archachatina* the cephalic aorta functions temporarily as a 'larval' heart (Brisson, 1968).

The kidney primordium becomes tubular. In *Achatina* this kidney tubule communicates from the outset with the pericardial cavity by way of a renopericardial aperture (Ghose, 1963c). In other pulmonates the kidney breaks through only secondarily into the pericardial cavity. The outer end of the renal tubule breaks through the ectoderm in the region of the mantle cavity. At this point an ectodermal invagination occurs, which gives rise to the primary ureter. In the Stylommatophora this is supplemented by a secondary ureter, arising as a groove in the wall of the mantle cavity, which closes to a tube from behind forwards.

The gonad as a rule becomes first recognizable as an isolated mesodermal cell mass in the body cavity (Ancel, 1903; Pabst, 1914; Fraser, 1946). A lumen appears in the gonad by separation of the cells, lined by a single-layered germinal epithelum. The gonad becomes lobate by local evaginations, which by further subdivision give rise to the acini.

The formation of the gonoducts is different in Basommatophora and Stylommatophora. In the Basommatophora (De Larambergue, 1939; Fraser, 1946) the efferent part of the reproductive system is derived from two separate primordia: 1, the rudiment of the primary gonoduct; and 2, the anlage of the copulatory organs and the distal part of the vas deferens. The primary gonoduct originates as an ectodermal invagination in the posterior part of the mantle cavity, which grows backwards and connects at an early stage with the rudiment of the gonad. A secondary outgrowth grows forwards toward the position of the future female genital pore. The deeper part of the mantle cavity of *Lymnaea*, according to Fraser, is constricted from the rest of the cavity, but remains in connection with the gonoduct by the original invagination opening; it becomes the receptaculum seminis. The proximal part of the gonoduct gives rise to the hermaphroditic duct, the seminal vesicles and the albumen gland. The rest of the gonoduct divides lengthwise into a male and female duct, which differentiate further into regions with different structure. The copulatory organs originate separately from an ectodermal invagination on the side of the head, and are secondarily connected with the anterior end of the male duct.

In the Stylommatophora, the gonoduct and copulatory organs arise from a common primordium (Pabst, 1914; Hoffmann, 1922; Laviolette, 1954). The original opening of the ectodermal invagination shifts forwards, ultimately reaching its definitive place on the side of the head. A swelling near the orifice forms the rudiment of the copulatory organs. The division of the gonoduct into male and female ducts is incomplete or lacking. The receptaculum seminis is formed from the gonoduct by constriction.

XII. References

Ancel, P. (1903). *Arch. Biol.* **19**, 389–652.

Andrew, A. (1959). *Proc. K. ned. Akad. Wet.*, C **62**, 68–75.

Beadle. L. C. (1969a). *J. exp. Biol.* **50**, 473–479.

Beadle, L. C. (1969b). *J. exp. Biol.* **50**, 491–499.

Beck, K. (1912). *Jena. Z. Naturw.* **48**, 187–262.

Berendsen, W. (1971). Thesis, University of Utrecht.

Bloch, S. (1938). *Rev. suisse Zool.* **45**, 157–220.

Bluemink, J. G. (1967). "The subcellular Structure of the Blastula of *Lymnaea stagnalis* L. (Mollusca) and the Mobilization of the Nutrient Reserve." University of Utrecht.

Bretschneider, L. H. (1948). *Proc. K. ned. Akad. Wet.* **51**, 358–363.

Bretschneider, L. H., and Raven, C. P. (1951). *Arch. Néerl. Zool.* **10**, 1–31.

Brisson, P. (1968). *Arch Anat. microsc. Morph. exp.* **57**, 345–368.

Byrnes, E. F. (1900) *J. Morph.* **16**, 201–236.

Camey, T., and Verdonk, N. H. (1970). *Neth. J. Zool.* **20**, 93–121.

Carrick, R. (1939). *Trans. R. Soc. Edinb.* **59**, 563–597.

Carriker, G. S. (1943). *Nautilus* **57**, 52–59.

Chétail, M. (1963). *Arch. Anat. microsc. Morphol. exp.* **52**, Suppl. 1, 129–203.

Clement, A. C. (1938). *J. exp. Zool.* **79**, 435–460.

Comandon, J., and De Fonbrune, P. (1935). *Arch. Anat. Microsc.* **31**, 79–100.

Crabb, E. D. (1927). *Biol. Bull.* **53**, 67–108.

Crampton, H. E. (1984). *Ann. N.Y. Acad. Sci.* **8**, 167–170.

De Larambergue, M. (1939). *Bull. biol. Fr. belg.* **73**, 21–231.

Eakin, R. M., and Brandenburger, J. L. (1967). *J. ultrastruct. Res.* **18**, 391–421.

Eckardt, E. (1914). *Jena. Z. Naturw.* **51**, 213–376.

Elbers, P. F. (1957). *Proc. K. ned. Akad. Wet.* C **60**, 96–98.

Elbers, P. F. (1959). "Over de Beginoorzaak van het Li-effect in de Morphogenese." University of Utrecht.

Elbers, P. F. (1966). *Biochim. Biophys. Acta* **112**, 318–329.

Elbers, P. F. (1969). *J. Embryol. exp. Morph.* **22**, 449–463.

Fahmy, O. G. (1949). *Q. Jl microsc. Sci.* **90**, 159–181.

Favard, P., and Carasso, N. (1958). *Arch Anat. microsc. Morph. exp.* **47**, 211–234.

Fioroni, P. (1971). *Z. wiss. Zool.* **182**, 263.

Fraser, L. A. (1946). *Trans. Am. microsc. Soc.* **65**, 279–298.

Fretter, V. (1943). *J. mar. biol. Ass. U.K.* **25**, 685–720.

Garnault, P. (1888). *Zool. Anz.* **11**, 731–736.

Garnault, P. (1889). *Zool. Anz.* **12**, 33–38.

George, J. C., and Jura, C. (1958). *Proc. K. Ned. Akad. Wet.*, C **61**, 598–603.

Ghose, K. C. (1962). *Proc. zool. Soc. Calcutta* **15**, 47–55.

Ghose, K. C. (1963a). *Proc. malac. Soc. Lond.* **35**, 119–126.

Ghose, K. C. (1963b). *Proc. R. Soc. Edinburgh* B **68**, 237–260.

Ghose, K. C. (1963c). *Proc. zool. Soc. Calcutta* **16**, 201.

Guerrier, P. (1968). *Ann. Embryol. Morphogen.* **1**, 119–139.

Guraya, S. S. (1969). *Acta Embryol. Morph. exp.* 1969, 91–96.

Hartung, E. W. (1947). *Biol. Bull. mar. biol. Lab. Woods Hole* **92**, 10–22.

Heyder, P. (1909). *Z. wiss. Zool.* **93**, 90–156.

Hoffmann, H. (1922). *Z. wiss. Zool.* **119**, 493–538.

Hoffmann, H. (1932). *Jen. Z. Naturwiss.* **67**, 535–553.

Holmes, S. J. (1900). *J. Morph.* **16**, 369–456.

Horstmann, H. J. (1956a). *Biochem. Z.* **328**, 342–347.

Horstmann, H. J. (1956b). *Biochem. Z.* **328**, 348–351.

Horstmann, H. J. (1958). *Z. vergl. Physiol.* **41**, 390–404.

Horstmann, H. J. (1964). *Hoppe-Seyler's Z. physiol. chem.* **337**, 57.

Hubendick, B. (1951). *Kungl. Svenska Vetensk. Hand.* **3**, 1–223.

Hunger, J., and Horstmann, H. J. (1968). *Z. Biol.* **116**, 90–104.

Ikeda, K. (1930). *Jap. J. Zool.* **3**, 89–94.

Joyeux-Laffuie, J. (1882). *Archs Zool. exp. gén.* **10**, 225–383.

Jura, C. (1960). *Zoologica Pol.* **10**, 95–128.

Jura, C. (1971). *Acta biol. cracov.*, Ser. Zool. **14**, 315–323.

Kapur, S. P., and Gibson, M. A. (1968). *Can. J. Zool.* **46**, 481–491.

Kerth, K., and Krause, G. (1969). *Wilhelm Roux Arch EntwMech. Org.* **164**, 48–82.

Kofoid, C. A. (1895). *Bull. Mus. comp. Zool. Harv.* **27**, 35–118.

Kostanecki, K., and Wierzejski, A. (1896). *Arch. mikrosk. Anat.* **47**, 309–386.

Labbé, A. (1933). *C. r. Séanc. Soc. Biol.* **114**, 1002–1003.

Lams, H. (1910). *Mém. Acad. r. Belg., Cl. Sci.* 4°, Ser. II, **2**, 142.

Laviolette, P. (1954). *Ann. Sci. nat. Zool. Biol. Anim.* (11) **16**, 427–535.

Linville, H. R. (1900). *Bull. Mus. comp. Zool. Harv.* **35**, 213.

McMahon, P., Von Brand, T., and Nolan, M. O. (1957). *J. cell. comp. Physiol.* **50**, 219–240.

Märkel, K. (1958). *Z. wiss. Zool.* **160,** 213–289.

Malhotra, S. K. (1960). *Cellule* **61**, 109–125.

May, F., and Weinland, H. (1953). *Z. Biol.* **105**, 339–347.

Meisenheimer, J. (1896). *Z. wiss. Zool.* **62**, 415–468.

Meisenheimer, J. (1898). *Z. wiss. Zool.* **63**, 573–664.

Meisenheimer, J. (1899). *Z. wiss. Zool.* **65**, 709–724.

Minganti, A. (1950). *Riv. Biol.* **42**, 295–319.

Morrill, J. B. (1963). *Acta Embry. Morph. exp.* **6**, 339–343.

Morrill, J. B. (1964). *Acta Embryol. Morph. exp.* **7**, 131–142.

Pabst, H. (1914). *Zool. Jb. (Anat. Ont.)* **38**, 465–508.

Parat, M. (1928). *Arch. Anat. microsc.* **24**, 73–357.

400   C. P. RAVEN

Pötzsch, O. (1904). *Zool. Jb.*, *Anat. Ont.* **20**, 409–438.
Quattrini, D., and Lanza, B. (1965). *Monit. Zool. Ital.* **73**, 3–60.
Rabl, C. (1879). *Morphol. Jb.* **5**, 562–660.
Raven, C. P. (1945). *Arch. néerl. Zool.* **7**, 91–121.
Raven, C. P. (1946). *Arch. néerl. Zool.* **7**, 353–434.
Raven, C. P. (1949). *Bijdr. Dierk.* **28**, 372–384.
Raven, C. P. (1952). *J. exp. Zool.* **121**, 1–77.
Raven, C. P. (1959). *J. Embryol. exp. Morph.* **7**, 344–360.
Raven, C. P. (1963). *Dev. Biol.* **7**, 130–143.
Raven, C. P. (1964). *J. Embryol. exp. Morph.* **12**, 805–823.
Raven, C. P. (1966). "Morphogenesis. The Analysis of Molluscan Development", 2nd ed. Pergamon Press, Oxford.
Raven, C. P. (1967). *Devl Biol.* **16**, 407–437.
Raven, C. P. (1970). *Int. Rev. Cytol.* **28**, 1–44.
Raven, C. P. (1974). *J, Embryol. exp. Morph.* **31**, 37–59.
Raven, C. P., and Klomp, H. (1946). *Proc. K. ned. Akad. Wet.* **49**, 101–109.
Raven, C. P., and Van Der Wal, U. P. (1964). *J. Embryol. Exp. Morph.* **12**, 123–139.
Raven, C. P., and Van Zeist, W. (1950). *Proc., K. ned. Akad. Wet.* **53**, 601–609.
Raven, C. P., Bezem, J. J. and Isings, J. (1952). *Proc. K. ned. Akad. Wet.* C **55**, 248–258.
Raven, C. P., Bezem, J. J. and Geelen, J. F. M. (1953). *Proc. K. ned. Akad. Wet.* C **56**, 409–417.
Raven, C. P., Escher, F. C. M., Herrebout, W. M. and Leussink, J. A. (1958). *J. Embryol. exp. Morph.* **6**, 28–51.
Recourt, A. (1961). "An Electron Microscopic Study of Oogenesis in *Limnaea stagnalis* L." University of Utrecht.
Régondaud, J. (1964). *Bull. Biol. France Belg.* **98**, 433–471.
Runham, N. W. (1963). *Q. Jl microsc. Sci.* **104**, 271–277.
Sathananthan, A. H. (1968). *Proc. Symp. Mollusca* 687–705.
Sathananthan, A. H. (1970). *J. Embryol. exp. Morph.* **24**, 555–582.
Schnabel, H. (1903). *Z. wiss. Zool.* **74**, 616–655.
Spek, J. (1921). *Z. wiss. Zool.* **118**, 313–363.
Timmermans, L. P. M. (1969). *Neth. J. Zool.* **19**, 417–523.
Ubbels, G. A. (1968). "A Cytochemical Study of Oogenesis in the Pond Snail *Limnaea stagnalis.*" University of Utrecht.
Ubbels, G. A., Bezem, J. J. and Raven, C. P. (1969). *J. Embryol. exp. Morph.* **21**, 445–466.
Van Der Wal, U. P. (1974). Thesis, University of Utrecht.
Verdonk, N. H. (1965). "Morphogenesis of the Head Region in *Limnaea stagnalis* L." University of Utrecht.
Waddington, C. H., Perry, M. M. and Okada, E. (1961). *Exp. Cell Res.* **23**, 631–633.
Weiss, M. (1968). *Rev. Suisse Zool.* **75**, 157–225.
Wierzejski, A. (1905). *Z. wiss. Zool.* **83**, 502–706.

Systematic Index

Names of phyla, classes and families are printed in roman type, genera and species in italic. Non-pulmonate species are indicated with an asterisk. Page numbers against the generic name include a non-specific reference to the genus. Page numbers in italic refer to a figure; in parentheses, to a table.

A

Acavus phoenix, 279
**Acer pseudoplatanus, (59)*
Achatina, 79, 335, 387, 390, 391, 393, 397
Achatina achatina, 131, 142, (112)
   fulica, 79, 84, 291, (65, 107, 112, 117)
Achatinella fulgens, 249, 251
Acteon, 311, 313, 314
**Acteon tornatilis, 310*
Acroloxidae, 320, 349
Acroloxus, 329, 330
Acroloxus lacustris, 319, 345
**Aegopodium podograria, 55, 56*
Agriolimax, 13, 16, 17, 55, 73, 75, 77, 79, 85, 86, 89, 90, 93, 95, 96, 97, 98, 99, 166, 205, 221, 224, 323, 327, 330, 336, 337, 343, 344, 346, 347, 387, 388, 15, 220, 221, 222, 223, 224
Agriolimax agrestis, 138, 374, (27, 125, 139)
   caruanae, (65)
   laevis, 64, 333
   reticulatus, 7, 13, 54, 56, 57, 60, 71, 73, 76, 79, 84, 87, 88, 111, 120, 136, 137, 138, 154, 211, 220, 288, 289, 290, 296, 297, 298, 321, 336, 349, 353, 355, 374, 14, 54, 83, 87, 220, 221, 222, 223, 224, (65, 74, 112, 114, 121, 125)
Allogona profunda, 109, 138, (135)
   ptychophora, 138
Amphibolidae, 386
Amphibola, 83
Amphibola crenata, 367
Ancylastrum fluviatilis, 155
Ancylidae, 83, 320, 329, 332, 349

Ancylus, 278, 282, 329, 332, 355
Ancylus fluviatilis, 24, 155, 275, 281, 348, 352, 355, 284, 319, 331, 345
Anguispira alternata, 358
**Anodonta, 144*
Aplexa hypnorum, 352
**Aplysia, 187, 188, 191*
**Aplysia californica, 248, 295*
Archachatina, 166, 175, 335, 387, 389, 391, 393, 397, 173, 174, 206
Archachatina (Calachatina) marginata, 173, 205
   ventricosa, 130, 131, 136, (112, 117)
Arianta, 136
Arianta arbustorum, 133, 136, 145, (119, 125, 126, 127)
Ariolimax, 342, 343
Ariolimax californicus, 291, 298
   columbianus, 347
Arion, 7, 35, 37, 217, 234, 235, 236, 316, 323, 329, 335, 336, 337, 346, 347, 368, 371, 386, 387, 388, 389, 393, 237
Arion ater, 12, 58, 59, 68, 77, 78, 97, 100, 136, 154, 289, 291, 321, 322, 330, 349, 353, 354, 11, (65, 107, 112, 119, 125, 135, 139)
   ater ater, 7
   ater rufus, 16, 59, 251, 292, 298, 377, (112)
   fasciatus, 138
   hortensis, 90, 137, 349, 353
   intermedius, 138
   rufus, 273, 275
   subfuscus, 291
Arionidae, 288, 332, 335, 386
Auricula, 315, 316

B

Basommatophora, 76, 82, 84, 95, 98, 116, 124, 153, 251, 253, 257, 259, 263, 265, 272, 273, 278, 279, 282, 287, 289, 313, 315, 318, 323, 332, 335, 337, 338, 344, 347, 348, 349, 351, 368, 386, 387, 389, 390, 391, 397, *319*

Biomphalaria, 48, 88, 89, 93, 119, 146, 153, 278, 282, 323, 329, 336, 338, 341, 343, 353, 357, 379, 381, 383, 384, *280, 281, 283*

Biomphalaria glabrata, 46, 48, 59, 118, 132, 133, 145, 146, 154, 254, 255, 257, 260, 263, 265, 273, 281, 287, 296, 336, *48*, (47, 48, 107, 134, 139)
  pfeifferi, 88
  sudanica, 153
Bithynia leachi, *49*
  tentaculata, *49*
Bivalvia, 46
Bradybaena similaris, 336
Buliminus detritus, (126)
Bulimulus dealbatus, 148
Bulinus, 337, 373
Bulinus contortus, 132, 333, 357
  natalensis tropicus, 63
  tropicus, 71
  truncatus, (134)
*Bullia, 29, 30, 31

C

*Calystegia sepium, 55
*Carpinus betulus, (59)
Carychium, 82, 278, 315, 316, 322, 338, 348, 349, *81*
Carychium tridentatum, 137, 350, *350*
*Cassidula, 315, 316
*Cassidula aurisfelis, 350
*Castanea sativa, (59)
Cepaea, 46, 133, 135, 136, 139, 354
Cepaea hortensis, 21, 249, (46, 125, 126, 127)
  nemoralis, 55, 57, 154, 249, 322, 323, 336, 349, (65, 107, 125, 126, 127)
  vindobonensis, (46)
*Cephalopoda, 46, 109
Cernuella variabilis, 145
*Chaerophyllum temulem, 57

Chilina, 166, *167, 310*
Chilinidae, 315, 318
*Circaea lutetiana, (59)
Cochlicella acuta, (139)
  ventrosa, (139)
Cochlicopa lubrica, 6

D

*Dendronotus, 205
Deroceras, 371
Discartemon, *310*
Discus, 24
Discus rotundatus, 22, 23, 120, (59, 121)
Drepanotrema, 118
Drepanotrema anatinum, 333

E

Ellobiidae, 2, 22, 82, 131, 156, 272, 313, 314, 315, 316, 328, 332, 335, 338, 344, 351
Ellobium, 338, 343, 348
Ellobium aurisjudae, 316, 332, *333*
*Elodea canadensis, 56
Eobania vermiculata, 249
Eremina desertorum, (135)
*Escherichia coli, 59
Eulota fruticum, (126)

F

*Fagus sylvatica, (59)
Ferrissia, 278
Ferrissia irrorata, 333
  shimekii, 246, 253, 263, 336

G

Gadinia, 338
Gadiniidae, 318
Gastropoda, 271
*Glechoma hederacea, 57
Gundlachia, 278

H

*Haliotis tuberculata, (46)
Hedleyella falconeri, 323
Helicella neglecta, (135, 139)
  obvia, 145, (46)
  virgata, 139, 141, (126, 135, 139)

Subject Index

Page numbers in italic refer to a figure; in parentheses, to a table.

H

Habituation, 207–210
  defined, 207
  cf. fatigue and sensitization, 207
Haemal pore, 30, 116
Haemocoel, 3, 13, 15, 16, 21, 28, 30, 31,
  90, 106, 283, 296, 311, *220, 224*
Haemocyanin
  α and β forms, 44
  chemical composition, 42, 43, 44, *43, 44*
  concentration of, 143, 144
  dissociation curve, 45, *45*
  pH and oxygen capacity, 45
Haemoglobin
  dissociation curve, 41
  Planorbarius corneus, 41–42
  molecular weight, 41
Haemolymph, 41, 106, 109, 111, 115,
  116, 118, 120, 130, 132, 142, 143, 147,
  156, 188, 282, 290
  variation in composition, 111, 115,
    144, 188
Head-foot, 2, 3, 4, 13, 20
  retraction and protrusion of, 29–31
Head vesicle, 386–387
Heart, 115, 116, 270
  development, 396–397
  hypertrophied tentacle and, 28, 29
  in terrestrial molluscs, 28
  pressure produced by, 28, (27)
Helicorubin, 94
Hemicholinium, 194
Hepatic ducts, 83, 85, 86, 89, *87*
  vestibule, 83, 85
Hibernation, 46, 67, 77, 129, 155
  arousal from, 138–140
  carbohydrate metabolism, 146–147
  death in, 140
  lack of, desert species, 139–140
  onset of, 137–138
  oxygen consumption and, 137–138
  water loss during, 138
Hindgut, 390
Homing behaviour, 138
Hydrostatic skeleton
  blood pressure and, 26, 28, (27)
  to support shell, 26
  to support tentacles, 26

I

Ibotenic acid, 201
ILD, 196
Inhibitory post synaptic potential
  (IPSP), 181–182, 186–187, *181*
  acetylcholine and, 194
Intestine, 46, 95ff
  bacterial action in, 99
  blood and nerve supply, 98
  enzymes in, 97, 98, 99
  groove, 82, 86, 96, *81, 83*
  peristalsis in, 97, 98
  sphincter of, 93

J

Jaw, 54, 60, 68, 70, 96
  construction, 61
  development, 396

K

Karyomeres, 373
Kidney, 46, 58, 95, 106, 115, 116, 130,
  131, 132, 137, 143, 148, 149, 265, 270,
  390, 391
  development, 397
  haemolymph concentration, 142
  sac, 130, 149, 270, (117)
  salt loss and, 142
Krogh's normal curve, 48, *48*

L

Lateral lobe, 272
Latero-dorsal bodies, 278, 279, *256*
Life cycles, 352–354, *351*
Lipid, 88, 146
Lips, 60, 68
Little hermaphrodite duct, 296, 311,
  313, 329, 336, 351, 397, *310, 312, 314,*
  317, 319, 327, 328
  species variation, 323, 327
Liver, 46
  string, 86, 90, 91
Locomotion
  ciliary, 17, 22-25
    environmental limiting factor, 23,
    24
    nervous control of, 24-25
    size and, 23